AF334608

Uniformly Accelerating Charged Particles

Fundamental Theories of Physics

*An International Book Series on The Fundamental Theories of Physics:
Their Clarification, Development and Application*

Editors:
GIANCARLO GHIRARDI, *University of Trieste, Italy*
VESSELIN PETKOV, *Concordia University, Montreal, Canada*
TONY SUDBERY, *University of York, U.K.*
ALWYN VAN DER MERWE, *University of Denver, U.S.A.*

Volume 158

Uniformly Accelerating Charged Particles
A Threat to the Equivalence Principle

by

Stephen N. Lyle
Alzen
France

 Springer

Stephen N. Lyle
Andébu
09240 Alzen
France
steven.lyle@wanadoo.fr

ISBN 978-3-540-68469-5   e-ISBN 978-3-540-68477-0

Library of Congress Control Number: 2008928513

Cover design: deblik, Berlin

Printed on acid-free paper

9 8 7 6 5 4 3 2 1

springer.com

To Joan M. Lazenby

Preface

Back in 1954, a paper [2] by Bondi and Gold was to pick up on a much older question and raise a new one that would trigger another long debate. The old question had been around since the beginning of the twentieth century, when Born first raised it [1] and others followed suit. This was the question of whether a uniformly accelerated charge (in flat spacetime) would radiate electromagnetic energy. The new question arose from the claim by Bondi and Gold that (in the context of general relativity now) a static charge in a static gravitational field cannot radiate energy. If this were the case, then a particular version of the equivalence principle would thereby be contradicted.

This book reviews the problem discovered by Bondi and Gold and discusses the ensuing debate as carried on by Fulton and Rohrlich [3], DeWitt and Brehme [4], Mould [5], Boulware [6], and Parrott [7]. Various solutions have been proposed by the above (and others who are not discussed here). One of the aims here will be to put forward a rather different solution to Bondi and Gold's radiation problem. So even though the papers cited are discussed to a large extent in chronological order, the reason for writing this is not just to produce an historical reference. And even though the version of general relativity applied here is entirely consensual, every one of these papers is criticised on at least one important count, so I suspect that the result as a whole should not be described as consensual.

Chronologically speaking, I began to read the above papers somewhere in the middle and moved randomly through them, ending with the first! But, of course, Bondi and Gold's paper is not the first in this debate. Here we have one of those situations where we expect the latest papers to reveal the state of the art, so that all previous ones are redundant, but in fact we find that there was something important in the paper just before the one we are reading, that requires more attention, and we end up delving further and further into the history of the problem. This was why I eventually got hold of Bondi and Gold's 1954 paper, and I am glad I did so. I think it is here that we find out why the equivalence principle came into the problem.

This is the kind of debate that one gets drawn into almost against one's better judgement. As I understood the notion of equivalence principle, it was one of the lynchpins of general relativity, and my whole understanding of this theory would have been in trouble without it. This is why I thought I had better have a look. Several years and hundreds of pages of calculations later, it seems to me that I have a clearer view of some of the pitfalls awaiting those who try to understand general relativity more deeply than the account in a standard introductory textbook.

Here is a well-known principle of learning: if you want to really understand something, you have to get involved with it, asking difficult questions and trying to answer them. So another aim here, in fact the main aim, is to look closely at what general relativity is telling us with regard to some difficult questions, and to emphasise those pitfalls. It is true that the specific question of electromagnetic radiation from charged particles dealt with in this book is very narrow, but it provides a stage on which one can exercise the art and become stronger.

So my idea was to take my own voluminous notes on this issue and extract a story with the chronology of the above sequence of papers, illustrating along the way this or that theoretical problem and forging my own conclusions about the notion of equivalence principle, the question of whether uniformly accelerating charges do or do not radiate electromagnetic energy, and the validity or otherwise of the Lorentz–Dirac equation. The logical structure of the result is not simpler for having taken this storyline as backbone, rather than a textbook approach where logic, or rather economy of logic, comes first. However, there is a detailed table of contents and an index, and there are many cross-references. And the reader interested in a non-linear theory is assumed to be thoroughly capable of reading a non-linear book.

But who is the reader? It is certainly someone who has read several of those standard textbooks, and on reading the second was a little disappointed to find that it recounted in the same way more or less the same problems, and with the same solutions as the first. It is also someone who is never happy with scientific theories and has a lot of determination. In brief, a scientist, I suppose. The calculations here are not always entirely self-contained, as happens when one leaves the world of textbooks and ventures into the journal literature. Even the notation is not always the same, as happens when one moves from one publication to the next. In any case, the mathematics is not the most important thing, it is only equally important, and the text is definitely intended to be user-friendly. This explains the frequent recycling and development of ideas. There is a deliberate redundancy in the text which means that you may find yourself reading almost the same thing several times, something I always find comforting in a book.

So who is the reader? An undergraduate who likes a challenge, or a graduate who finds general relativity more difficult than the textbooks suggest with their glib solutions to Einstein's equations. Anybody who wants to know more about the way electromagnetism can be done in a curved spacetime, or the way the notion of equivalence principle can subtly change from one author to another. And here is another problem when one turns to the literature. Often abbreviated to get it into limited journal space, there is not always enough room to state all one's principles carefully. The potential reader is invited to get hold of the papers cited above and scan them for statements of the equivalence principle, bearing in mind the question: is that useable in practice? In the present book, it is hoped that things have been stated clearly. For example, the easy reference in the first paragraph of this preface to a static charge in a static spacetime is spelt out and given due attention. The potential reader is therefore someone who baulks at theory-laden statements when they are not treated with care, and someone who does not mind a slightly bulkier paraphrase that reminds us of the full content of a claim when that is useful. Some parts of the book look rather pedantic as a result, and I apologise for this in advance, although I suspect that it will not be disagreeable to the veterans of this debate.

Naturally, it is hoped that anyone who reads this book will come out stronger. It is hoped that they will find the weak points and improve things or even reject something where necessary. My intention is certainly not to consolidate some widely accepted theory or other, only to put it to the test, to see whether it can hold its ground on these particular issues. There are many other conflicts between general relativity and the rest of theoretical physics which should make us continue to be very tough on *all* our theories, as befits scientific endeavour.

Chapter 1 is a brief introduction to the question raised by Bondi and Gold, which sets the scene as it were. Chapter 2 then goes straight to the heart of the matter, because it discusses semi-Euclidean coordinate systems, sometimes advocated for accelerating observers in Minkowski spacetime, and their interpretation after taking the step to general relativity as describing static and homogeneous gravitational fields. This chapter launches the discussion about the notion of equivalence principle, a notion which turns out to look upon the world in many disparate ways. It is one of the key chapters. Chapter 3 then recalls how one might treat gravity as a force in the framework of special relativity, drawing attention to an amusing fact, viz., it is easy to get uniform acceleration in this scenario.

Chapter 4 introduces what has become known as the strong equivalence principle, which allows one to do electromagnetism in curved spacetimes, and indeed to transpose other bits of non-gravitational physics to this context. But in fact this principle does far more than that. It is a *sine qua non* for relating coordinates on curved spacetimes to length and time measurements one might make in the real world. From the historical angle, this chapter returns to Bondi and Gold's paper and spells out their problem.

Chapter 5 looks at the way Fulton and Rohrlich took up the debate, discussing the problem of electromagnetic radiation from an eternally uniformly accelerating charged particle in some detail. Chapter 6 then takes a first glance at Parrott's excellent review of this question, which is much more recent, so that the chronology of the story is broken (a kind of flash-forward). Chapters 7 and 8 return to the main time line and get to grips with the the problem of defining radiation in the context of Fulton and Rohrlich's paper. They also reveal some of the subtleties of the fields due to an eternally uniformly accelerating charged particle. Chapter 9 analyses Fulton and Rohrlich's view on the threat to the equivalence principle. This illustrates just how difficult it can be to interpret an author when the main claims have not been unambiguously stated. One possible interpretation of Fulton and Rohrlich's view is that they confuse some features of special and general relativity, and Chap. 10 aims to spell out the differences between the two theories with regard to radiation predictions.

Chapter 11 comes back to the Lorentz–Dirac equation, discussing the seminal Dirac paper rather briefly and exposing the much more recent account by Parrott, which is taken as definitive here. A brief mention is made of self-force calculations for spatially extended charge sources and their relevance to the Lorentz–Dirac equation. This is one of the key chapters for anyone wishing to find out about this much-discussed equation of motion. Chapter 12 moves on to DeWitt and Brehme's famous extension of the Lorentz–Dirac equation to curved spacetimes, a notoriously difficult paper which is essential reading for anyone who needs to sharpen their teeth on relativity theory. I mention some of the problems with this calculation and also with the later paper by Hobbs, which attempts to transpose the method to a more modern notation. It then reviews what DeWitt and Brehme have to say concerning the equivalence principle and the possibility of radiation from static charges in static spacetimes. Chapter 13 focuses more closely on the latter question, in a highly critical way, and is perhaps the key chapter for understanding the solution to the radiation problem that I am advocating here.

Chapter 14 discusses an entirely theoretical radiation detector proposed by Mould in 1964. The discussion here mainly concerns how the theoretical predictions for what this detector will detect affect or do not affect conclusions about the equivalence principle. But there is some analysis of the theory itself since it does illustrate the difficulties involved in interpreting results expressed relative to general coordinate systems, and it raises the question of how one should define radiation.

Chapter 15 is a detailed criticism of the paper by Boulware that is now often taken as definitive on these issues (although not by Parrott). Indeed one might describe it as the definitive mathematical analysis of the electromagnetic fields due to an eternally uniformly accelerating charged particle. However, there is a great deal of criticism here of the way the mathematics is interpreted, and especially the conclusions regarding equivalence principles. Parrott's arguments against Boulware are exposed in some detail in Chap. 16. The aim once again is to highlight the dangers of naive interpretation of tensor quantities when coordinates are not (locally) inertial. Parrott's neat analysis of the charged rocket, so useful for revealing the strangeness of the Lorentz–Dirac equation that it seems essential reading at some point, is recounted in detail in Chap. 17.

Chapter 18 beginning on p. 307 is effectively an abstract and could be read right at the beginning. It is long for a summary. I felt it was useful to recycle and repeat some of the main themes, given the logical complexity of the layout.

Acknowledgements

My first duty is to all those authors cited in this book, without whom it could not exist. Many thanks to Vesselin Petkov in Concordia, Montreal, for his constant interest, even when ideas conflict with his own. Also to Angela Lahee at Springer, who is much, much more than an editor. And I must not forget the unerring support of those close to me.

Alzen, France   *Stephen N. Lyle*
April 2008

Contents

Acronyms

Here are some of the abbreviations used throughout the book:

AO	accelerating observer
EM	electromagnetic
GR	general relativity
HOS	hyperplane of simultaneity
ICIO	instantaneously comoving inertial observer
LD	Lorentz–Dirac
MEME	minimal extension of Maxwell's equations
NG	Newtonian gravitation
SE	semi-Euclidean
SEP	strong equivalence principle
SHGF	static homogeneous gravitational field
SR	special relativity
WEP	weak equivalence principle

Chapter 1
A Doubt about the Equivalence Principle

In 1954, Bondi and Gold published a paper [2] in which they found the electromagnetic fields that would be generated by a charged particle that had been uniformly accelerating forever. This was a subject that had already been treated by other authors but without finding the complete solution. They also raised a new question about this rather idealistic situation: could the equivalence principle apply to such a scenario?

The idea here is not to go into all the technical details of this paper, but rather to provide a qualitative description of the issues. The main part of the paper concerns a charged particle undergoing uniform acceleration along a straight line in a flat spacetime for all eternity. This means that the four-acceleration a^μ of the charge is such that $a^2 := a^\mu a_\mu$ is constant, and the metric is the Minkowski metric

$$
\eta = \begin{pmatrix} 1 & 0 & 0 & 0 \\ 0 & -1 & 0 & 0 \\ 0 & 0 & -1 & 0 \\ 0 & 0 & 0 & -1 \end{pmatrix},
\tag{1.1}
$$

where the speed of light c has been set equal to unity. Uniform acceleration can also be characterised by saying that the four-acceleration of the charge always looks exactly the same in its instantaneous rest frame. It is a motion that is a priori difficult to envisage in any real physical context, because it involves the particle asymptotically approaching the speed of light in the distant past and the distant future. It is also a very special type of motion for various purely theoretical reasons to be explained shortly.

Born derived the electromagnetic fields that he thought would be generated by such a charge, using the so-called method of retarded potentials (the Lienard–Wiechert retarded potentials) [1]. However, his solution is only correct in a certain region of spacetime. Let us just see qualitatively what went wrong.

We imagine the charge coming down one of the space axes of an inertial frame from large values of the corresponding coordinate toward the origin, stopping at some distance from the origin, and then accelerating back up that same axis, as

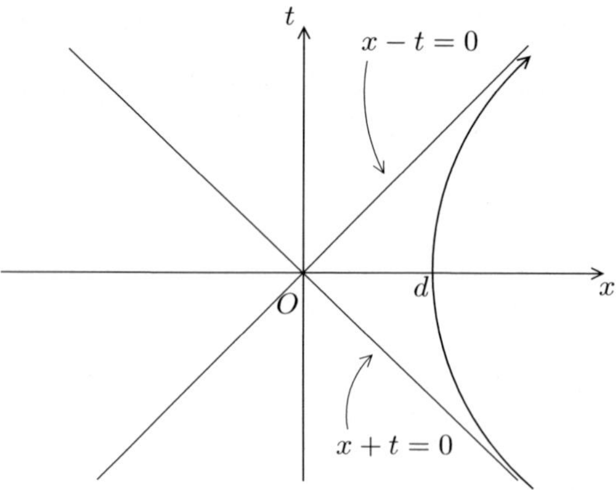

Fig. 1.1 A charged particle arrives from large positive x (*bottom right of diagram*), slows down to a halt at $x = d$ (in this frame), then accelerates back up the x axis. The worldline is asymptotic to the null cones at the origin, i.e., it is asymptotic to $x + t = 0$ for large negative times, and $x - t = 0$ for large positive times. Naturally, it never actually reaches the speed of light

shown in Fig. 1.1. In fact, the worldline of the charge will look exactly the same to any inertial observer moving along the axis of motion of the charge. Indeed, since uniformly accelerated motion leads to a situation where the charge has all possible speeds at different times throughout the motion, in any inertial frame S, there is an instant of time when the charge is at rest in that frame. It turns out that the equation for the worldline will be the same for all frames S' that can be reached from S by a Lorentz transformation, provided the spacetime origin is suitably chosen. This is one very special property of this kind of motion.

Now in any such inertial frame, the charge worldline will be asymptotic to the past and future null cones through the spacetime origin. If the charge comes down the x axis to some value $x = d$ where it stops before accelerating back out along the same axis, then Born got the usually accepted (Lienard–Wiechert) solution everywhere in the region $x + t > 0$. Later Schott [8] added the obvious restriction that the fields should be zero in the region $x + t < 0$, but kept Born's fields in the region $x + t > 0$. Maxwell's equations could not then be satisfied in the region $x + t = 0$ of the null cones.

The main part of Bondi and Gold's paper is thus concerned with obtaining a proper solution of Maxwell's equations, even in the delicate region $x + t = 0$, where one must include distributions (Dirac delta functions). Their method involves imagining a physical process and considering the idealistic uniformly accelerated motion as a limit. Like so many calculations connected with these questions, it is somewhat heavy and relegated to an appendix of their paper. Indeed, the details and even the method itself are not particularly enlightening.

But they did obtain what is now generally considered to be the right expression for the electromagnetic fields. So we may now turn to that old question mentioned at the beginning of this discussion: does such a uniformly accelerated charge radiate electromagnetic energy?

Wolfgang Pauli said it would not [9] back in 1920. Using Born's original solution, he observed that at $t = 0$, when the charge is instantaneously at rest, the magnetic field is zero everywhere in the corresponding inertial frame. This means that the Poynting vector is zero everywhere in the charge rest frame at that instant of time. One then argues that a Lorentz transformation can reduce any point of the worldline to rest, whence the magnetic field and Poynting vector are always zero everywhere provided one keeps changing inertial frame, so that one is always using the inertial frame instantaneously comoving with the charge.

This can be read directly from Born's original solution in the region where it is correct, but it is no longer valid when the fields are fixed up on the null cones. In Bondi and Gold's view, all the travelling energy at $t = 0$ is due to the delta function. But it is not difficult to see another weakness in Pauli's argument. In any given inertial frame the Poynting vector is going to change from zero as time goes by. Is there any relevance in what one would observe by continually changing inertial frame? On the other hand, we shall see later that, in a well known non-inertial frame adapted to the motion of the charge in a certain well-defined sense, the magnetic field components are indeed identically zero (see p. 231). This too has kept the debate alive.

There is another reason for thinking that a uniformly accelerating charge will not radiate, which is not mentioned by Bondi and Gold. This is the question of the radiation reaction. When a charged particle does radiate, one expects a reaction force on it, just as one would in the case of an emitting radio antenna. This reaction force, which will be discussed in more detail later, can be obtained either by assuming that the charge is spatially extended and doing a self-force calculation (see p. 110), or by treating it as a mathematical point and carrying out integrations of energy and momentum transfer across a tube containing the worldline and invoking energy and momentum conservation for the system as a whole. Within their limitations, the two methods agree on what the radiation reaction force should be, and if they are applicable to the uniformly accelerating charge, they both give a value of zero! This is discussed further in later chapters.

In any case, the conclusion in [2] was that, due to the presence of the delta function terms on the null cone, there would be radiation from the charge. So much for the older problem, which was clearly still evolving at the time of Bondi and Gold's contribution. But now, what of the equivalence principle? Let us quote from [2]:

> The discussion given in the present paper has an interesting application to a problem in the theory of gravitation. The principle of equivalence states that it is impossible to distinguish between the action on a particle of matter of a constant acceleration or of static support in a gravitational field (Einstein 1923). This might be thought to raise a paradox when a charged particle, statically supported in a gravitational field, is considered, for it might be thought that a radiation field is required to assure that no distinction can be made between the cases of gravitation and acceleration. But if the charge had an associated radiation field, this would reach distant regions unaffected by the gravitating body, where Maxwell's equations apply without any modification due to gravitational effects. According to these equations there

can be no static radiation field; and as the whole system is static the electromagnetic field
cannot depend upon time.

The Einstein citation is [10]. We are clearly referring to general relativity theory
here. This means that there has been a certain intellectual leap from the case of
a charge uniformly accelerating in Minkowski spacetime to one that is statically
supported in a gravitational field. The next chapter spells out the details of this
intellectual leap, before returning to the issues raised in the quote.

Chapter 2
From Minkowski Spacetime
to General Relativity

The aim here is to see how one can infer something about a statically supported charge in a gravitational field from the description, entirely within the special theory of relativity, of a uniformly accelerating charge in a spacetime with no gravitational fields. This will be an opportunity to introduce some of the mathematical machinery required to carry out calculations in this area. Section 2.1 describes coordinate systems that might be adopted by accelerating observers in flat spacetimes. Section 2.2 then proves that only when the observer has an eternally uniformly accelerating motion will the metric components in these coordinate systems have a static form. But the key part of the chapter is Sect. 2.3, which discusses different versions of the weak equivalence principle.

Section 2.4 is a detailed review of the weak field approximation usually used to make a connection between general relativity and Newtonian theory as an approximation to it. The aim here is to examine how one might interpret the GR view of a static and homogeneous gravitational field. Finally, Sect. 2.5 gives a critical review of the geodesic principle, which says that a particle left to its own devices in a curved spacetime, i.e., not subject to any external fields, will follow a geodesic. We discuss the demonstration of the latter 'principle' as a theorem of general relativity, and its limitations when the particle in question is not a point particle.

2.1 Semi-Euclidean Coordinate Systems

We begin therefore in Minkowski spacetime. We introduce a little extra generality by considering an accelerating observer AO with arbitrary acceleration along a straight line, specialising to the case of uniform acceleration later. This will allow us to bring out some of the particularities of the special case.

It is well known [11, p. 133] that an observer like AO can find well-adapted coordinates y^μ with the following properties (where the Latin index runs over $\{1, 2, 3\}$):

- First of all, any curve with all three y^i constant is timelike and any curve with y^0 constant is spacelike.

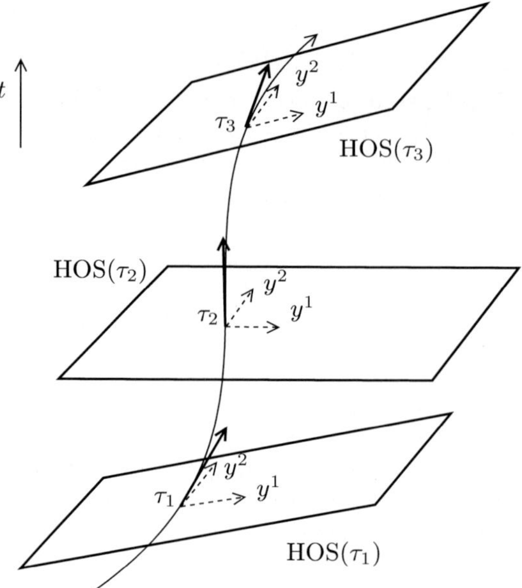

Fig. 2.1 Constructing a semi-Euclidean (SE) frame for an accelerating observer. View from an inertial frame with time coordinate t. The curve is the observer worldline given by (2.1). Three hyperplanes of simultaneity (HOS) are shown at three successive proper times τ_1, τ_2, and τ_3 of the observer. These hyperplanes of simultaneity are borrowed from the instantaneously comoving inertial observer, as are the coordinates y^1, y^2, and y^3 used to coordinatise them. Only two of the latter coordinates can be shown in the spacetime diagram

- At any point along the worldline of AO, the zero coordinate y^0 equals the proper time along that worldline.
- At each point of the worldline of AO, curves with constant y^0 which intersect it are orthogonal to it where they intersect it.
- The metric has the Minkowski form along the worldline of AO.
- The coordinates y^i are Cartesian on every hypersurface of constant y^0.
- The equation for the worldline of AO has the form $y^i = 0$ for $i = 1, 2, 3$.

Such coordinates could be called semi-Euclidean.

Let us suppose that the worldline of the accelerating observer is given in inertial coordinates in the inertial frame $\mathscr{I}$ by

$$t = \sigma \,, \qquad x = x(\sigma) \,, \qquad \frac{\mathrm{d}x}{\mathrm{d}\sigma} = v(\sigma) \,, \tag{2.1}$$

$$\frac{\mathrm{d}^2 x}{\mathrm{d}\sigma^2} = a(\sigma) \,, \qquad y(\sigma) = 0 = z(\sigma) \,, \tag{2.2}$$

using the time t in $\mathscr{I}$ to parametrise. The proper time $\tau(\sigma)$ of AO is given by

$$\frac{\mathrm{d}\tau}{\mathrm{d}\sigma} = (1 - v^2/c^2)^{1/2} \,. \tag{2.3}$$

The coordinates y^μ are constructed on an open neighbourhood of the AO worldline as follows (see Fig. 2.1). For an event (t,x,y,z) not too far from the worldline, there is a unique value of τ and hence also the parameter σ such that the point lies in the hyperplane of simultaneity (HOS) of AO when its proper time is τ. This hyperplane of simultaneity is given by

$$t - \sigma(\tau) = \frac{v\big(\sigma(\tau)\big)}{c^2} \big[x - x\big(\sigma(\tau)\big)\big] \,, \tag{2.4}$$

which solves, for any x and t, to give $\sigma(\tau)(x,t)$.

The semi-Euclidean coordinates attributed to the event (t,x,y,z) are, for the time coordinate y^0, (c times) the proper time τ found from (2.4) and, for the spatial coordinates, the spatial coordinates of this event in an instantaneously comoving inertial frame at proper time τ of AO. In fact, every other event in this instantaneously comoving inertial frame is attributed the same time coordinate $y^0 = c\tau$ and the appropriate spatial coordinates borrowed from this frame. Of course, the HOS of AO at time τ is also the one borrowed from the instantaneously comoving inertial frame.

There is just one detail to get out of the way: there are many different instantaneously comoving inertial frames for a given τ, and there are even many different ways to choose these frames as a smooth function of τ as one moves along the AO worldline, rotating back and forth around various axes in the original inertial frame $\mathscr{I}$ as τ progresses. We choose a sequence with no rotation about any space axis in the local rest frame. It can always be done by solving the Fermi–Walker transport equations. The semi-Euclidean coordinates are then given by

$$\begin{cases} y^0 = c\tau \,, \\[2ex] y^1 = \dfrac{[x - x(\sigma)] - v(\sigma)(t - \sigma)}{\sqrt{1 - v^2/c^2}} \,, \\[2ex] y^2 = y \,, \\[2ex] y^3 = z \,, \end{cases} \tag{2.5}$$

where $\sigma = \sigma(t,x)$ is found from (2.4). The inverse transformation, from semi-Euclidean coordinates to inertial coordinates, is given by

$$\begin{cases} t = \sigma(y^0) + \dfrac{v(y^0)}{c^2} y^1 \left[1 - \dfrac{v(y^0)^2}{c^2}\right]^{-1/2} , \\[2ex] x = x(y^0) + y^1 \left[1 - \dfrac{v(y^0)^2}{c^2}\right]^{-1/2} , \\[2ex] y = y^2 , \\[2ex] z = y^3 , \end{cases} \tag{2.6}$$

where the function $\sigma(y^0)$ is just the expression relating inertial time to proper time for the accelerating observer, and the functions $x(y^0)$ and $v(y^0)$ should really be written $x(\sigma(y^0))$ and $v(\sigma(y^0))$, respectively.

The above relations are not very enlightening. They are only displayed to show that the idea of such coordinates can be made perfectly concrete. One calculates the metric components in this frame, viz.,

$$g_{00} = 1/g^{00} = \left[1 + \frac{a(\sigma)[x - x(\sigma)]/c^2}{1 - v(\sigma)^2/c^2}\right]^2 , \tag{2.7}$$

where $\sigma = \sigma(t,x)$ as found from (2.4), and

$$g_{i0} = 0 = g_{0i} , \qquad g_{ij} = -\delta_{ij} , \qquad i,j \in \{1,2,3\} , \tag{2.8}$$

and checks the list of requirements for the coordinates to be suitably adapted to the accelerating observer.

Although perfectly concrete, the coordinates are not perfectly explicit: the component g_{00} of the semi-Euclidean metric has been expressed in terms of the original inertial coordinates! To obtain explicit formulas, one needs to consider a specific motion $x(\sigma)$ of AO, the classic example being uniform acceleration:

$$x(\sigma) = \frac{c^2}{g} \left[\left(1 + \frac{g^2 \sigma^2}{c^2}\right)^{1/2} - 1\right] , \qquad t = \sigma , \tag{2.9}$$

where g is some constant with units of acceleration. This does not look like a constant acceleration in the inertial frame:

$$\frac{dx}{d\sigma} = \frac{g\sigma}{(1 + g^2\sigma^2/c^2)^{1/2}} , \qquad \frac{d^2x}{d\sigma^2} = \frac{g}{(1 + g^2\sigma^2/c^2)^{3/2}} . \tag{2.10}$$

However, the 4-acceleration defined in the inertial frame $\mathscr{I}$ by

$$a^\mu = \frac{d^2 x^\mu}{d\tau^2} , \tag{2.11}$$

where τ is the proper time, has constant magnitude. It turns out that

$$a^2 := a_\mu a^\mu = -g^2 ,$$

with a suitable convention for the signature of the metric.

In this case, the transformation from inertial to semi-Euclidean coordinates is

$$y^0 = \frac{c^2}{g} \tanh^{-1} \frac{ct}{x + c^2/g} , \tag{2.12}$$

$$y^1 = \left[\left(x + \frac{c^2}{g} \right)^2 - c^2 t^2 \right]^{1/2} - \frac{c^2}{g} , \qquad y^2 = y , \qquad y^3 = z , \tag{2.13}$$

and the inverse transformation is

$$t = \frac{c}{g} \sinh \frac{g y^0}{c^2} + \frac{y^1}{c} \sinh \frac{g y^0}{c^2} , \tag{2.14}$$

$$x = \frac{c^2}{g} \left(\cosh \frac{g y^0}{c^2} - 1 \right) + y^1 \cosh \frac{g y^0}{c^2} , \qquad y = y^2 , \qquad z = y^3 . \tag{2.15}$$

One finds the metric components to be

$$g_{00} = \left(1 + \frac{g y^1}{c^2} \right)^2 , \qquad g_{0i} = 0 = g_{i0} , \qquad g_{ij} = -\delta_{ij} , \tag{2.16}$$

for $i, j \in \{1, 2, 3\}$, in the semi-Euclidean frame. Interestingly, this metric is static, i.e., g_{00} is independent of y^0. It is the only semi-Euclidean metric that is, and this is one of the theoretical particularities of uniform acceleration.

Before proving this, it is worth pausing to wonder why AO should adopt semi-Euclidean coordinates. It must be comforting to attribute one's own proper time to events that appear simultaneous. But what events are simultaneous with AO? In the above construction, AO borrows the hyperplane of simultaneity of an inertially moving observer, who does not have the same motion at all. AO also borrows the lengths of this inertially moving observer. It is not obvious that an accelerating observer would set up such coordinates, or indeed that they have any significance whatever! A well-informed observer, who knew the theory of relativity, might well prefer to use the globally Minkowskian frame that happens to be available here, even though it is not adapted to her motion.

2.2 The SE Metric for Uniform Acceleration
Is the Only Static SE Metric

We shall now show that a semi-Euclidean frame is static iff the acceleration of the observer is uniform in the sense of having constant pseudolength. We already have this in one direction: we have shown that the semi-Euclidean frame is static if the 4-acceleration is uniform. We now examine the metric for a general acceleration in one dimension and show that the condition for it to be static implies that the acceleration is uniform.

For this exercise, we shall need the partial derivatives

$$c\frac{\partial x}{\partial y^0} = \frac{v(\sigma)}{(1 - v^2/c^2)^{1/2}} + \frac{y^1 a(\sigma)v(\sigma)/c^2}{(1 - v^2/c^2)^2} \,, \tag{2.17}$$

$$\frac{\partial x}{\partial y^1} = \left[1 - \frac{v(\sigma)^2}{c^2}\right]^{-1/2} \,, \tag{2.18}$$

$$c\frac{\partial t}{\partial y^0} = \frac{1}{(1 - v^2/c^2)^{1/2}} + \frac{a(\sigma)y^1}{c^2(1 - v^2/c^2)^2} \,, \tag{2.19}$$

and

$$\frac{\partial t}{\partial y^1} = \frac{v(\sigma)}{c^2}\left[1 - \frac{v(\sigma)^2}{c^2}\right]^{-1/2} \,, \tag{2.20}$$

where $\sigma = \sigma(y^0)$, giving expressions entirely in terms of semi-Euclidean coordinates. These are obtained by differentiating (2.6) and using

$$\frac{\mathrm{d}\sigma(y^0)}{\mathrm{d}y^0} = \frac{1}{c(1 - v^2/c^2)^{1/2}} \,. \tag{2.21}$$

Recall from (2.7) that the metric components are

$$g_{00} = 1/g^{00} = \left[1 + \frac{a(\sigma)[x - x(\sigma)]/c^2}{1 - v^2/c^2}\right]^2 \,, \tag{2.22}$$

$$g_{01} = g_{10} = 0 \,, \qquad g_{11} = -1 \,, \tag{2.23}$$

with the rest also diagonal, and in the Minkowski way. So the semi-Euclidean metric is static iff

$$\frac{\partial}{\partial y^0}\left\{\frac{a(\sigma)[x - x(\sigma)]/c^2}{1 - v^2/c^2}\right\} = 0 \,. \tag{2.24}$$

Note that x is one of the inertial coordinates, while σ refers to $\sigma(\tau)(x,t)$, the solution of

$$t - \sigma(\tau) = \frac{v\big(\sigma(\tau)\big)}{c^2}\big[x - x(\sigma(\tau))\big]\,, \tag{2.25}$$

giving the HOS through the field point (x,t). Note, for example, that $x\big(\sigma(\tau)\big)$ is not just x.

The aim now is to reexpress (2.24) as a condition on the trajectory $x(\sigma)$ of the observer, and show that the corresponding worldline is uniformly accelerated. To begin with,

$$\frac{\partial}{\partial y^0}\left\{\frac{a(\sigma)[x - x(\sigma)]}{c^2 - v^2}\right\} = \frac{\mathrm{d}}{\mathrm{d}\sigma}\left\{\frac{a(\sigma)[x - x(\sigma)]}{c^2 - v^2}\right\}\frac{\partial\sigma}{\partial y^0} + \frac{a(\sigma)}{c^2 - v^2}\frac{\partial x}{\partial y^0}\,. \tag{2.26}$$

Now we have both $\partial x/\partial y^0$ and $\partial t/\partial y^0$ from (2.17) and (2.19). Note that, by (2.6),

$$y^1 = [x - x(\sigma)](1 - v^2/c^2)^{1/2}\,, \tag{2.27}$$

whence

$$\frac{\partial x}{\partial y^0} = \frac{v(\sigma)}{(c^2 - v^2)^{1/2}} + \frac{a(\sigma)v(\sigma)}{c^3(1 - v^2/c^2)^2}[x - x(\sigma)](1 - v^2/c^2)^{1/2}$$

$$= \frac{v(\sigma)}{(c^2 - v^2)^{1/2}}\left\{1 + \frac{a(\sigma)[x - x(\sigma)]}{c^2 - v^2}\right\}. \tag{2.28}$$

Furthermore, from the fact that $\sigma(x,t) = \sigma(y^0)$ and (2.21), we know that, when we put $x(y^0,y^1)$ and $t(y^0,y^1)$ into $\sigma(x,t)$, we get $\sigma(y^0)$, the function giving the Minkowski time corresponding to the proper time y^0. The latter has no dependence on y^1, and (2.21) gives

$$\frac{\mathrm{d}\sigma(y^0)}{\mathrm{d}y^0} = \frac{1}{(c^2 - v^2)^{1/2}}\,. \tag{2.29}$$

We can now proceed from (2.26), noting that

$$\frac{\mathrm{d}}{\mathrm{d}\sigma}\left\{\frac{a(\sigma)[x - x(\sigma)]}{c^2 - v^2}\right\} = \frac{a'(\sigma)[x - x(\sigma)]}{c^2 - v^2} - \frac{a(\sigma)v(\sigma)}{c^2 - v^2} + \frac{2a^2 v[x - x(\sigma)]}{(c^2 - v^2)^2}\,, \tag{2.30}$$

whence (2.24) is equivalent to

$$\left\{\frac{a'(\sigma)[x - x(\sigma)]}{c^2 - v^2} - \frac{a(\sigma)v(\sigma)}{c^2 - v^2} + \frac{2a^2 v[x - x(\sigma)]}{(c^2 - v^2)^2}\right\}\frac{1}{(c^2 - v^2)^{1/2}}$$

$$+ \frac{a(\sigma)}{c^2 - v^2}\left\{1 + \frac{a(\sigma)[x - x(\sigma)]}{c^2 - v^2}\right\}\frac{v(\sigma)}{(c^2 - v^2)^{1/2}} = 0\,.$$

There is now some cancellation. We find that

$$\left[a'(\sigma) + \frac{3a(\sigma)^2 v(\sigma)}{c^2 - v^2} \right] [x - x(\sigma)] = 0 \,, \tag{2.31}$$

and since we are not allowing $x = x(\sigma)$ as an interesting solution, we conclude that

$$\text{metric static} \quad \Longleftrightarrow \quad a'(\sigma) + \frac{3a(\sigma)^2 v(\sigma)}{c^2 - v^2} = 0 \,. \tag{2.32}$$

The next step is to look at the solutions to this equation.

As an aside, it is interesting to rederive (2.29) by this approach. We begin with the relation (2.25) that defines $\sigma(\tau)(x,t)$, thinking of it when x and t are written in their turn as functions of y^0 and y^1. Differentiating both sides with respect to y^0, we obtain

$$\frac{\partial t}{\partial y^0} - \frac{\partial \sigma}{\partial y^0} = \frac{a}{c^2} \frac{\partial \sigma}{\partial y^0} \big[x - x(\sigma(\tau))\big] + \frac{v(\sigma(\tau))}{c^2} \left[\frac{\partial x}{\partial y^0} - v(\sigma(\tau)) \frac{\partial \sigma}{\partial y^0} \right] . \tag{2.33}$$

We solve this for $\partial \sigma / \partial y^0$, obtaining

$$\frac{\partial \sigma}{\partial y^0} = \frac{\dfrac{\partial t}{\partial y^0} - \dfrac{v(\sigma)}{c^2} \dfrac{\partial x}{\partial y^0}}{1 - \dfrac{v^2}{c^2} + \dfrac{a[x - x(\sigma)]}{c^2}} \,. \tag{2.34}$$

Likewise, by the same algebra,

$$\frac{\partial \sigma}{\partial y^1} = \frac{\dfrac{\partial t}{\partial y^1} - \dfrac{v(\sigma)}{c^2} \dfrac{\partial x}{\partial y^1}}{1 - \dfrac{v^2}{c^2} + \dfrac{a[x - x(\sigma)]}{c^2}} \,. \tag{2.35}$$

Substituting (2.17) and (2.19) into (2.34) and doing a little algebra, we do indeed find the result (2.29) above, written here with the partial derivative, viz.,

$$\frac{\partial \sigma(y^0)}{\partial y^0} = \frac{1}{(c^2 - v^2)^{1/2}} \,. \tag{2.36}$$

But even more convincingly, if we substitute (2.18) and (2.20) of p. 10, viz.,

$$\frac{\partial x}{\partial y^1} = \left[1 - \frac{v(\sigma)^2}{c^2} \right]^{-1/2} , \tag{2.37}$$

and

$$\frac{\partial t}{\partial y^1} = \frac{v(\sigma)}{c^2}\left[1 - \frac{v(\sigma)^2}{c^2}\right]^{-1/2} ,$$

(2.38)

into (2.35), we obtain zero! This is just because

$$\frac{\partial t}{\partial y^1} = \frac{v(\sigma)}{c^2}\frac{\partial x}{\partial y^1} .$$

Solving the Equation for a Static Metric

We now return to (2.32) and consider solutions. It is instructive to begin by checking that the uniform acceleration does solve the equation, since the calculation is not quite as simple as we might expect. We have the formulas for such an acceleration on p. 8, viz.,

$$v(\sigma) = \frac{g\sigma}{(1+g^2\sigma^2/c^2)^{1/2}} , \qquad a(\sigma) = \frac{g}{(1+g^2\sigma^2/c^2)^{3/2}} ,$$

(2.39)

whence also

$$a'(\sigma) = -\frac{3g^3\sigma/c^2}{(1+g^2\sigma^2/c^2)^{5/2}} .$$

(2.40)

Then

$$\frac{3a(\sigma)^2 v(\sigma)}{c^2 - v^2} = \frac{3g^2}{(1+g^2\sigma^2/c^2)^3}\frac{g\sigma}{(1+g^2\sigma^2/c^2)^{1/2}}\frac{1+g^2\sigma^2/c^2}{c^2}$$

$$= \frac{3g^3\sigma/c^2}{(1+g^2\sigma^2/c^2)^{5/2}} .$$

Equation (2.32) is therefore satisfied, which we knew it should be, and that is encouraging.

But are there other solutions? The last calculation can guide us intuitively. We observe that

$$\frac{d}{d\sigma}\left\{\frac{a(\sigma)}{[1-v(\sigma)^2/c^2]^{3/2}}\right\} = \frac{1}{[1-v(\sigma)^2/c^2]^{3/2}}\left[a'(\sigma) + \frac{3a(\sigma)^2 v(\sigma)}{c^2 - v^2}\right] ,$$

so that

$$\text{metric static} \quad \Longleftrightarrow \quad \frac{d}{d\sigma}\left\{\frac{a(\sigma)}{[1-v(\sigma)^2/c^2]^{3/2}}\right\} = 0 .$$

(2.41)

Put another way,

$$\text{metric static} \quad \Longleftrightarrow \quad \frac{a(\sigma)}{[1-v(\sigma)^2/c^2]^{3/2}} = \kappa\,, \tag{2.42}$$

for some constant κ. Once again, we can check that this is satisfied by the uniform acceleration. We note that

$$1 - \frac{v(\sigma)^2}{c^2} = 1 - \frac{g^2\sigma^2/c^2}{1+g^2\sigma^2/c^2} = \frac{1}{1+g^2\sigma^2/c^2}\,,$$

so that the above uniform acceleration satisfies the condition iff $\kappa = g$.

We can completely solve the above differential equation, which is rather surprising in itself. We have

$$\frac{\mathrm{d}v}{[1-v(\sigma)^2/c^2]^{3/2}} = \kappa\,\mathrm{d}\sigma\,.$$

The left-hand side is integrated by first substituting $v = c\sin\theta$, so that

$$\mathrm{d}v = c\cos\theta\,\mathrm{d}\theta\,,$$

and

$$\int \frac{\mathrm{d}v}{[1-v(\sigma)^2/c^2]^{3/2}} = c\int \frac{\mathrm{d}\theta}{\cos^2\theta}$$

$$= c\int (1+\tan^2\theta)\,\mathrm{d}\theta$$

$$= c\tan\theta\,.$$

Of course,

$$\tan\theta = \frac{\sin\theta}{\cos\theta} = \frac{v/c}{(1-v^2/c^2)^{1/2}}\,,$$

and the final, most general solution is

$$\kappa\sigma + \kappa' = \frac{v}{(1-v^2/c^2)^{1/2}}\,, \tag{2.43}$$

where κ and κ' are constants. We can rearrange this to obtain $v(\sigma)$. After a little manipulation,

$$v(\sigma) = \frac{\kappa\sigma + \kappa'}{\left[1+(\kappa\sigma+\kappa')^2/c^2\right]^{1/2}}\,. \tag{2.44}$$

This is to be compared with the uniform acceleration

$$v(\sigma) = \frac{g\sigma}{(1+g^2\sigma^2/c^2)^{1/2}}\,,$$

which is the solution when $\kappa = g$ and $\kappa' = 0$. Since κ' in (2.44) is just a readjustment of the zero of the variable σ, we see that all solutions of the static metric equation are in fact uniform accelerations, which is what had to be proved.

2.3 The Step to General Relativity

So what does all this tell us about gravitational fields? Suppose we are doing general relativity and we are given a spacetime with coordinates $\{y^\mu\}$ and interval

$$\mathrm{d}s^2 = \left(1 + \frac{gy^1}{c^2}\right)^2 (\mathrm{d}y^0)^2 - (\mathrm{d}y^1)^2 - (\mathrm{d}y^2)^2 - (\mathrm{d}y^3)^2 \, , \qquad (2.45)$$

where g is a constant. We might begin by calculating the connection coefficients. They turn out to be fairly simple functions of the coordinates. In fact using the usual formula

$$\Gamma^\mu_{\sigma\tau} = \frac{1}{2}g^{\mu\nu}\left(g_{\nu\sigma,\tau} + g_{\nu\tau,\sigma} - g_{\sigma\tau,\nu}\right) \qquad (2.46)$$

for the Levi-Civita connection coefficients, where commas denote ordinary coordinate derivatives, we find that all the coefficients are zero except

$$\Gamma^1_{00} = \frac{g}{c^2}\left(1 + \frac{gy^1}{c^2}\right) \, , \qquad \Gamma^0_{01} = \Gamma^0_{10} = \frac{g}{c^2}\left(1 + \frac{gy^1}{c^2}\right)^{-1} \, . \qquad (2.47)$$

We would then calculate the Riemann curvature tensor, finding it to be identically zero. This would immediately tell us that we were dealing with some non-inertial coordinatisation of a flat spacetime, in which there were no tidal effects.

One ought to point out a somewhat delicate matter regarding the above interval, which comes from the metric (2.16): when $y^1 = -c^2/g$, the metric is degenerate because $g_{00} = 0$, and g^{00} is not defined. However, one would soon observe that this was due to the choice of coordinates, rather like the singularity at the event horizon in the Schwarzschild metric. Such singularities can be removed by a better choice of coordinates. In the present case, one would transform by (2.14) and (2.15) and the metric would reduce to the Minkowski form η_{ij} of (1.1).

According to the usual rules of general relativity, one would then argue as follows: the coordinates (t,x,y,z) are those that would be set up in a freely falling frame, in which the connection coefficients are zero and freely falling bodies not subjected to any forces follow straight lines. There are two things to note:

- In a general, curved spacetime, one can always find such coordinates locally in the neighbourhood of any given event. This is an application of what is sometimes called the weak equivalence principle WEP and it is inherent in the manifold formulation of general relativity when one adopts the Levi-Civita connection (zero torsion). In that case, for any event (point) P in spacetime, there is always a choice of coordinates in some neighbourhood of that event for which the connection coefficients are zero at P and the metric takes the Minkowski form at P. By continuity, the connection remains close to zero and the metric close to η over some small enough neighbourhood of P.

- In general relativity, one adopts the idea that free fall is not due to a force, i.e., gravity is no longer a force in this view. This is a linguistic adjustment made to accord with the idea that forces and accelerations are intimately linked. In general relativity, freely falling bodies follow geodesics (with some provisos, as explained in Sect. 2.5), and the geodesic equation says precisely that their four-acceleration is zero.

We have already encountered our first equivalence principle here, and it is so deeply embedded in the theory that it would be disastrous if the theory itself contradicted it. Fortunately, this does not happen, at least, not on the basis of the discussion in [2]. For one thing, the equivalence principle cited in the quotation from Bondi and Gold on p. 3 refers to electromagnetic radiation fields, whereas the weak equivalence principle WEP mentioned here does not yet have anything to say about electromagnetism.

Note also that, since the proof concerning the existence of this convenient form of the metric and connection coefficients only requires one to assume that the connection is the usual Levi-Civita connection entirely determined by the metric, and this follows if one assumes zero torsion, i.e., that the connection coefficients are symmetric in their two lower indices, one might say that this equivalence principle hardly looks like a principle at all: the whole structure of the theory as it is usually formulated implies the weak principle of equivalence. On the other hand, one might also say that it is precisely the prerogative of a principle to be inescapable.

So what would the general relativist deduce from the discovery of a global inertial frame? There would appear to be several possibilities, all of which are equivalent as far as general relativity is concerned. One might say that one was in a perfectly flat region of spacetime far from any source of curvature (gravity), and that the coordinates $\{y^\mu\}$ were simply those adopted by a uniformly accelerating observer clever enough to set up a system so well adapted to her motion. This is what we were considering before. But one might also say that the coordinates (t,x,y,z) are those of an observer freely falling in a static and homogeneous gravitational field (SHGF). The fact that there are no tidal effects, i.e., the curvature is zero, is what inspires us to say that the field is static and homogeneous.

Presumably, in the second case, if the observer looked around, she would find some source, i.e., some distribution of energy (e.g., mass energy) to which one could attribute the presence of this SHGF. It is hard to imagine what it might be and we shall discuss this below (see Sect. 2.4). However, a practically-minded observer with knowledge of physics would probably just say that, in reality, the curvature is not quite zero but that this is a good approximation over some region of spacetime, of the kind usually made in Earth-based laboratories, given the accuracy with which measurements can be made.

But the Earth-based observer would presumably have a considerable advantage over one who was simply presented with the interval (2.45) and could not see the source. In the laboratory, one can measure the acceleration relative to the floor, for example, assuming that it is at rest in some sense (or can be treated as being at rest) relative to the gravitational source. At some risk of confusion, let us call this the acceleration due to gravity, but bearing in mind that there is no acceleration due to gravity in general relativity (there is only acceleration when one is not allowed to free-fall). Without sight of the gravitational source or other reference point, our relativist in possession of the above interval could not even determine this acceleration due to gravity. Looking at the expression in (2.45), one might want to say that it was equal to g, because the interval does indeed single out this value. But in a certain sense, the value g is just an artefact of the choice of coordinates.

One should have no doubt about this. Starting with the Minkowski frame, one could have chosen any uniformly accelerating observer, with value g' say, and obtained new coordinates $\{y'^\mu\}$ such that the interval takes the form

$$ds^2 = \left(1 + \frac{g'y'^1}{c^2}\right)^2 (dy'^0)^2 - (dy'^1)^2 - (dy'^2)^2 - (dy'^3)^2 \ . \tag{2.48}$$

The fact is, of course, that in this specific context, one cannot say how much of the acceleration is due to gravity and how much is due to some other effect, e.g., the kind of accelerating effect we were presumably imagining at the outset when we considered a uniformly accelerating body in a flat spacetime (with no gravity).

One might appeal to the well known weak field approximation in which the theory begins to look like Newton's theory. Could this give us some reason to think that g rather than g' was the acceleration due to gravity in a given context? We shall discuss this in some detail in Sect. 2.4, to bring out another rather subtle point, but let us just see globally what is involved. One writes the metric in the form

$$g_{\mu\nu} = \eta_{\mu\nu} + h_{\mu\nu} \ ,$$

where $h_{\mu\nu}$ is assumed to be small. A range of other assumptions is usually made about the metric components and allowed coordinate transformations which maintain this range of assumptions, but there is no need to go into too much detail here. In the case of (2.45), one reads off

$$h_{00} \approx 2gy^1/c^2 \ , \tag{2.49}$$

in a region where $gy^1/c^2 \ll 1$. One then considers the geodesic equation giving the worldline of a freely falling object relative to these coordinates and concludes that, if the motion is indeed entirely due to a gravitational potential Φ, then one must have

$$h_{00} = 2\Phi/c^2 \ . \tag{2.50}$$

This would give

$$\Phi \approx gy^1 \ . \tag{2.51}$$

In this Newtonian view, the potential causes a force per unit mass of

$$(-\Phi_{,1}, -\Phi_{,2}, -\Phi_{,3}) = (-g, 0, 0) \,, \tag{2.52}$$

so that we are indeed explaining the quantity g appearing in the metric by the effects of gravity, on this interpretation.

There is no surprise here, but of course the metric components (2.48) would give us a gravitational potential Φ' such that

$$(-\Phi'_{,1}, -\Phi'_{,2}, -\Phi'_{,3}) = (-g', 0, 0) \,, \tag{2.53}$$

if we could assume that the motion were indeed entirely due to a gravitational potential Φ'. And this is precisely what we do not know: how much of g or g' comes from a gravitational effect?

This highlights a fundamental point about general relativity which might be stated as a form of the equivalence principle WEP2: from local observations alone, one cannot tell whether the fields one observes via the spacetime metric are due to one's own motion or a gravitational potential, or both, and in the latter case, in what proportions. The word 'local' is essential here, for two reasons:

- If one could look around, one would presumably see any sources of gravitational potential operating in the neighbourhood. One could then apply Einstein's equations to try to understand the metric and one could perhaps try to measure the 'acceleration due to gravity' by measuring a rate of change of proper distance to the gravitational source for objects in free fall.
- An acceleration of the observer which changes the metric components via an appropriate choice of the observer's coordinates cannot introduce curvature, so if there is curvature, one can attribute it to a gravitational potential. However, one must be able to survey a large enough region of spacetime to be able to determine the curvature, if there is any.

Considering the second point, it might seem that data from any neighbourhood of a given event, no matter how small, would suffice to establish the curvature if there were any. The reason why a certain size of neighbourhood is required is just that one is always restricted by measurement accuracy. For any given level of measurement accuracy, supposing that one has arranged by coordinate choice for the metric to be Minkowskian and the connection components zero at some given point, there will be some neighbourhood of the event in which the metric appears to be Minkowskian and the connection components zero. More about this in a moment.

Referring to the first of the above points, if one has the distribution of energy and momentum relative to some coordinate system in the spacetime, one then knows the metric, connection, and curvature relative to those coordinates by solving Einstein's equations (at least insofar as they and the relevant boundary conditions are determinate). On this basis one would then like to say which aspects of a motion were due to gravity and which were an artefact of the coordinates. For example, within the context of the weak field approximation, one can consider the effects of a massive point particle, with energy–momentum distribution

$$T^{00} = M\delta(\mathbf{x}) \,, \tag{2.54}$$

where M is its mass and δ is a point-support distribution on the spacelike hypersurface with coordinates $\mathbf{x}$, whence Einstein's equations reduce to something like

$$h_{00} = -\frac{2MG}{|\mathbf{x}|} \,,$$

and one deduces the form of Φ, viz.,

$$\Phi = -\frac{GM}{|\mathbf{x}|} \,.$$

This is discussed more carefully in Sect. 2.4. The present scenario would appear to escape somewhat from the kind of constraint imposed by Einstein's equations. For one thing, it is not obvious that any energy–momentum distribution would lead via Einstein's equations to the static, homogeneous gravitational field proposed to explain the intervals (2.45) or (2.48).

Of course, as mentioned above, one can easily imagine an approximate context, and this context is precisely the one in which the neighbourhood over which one can carry out measurements of the gravitational potential as embodied by the metric is too small to allow one to detect the curvature. This is what allows the above version WEP2 of the weak equivalence principle: if one is restricted to a small enough neighbourhood, where 'small enough' is dictated by one's measurement accuracy, then one is forced into the situation of a spacetime that could be attributed the intervals (2.45) or (2.48).

This follows from the weak equivalence principle WEP first mentioned on p. 15. At a given event in spacetime, one arranges for the metric to have Minkowski form and the Levi-Civita connection components to be zero. But the latter are simply related to the coordinate derivatives of the metric components and the fact that they are zero at the given event means that the coordinate derivatives of the metric components are also zero there. This in turn means that the metric components relative to these coordinates will change very slowly from the Minkowski form in any small neighbourhood of the event. However, it should be noted that the coordinate derivatives of the connection components, which depend on the curvature components, are not generally zero, so the connection components may be changing rapidly from zero over any neighbourhood of the event.

Since the above statement of WEP2 refers to an acceleration of the observer which changes the metric components via an appropriate choice of the observer's coordinates, one ought to say more about what might be an appropriate choice of the observer's coordinates. We already encountered this question at the beginning of this chapter for an observer accelerating in a flat spacetime. With coordinates satisfying the conditions listed on p. 5, the equation for a geodesic that happens to intersect the observer worldline at some event, i.e., describing a free-falling object encountered by the observer, takes the form [11, p. 135].

$$\frac{\mathrm{d}^2 y^i}{\mathrm{d}y^{0\,2}} + \underset{\substack{\text{inertial}\\ \text{force}}}{a^i} + \underset{\substack{\text{Coriolis}\\ \text{force}}}{2\Omega^i{}_j \frac{\mathrm{d}y^j}{\mathrm{d}y^0}} - \underset{\substack{\text{relativistic}\\ \text{correction}}}{a^j \frac{\mathrm{d}y^j}{\mathrm{d}y^0}\frac{\mathrm{d}y^i}{\mathrm{d}y^0}} = 0 , \tag{2.55}$$

with suitable interpretation of the terms, where a^i and $\Omega^i{}_j$ are determined by the connection coefficients according to

$$a^i := \Gamma^i_{00} , \qquad \Gamma^0_{0i} = \Gamma^0_{i0} = a^i \quad (i = 1,2,3) , \tag{2.56}$$

$$\Omega^i{}_j := \Gamma^i_{j0} = \Gamma^i_{0j} = -\Gamma^j_{0i} \quad (i,j = 1,2,3) . \tag{2.57}$$

Furthermore, one can arrange for the antisymmetric rotation matrix $\Omega^i{}_j$ to be zero, viz., $\Omega^i{}_j = 0$, for $i,j = 1,2,3$, by suitable choice of coordinates (basically, using Fermi–Walker transport). It is important to note that (2.55) holds exactly only on the observer worldline, although it will hold approximately nearby.

One can do something very similar even in a curved spacetime [11, p. 181]. An observer with quite arbitrary motion can find well-adapted coordinates y^μ with the following properties (where the Latin index runs over $\{1,2,3\}$):

- First of all, any curve with all three y^i constant is timelike and any curve with y^0 constant is spacelike.
- At any point along the observer worldline, the zero coordinate y^0 equals the proper time along that worldline.
- At each point of the observer worldline, curves with constant y^0 which intersect it are orthogonal to it where they intersect it.
- The metric has the Minkowski form along the observer worldline.
- Any purely spatial geodesic through the observer worldline on a hypersurface $y^0 = $ constant satisfies

$$\frac{\mathrm{d}^2 y^i}{\mathrm{d}s^2} = 0 , \tag{2.58}$$

  where s is an affine parameter for the geodesic.
- The equation for the observer worldline has the form $y^i = 0$ for $i = 1,2,3$.

Such coordinates are said to be normal. Note the difference with the flat spacetime case: the coordinates y^i are not necessarily Cartesian on every hypersurface of constant y^0 as they were on p. 5, but we have $\mathrm{d}^2 y^i / \mathrm{d}s^2 = 0$ for any purely spatial geodesic through the observer worldline on a hypersurface $y^0 = $ constant. Put another way, in a semi-Euclidean frame (for flat spacetime), the fifth condition (2.58) is true for any purely spatial geodesic, not just for spatial geodesics through the observer worldline.

Carrying out the above construction, one obtains precisely the same relations (2.55)–(2.57) for the worldline of any freely falling object where it encounters the observer worldline.

In the last two examples, the observer chose coordinates in such a way that she always remained at the origin of the spatial hypersurfaces given by keeping her time coordinate constant. In this situation, even though the connection components associated with the rotation matrix $\Omega^i{}_j$ can be made equal to zero, those related to a^i cannot. So with this appropriate choice of coordinates, said to be adapted to the observer's worldline, one cannot make the connection components equal to zero on the observer worldline, unless the latter happens to be a geodesic, i.e., unless the observer happens to be in free fall.

Of course, as mentioned before, the observer can chose any coordinates she likes in general relativity. In a flat spacetime, the best choice may well be some set of globally inertial coordinates, and in a curved spacetime, some set of locally inertial coordinates, even though the observer then has motion relative to those coordinates. But there is a reason for looking at the worldline of a freely falling object relative to the kind of coordinates an observer might consider to be adapted to her own worldline, and this is that it brings out the idea that an object can appear to be accelerating even when it is not, in particular, when it is the observer that is accelerating rather than the observed object. Note how the components a^i of the observer 4-acceleration become interpreted as the components of the inertial acceleration of the test particle relative to such coordinates.

This leads to the idea that one cannot then say how much of the motion of an object relative to one's coordinates can be attributed to gravitational effects and how much to one's choice of coordinates. Let us refer to this as WEP3, another contender for the post of equivalence principle. One is familiar with this idea from the Newtonian theory of gravity in the pre-relativistic context: because inertial mass and passive gravitational mass are equal (an experimental result, always open to further testing as techniques improve), an observer cannot tell whether the acceleration of a test particle relative to her Euclidean coordinate system is due to some gravitational field or due to her own acceleration and the consequent acceleration of her comoving Euclidean coordinate system. (The acceleration of that Euclidean system must not be rotational here.) In pre-relativistic theory, one has no difficulty setting up Euclidean coordinate systems in spatial hypersurfaces and time is the same for everyone. What becomes of WEP3 in GR?

The Newtonian idea of a test particle moving only due to gravitational forces (a motion called free fall) is replaced in GR by the principle that such a test particle will follow a timelike geodesic with the proper time of the particle as affine parameter. In fact this 'principle' can be deduced from Einstein's equations, a point which deserves its own short section at the end of this chapter. This so-called geodesic principle immediately builds in the idea that the purely gravitational motions of any two objects are going to be the same if they start at the same spacetime event and with the same 4-velocity, regardless of their inertial masses, or indeed any detail of their internal physical makeup. Their motions depend only on the geometrical structure of the spacetime as given by the metric (which itself determines the connection and curvature), and of course initial conditions. We shall see in Sect. 2.5 that this is only an approximation, a point to be confirmed in Chap. 12. Anyway, such motion is described as inertial, or non-accelerational, and herein lies a difficulty for the pre-relativistic equivalence principle mentioned in the last paragraph.

The problem is this. The pre-relativistic equivalence principle can compare a test particle that is freely falling under the effect of a gravitational force (and gravity is a force in this theory) with one that has the special property of moving inertially (and appears to accelerate because the observer is accelerating relative to the special inertial frames that exist in this theory). In this theory one does of course have a difficulty in saying why some frames should be inertial while others are not, but this is another matter. The point is that, apart from the freely falling locally inertial frames of GR, there are no other, distinct inertial frames. Put another way, if one were to say in GR that one could not distinguish the motion of a test particle subject only to gravitational effects from one with 'inertial acceleration', one would be committing tautology: the test particle with 'inertial acceleration' would presumably be one subject only to gravitational effects, i.e., with no acceleration, but appearing to accelerate as in equations like (2.55). One would be talking about the very same test particle on each side of the comparison.

So is there nothing left of WEP3 in general relativity? Could an accelerating observer (on a non-geodesic worldline) not at least mistake the motion of a freely falling test object as an acceleration by setting up the coordinates described above and obtaining an equation of motion like (2.55) for the worldline of the test object relative to her coordinates? If such were the case, one could then say that acceleration of the observer and 'gravitational acceleration' of a test particle were indistinguishable. This certainly looks to be the case. On the other hand, we have to ask whether the observer would not realise in some direct way that she was accelerating. For example, if she knew the metric, connection and curvature components relative to her adapted coordinates, and it suffices to know only the metric components over some neighbourhood in order to achieve this, would she not realise that she was accelerating?

Of course, she would. Although the metric components are zero actually on the observer worldline, they are not zero in any neighbourhood of the worldline, even if it happens to be a geodesic. Worse, the connection components, and hence the coordinate derivatives of the metric components, are not even zero on the worldline, unless it is a geodesic. So in the case of an accelerating observer, the metric will not generally have the Minkowski form in these coordinates as soon as one leaves the observer worldline. If the observer can measure the metric accurately enough, she will know that she is accelerating and the apparent acceleration of the freely falling object can be attributed to coordinate choice.

The point about the above tautology is that, in the pre-relativistic case one was comparing two different test particles, one undergoing gravitational acceleration and the other in inertial motion (subject to no forces), whereas in the GR version of WEP3 one considers just one test particle in free fall, i.e., unaccelerating, and notes that it may appear to be accelerating relative to suitable coordinates. So this idea has changed somewhat, but it survives. It becomes something like this: locally, an observer cannot tell whether a test particle is freely falling or accelerating. The word 'locally' is there to remind us that the observer must not be able to survey a sufficiently wide neighbourhood of her worldline to detect the fact that it is really herself who accelerates, if that be the case.

We need to consider one more claim, which is in fact made by Bondi and Gold [2]: that it is impossible to distinguish between the action on a particle of matter of a constant acceleration or of static support in a gravitational field. This looks very like WEP2, except that it mentions a test particle. WEP2 says that, in a flat spacetime, an observer cannot decide whether there is a non-tidal gravitational field or not. So if the observer is moving with the test particle, she will not be able to say whether the particle is supported, with her, in a gravitational field, or whether the particle is undergoing a uniform acceleration, with her, depending on whether there is a gravitational field or not, respectively. Of course, one could say more simply that, in the GR understanding of Minkowski spacetime, static support in an SHGF interpretation of the spacetime is in fact just a uniform acceleration. If this can be viewed as a version of the weak equivalence principle, let us label it WEP4.

The measurements here refer to determination of the metric. If the test particles one is observing are charged, one expects to find EM fields and we have not yet addressed the question as to whether these fields could tell us more. This is in part the subject of the present book. It should be noted that the above versions WEP2 and WEP3 of the weak equivalence principle are not as fundamental as WEP on p. 15. When we come to examine EM fields, there is no reason a priori to suppose that the statements WEP2 or WEP3 should extend to cover such effects.

In fact one should ask what really is the utility of these ideas. Once GR has been constructed with WEP as an immediate consequence of the theory, and deployed physically with the help of the strong equivalence principle SEP (discussed further in Chap. 4), what need has one for any other consequences they may have? This will become a more pressing point in Chap. 12 which discusses the worldlines that charged particles are expected to have when falling 'freely' in curved spacetimes.

But to return to the main issue of the present chapter, as a flat spacetime, the spacetime with interval (2.45) or (2.48) escapes totally from the constraints imposed by Einstein's equations: the gravitational effects one might attribute to this scenario are non-tidal, i.e., the curvature is in fact zero. Put another way, although there might be some energy–momentum distribution that would lead via Einstein's equations to the SHGF proposed to explain the intervals, the completely empty energy–momentum distribution will also lead to this solution.

So the main point to bear in mind as we proceed is this: in the spacetime described by (2.45) or (2.48), without seeing any sources for gravitational effects, an observer cannot say whether she is uniformly accelerating in a flat spacetime (and using semi-Euclidean coordinates) or sitting still relative to coordinates that describe a spacetime containing a static, homogeneous gravitational field, or even a mixture of both. Note that sitting still in coordinates that describe such a spacetime is indeed an accelerating motion according to general relativity because one does not have geodesic motion. One might say that an observer cannot distinguish whether she has a uniform acceleration in a flat spacetime or a stationary state in a static, homogeneous gravitational field. The proviso is of course that one cannot see other things in the neighbourhood, such as sources of gravitation, or possibly other things, to be discussed in this book, viz., electromagnetic fields.

Furthermore, in the context of the spacetime described by (2.45) or (2.48), the weak principle of equivalence WEP as stated on p. 15 plays a key role. This principle is a fundamental building block in the general theory of relativity (without torsion). To recapitulate, it states that, given any spacetime event, one can always find coordinates in some neighbourhood of that event for which the metric has approximately the Minkowski form and the connection coefficients are approximately zero. But in the spacetime discussed here, everything works out exactly, i.e., there are coordinates for which the metric is everywhere exactly the Minkowski metric and the connection coefficients are everywhere exactly zero. This means that there are not even any higher order effects to help one make the kind of distinction described above.

And it will be worth remembering for future discussion that the gravitational field described by (2.45) or (2.48), if there is one, is necessarily static and homogeneous, for there are no tidal effects. One would like to say that this precludes any spatiotemporal variation in the field. However, it turns out that this conclusion is not so obvious as one would think, and this will lead to more subtleties in Sect. 15.7.

2.4 Weak Field Approximation

In the last section, we mentioned the weak field approximation, because it provides a point of contact between theory and observation that might help one to analyse the motions of test particles. The reason this gets a further section here is that there is a point to be noted about the way coordinates, and objects expressed relative to coordinates, are often somewhat naively interpreted. There are two parts to this:

- The first is rather standard, obtaining the approximate relation (2.50) between the metric and a Newtonian gravitational potential.
- The second concerns the relation between gravitational source and metric via Einstein's equation in this approximate context.

The following is adapted from DeWitt's Stanford lectures [24].

We consider a situation in which coordinates, called quasi-canonical coordinates, can be found, defined by the property that

$$g_{\mu\nu} = \eta_{\mu\nu} + h_{\mu\nu}, \qquad \text{where } |h_{\mu\nu}| \lesssim \varepsilon \ll 1, \quad \forall \mu, \nu. \qquad (2.59)$$

Such a coordinate system is not unique. It remains quasi-canonical under rotations and low-velocity Lorentz boosts. (Note that the $\eta_{\mu\nu}$ term in the metric would be fixed even by high-velocity boosts, but the $h_{\mu\nu}$ might be made large by such a transformation.) It also remains quasi-canonical under general coordinate transformations $\bar{x}^{\mu} = x^{\mu} + \xi^{\mu}$ restricted only by the requirement

$$|\xi^{\mu}{}_{,v}| \lesssim \varepsilon \qquad \text{and} \qquad |h_{\mu v,\sigma}\xi^{\sigma}| \lesssim \varepsilon^2 , \qquad \forall \mu, v .$$ (2.60)

One then finds a new $\bar{h}_{\mu v}$ defined by $\bar{h}_{\mu v} := \bar{g}_{\mu v} - \eta_{\mu v}$, whence

$$\begin{aligned}
\eta_{\mu v} + \bar{h}_{\mu v} = \bar{g}_{\mu v} &= \frac{dx^{\sigma}}{d\bar{x}^{\mu}}\frac{dx^{\tau}}{d\bar{x}^{v}} g_{\sigma\tau} \\
&= (\delta^{\sigma}{}_{\mu} - \xi^{\sigma}{}_{,\mu})(\delta^{\tau}{}_{v} - \xi^{\tau}{}_{,v})(\eta_{\sigma\tau} + h_{\sigma\tau}) \\
&= \eta_{\mu v} + h_{\mu v} - \xi_{v,\mu} - \xi_{\mu,v} + O(\varepsilon^2) ,
\end{aligned}$$ (2.61)

using the conditions in (2.60). So $h_{\mu v}$ undergoes the transformation

$$\bar{h}_{\mu v} = h_{\mu v} - \xi_{\mu,v} - \xi_{v,\mu} , \qquad \xi_{\mu} := \eta_{\mu v}\xi^{v} ,$$ (2.62)

for such a change of coordinates. Note that in these approximations, one can always use the Minkowski metric to raise and lower indices.

The choice of quasi-canonical coordinate system can be restricted by imposing a supplementary condition

$$l^{\mu v}{}_{,v} = 0 ,$$ (2.63)

where

$$l_{\mu v} := h_{\mu v} - \frac{1}{2}\eta_{\mu v}h , \qquad h := h_{\sigma}{}^{\sigma} ,$$ (2.64)

$$h_{\mu v} = l_{\mu v} - \frac{1}{2}\eta_{\mu v}l , \qquad l := l_{\sigma}{}^{\sigma} = -h .$$ (2.65)

Now $l_{\mu v}$ transforms by

$$\begin{aligned}
\bar{l}_{\mu v} = \bar{h}_{\mu v} - \frac{1}{2}\eta_{\mu v}\bar{h} \\
= l_{\mu v} - \xi_{\mu,v} - \xi_{v,\mu} + \eta_{\mu v}\xi^{\sigma}{}_{,\sigma} .
\end{aligned}$$ (2.66)

It follows that

$$\begin{aligned}
\bar{l}^{\mu v}{}_{,v} &= l^{\mu v}{}_{,v} - \xi^{\mu}{}^{,v}{}_{v} - \xi^{v}{}^{,\mu}{}_{v} + \xi^{\sigma}{}_{,\sigma}{}^{\mu} \\
&= l^{\mu v}{}_{,v} - \xi^{\mu}{}^{,v}{}_{v} .
\end{aligned}$$

So $\bar{l}_{\mu v}$ can be made to satisfy the supplementary condition (2.63) by choosing ξ^{μ} as a solution of the wave equation with a source:

$$\Box^2\xi^{\mu} := \xi^{\mu}{}^{,v}{}_{v} = l^{\mu v}{}_{,v} .$$ (2.67)

The supplementary condition (2.63) does not fix the coordinate system completely, because it is still possible to carry out rotations, low-velocity Lorentz boosts, and coordinate transformations satisfying the homogeneous wave equation

$$\Box^2 \xi^\mu = 0 \,. \tag{2.68}$$

For gravitational sources with low velocities and when there is no gravitational radiation around, one can impose another condition on the above quasi-canonical coordinate systems. Indeed, one expects there to exist quasi-stationary coordinates. These are quasi-canonical coordinate systems, fixed in some intuitive sense with respect to the gravitational sources, in which all time derivatives $h_{\mu\nu,0}$ may be neglected compared with the spatial derivatives $h_{\mu\nu,i}$ ($i = 1,2,3$). Note the interpretation of the zero coordinate as time and the other three coordinates as space, on the basis of the almost Minkowskian form of the metric. If such coordinates exist, the gravitational field itself is said to be quasi-stationary.

Any two quasi-stationary coordinate systems can be related by a rotation or a general transformation $\bar{x}^\mu = x^\mu + \xi^\mu$ in which the ξ^μ satisfy the previous conditions and in addition have negligible time derivatives. Under the latter, the components of $h_{\mu\nu}$ and $l_{\mu\nu}$ transform according to

$$\bar{h}_{00} = h_{00} \,, \qquad \bar{h}_{0i} = h_{0i} - \xi_{0,i} \,, \qquad \bar{h}_{ij} = h_{ij} - \xi_{i,j} - \xi_{j,i} \,, \tag{2.69}$$

$$\bar{l}_{00} = l_{00} - \xi_{i,i} \,, \qquad \bar{l}_{0i} = l_{0i} - \xi_{0,i} \,, \qquad \bar{l}_{ij} = l_{ij} - \xi_{i,j} - \xi_{j,i} + \delta_{ij}\xi_{k,k} \,. \tag{2.70}$$

The supplementary condition (2.63) is then imposed on $\bar{l}_{\mu\nu}$ by choosing ξ^μ to satisfy

$$\nabla^2 \xi^\mu := \xi^\mu{}_{,ii} = l^{\mu i}{}_{,i} \,. \tag{2.71}$$

Now h_{00} is the same for all these quasi-stationary coordinate systems, since it is not affected by the general coordinate transformations $\bar{x}^\mu = x^\mu + \xi^\mu$ and it is not affected by rotations. It is this uniqueness that allows one to make contact between the formalism of general relativity and observation, or at least, between Einstein's theory of gravity and Newton's.

The idea is to consider a freely falling particle, assumed to be moving slowly compared with light. It has worldline $z^\mu(t)$, where $t = x^0$, and this is assumed to satisfy the geodesic equation (but see Sect. 2.5 for more on that). The slow motion implies that

$$\left| \frac{\mathrm{d}z^i}{\mathrm{d}t} \right| \ll \left| \frac{\mathrm{d}z^0}{\mathrm{d}t} \right| = 1 \,. \tag{2.72}$$

The geodesic equation [see (2.95) on p. 36] becomes

$$\frac{\mathrm{d}^2 z^\mu}{\mathrm{d}t^2} + \Gamma^\mu_{\nu\sigma} \frac{\mathrm{d}z^\nu}{\mathrm{d}t} \frac{\mathrm{d}z^\sigma}{\mathrm{d}t} = h(t) \frac{\mathrm{d}z^\mu}{\mathrm{d}t} \,, \tag{2.73}$$

where

$$h(t) = -\frac{\mathrm{d}^2 t/\mathrm{d}\tau^2}{(\mathrm{d}t/\mathrm{d}\tau)^2} = \frac{\mathrm{d}^2 \tau/\mathrm{d}t^2}{\mathrm{d}\tau/\mathrm{d}t} \,,$$

using t as parameter rather than the proper time τ of the test particle. Note that t is not an affine parameter and this explains the term in h. The spatial part of this equation is

$$\frac{\mathrm{d}^2 z^i}{\mathrm{d}t^2} + \Gamma^i_{00} + 2\Gamma^i_{0j}\frac{\mathrm{d}z^i}{\mathrm{d}t} + \Gamma^i_{jk}\frac{\mathrm{d}z^j}{\mathrm{d}t}\frac{\mathrm{d}z^k}{\mathrm{d}t} = h(t)\frac{\mathrm{d}z^i}{\mathrm{d}t} \,,$$

in which we neglect the last term on the left. We have

$$h^{\mu\nu} = \eta^{\mu\sigma}\eta^{\nu\rho}h_{\sigma\rho} \,,$$

where $\eta^{\mu\sigma} = \eta_{\mu\sigma}$. Then note that, from

$$\Gamma^\mu_{\nu\sigma} = \frac{1}{2}g^{\mu\rho}\left(\partial_\nu g_{\sigma\rho} + \partial_\sigma g_{\nu\rho} - \partial_\rho g_{\nu\sigma}\right) ,$$

our approximation gives

$$\Gamma^\mu_{\nu\sigma} \approx \frac{1}{2}(\eta^{\mu\rho} - h^{\mu\rho})\left(\partial_\nu h_{\sigma\rho} + \partial_\sigma h_{\nu\rho} - \partial_\rho h_{\nu\sigma}\right) .$$

When we insert this into the above formula, we will be able to use

$$\Gamma^\mu_{\nu\sigma} \approx \frac{1}{2}\eta^{\mu\rho}\left(\partial_\nu h_{\sigma\rho} + \partial_\sigma h_{\nu\rho} - \partial_\rho h_{\nu\sigma}\right) .$$

Note that we have used the approximation

$$g^{\mu\rho} \approx \eta^{\mu\rho} - h^{\mu\rho} \,,$$

which follows from

$$(\eta^{\mu\rho} - h^{\mu\rho})(\eta_{\rho\nu} + h_{\rho\nu}) \approx \delta^\mu_\nu \,.$$

Now

$$\Gamma^i_{00} \approx \frac{1}{2}\eta^{i\rho}\left(\partial_0 h_{0\rho} + \partial_0 h_{0\rho} - \partial_\rho h_{00}\right) \approx -\frac{1}{2}\eta^{ij}\partial_j h_{00} \,,$$

so we can conclude that

$$\Gamma^i_{00} \approx \frac{1}{2}\delta^{ij}\partial_j h_{00} \,.$$

We have neglected $\partial_0 h_{\mu\nu}$ in comparison to $\partial_i h_{\mu\nu}$, $i = 1, 2, 3$. Also,

$$\Gamma^i_{0j} \approx \frac{1}{2}\eta^{i\rho}\left(\partial_0 h_{j\rho} + \partial_j h_{0\rho} - \partial_\rho h_{0j}\right) \approx -\frac{1}{2}\delta^{ik}\left(\partial_j h_{0k} - \partial_k h_{0j}\right) ,$$

once again neglecting $\partial_0 h_{\mu\nu}$.

We must now approximate the right-hand side of the geodesic equation (2.73). Since

$$\left(\frac{\mathrm{d}\tau}{\mathrm{d}t}\right)^2 = g_{\mu\nu}\frac{\mathrm{d}z^\mu}{\mathrm{d}t}\frac{\mathrm{d}z^\nu}{\mathrm{d}t} \,,$$

we have

$$\frac{\mathrm{d}\tau}{\mathrm{d}t} \approx (1+h_{00})^{1/2} \approx 1 + \frac{1}{2}h_{00} \,.$$

Consequently,

$$\frac{\mathrm{d}^2\tau}{\mathrm{d}t^2} \approx \frac{1}{2}h_{00,0} \,.$$

We now have a formula for the function $h(t)$:

$$h(t) \approx \frac{1}{2}h_{00,0}\left(1 - \frac{1}{2}h_{00}\right) \approx \frac{1}{2}h_{00,0} \,.$$

We can therefore neglect the term on the right-hand side of the geodesic equation (2.73).

Finally, the geodesic equation becomes approximately

$$m\frac{\mathrm{d}^2z^i}{\mathrm{d}t^2} \approx -m\delta^{ij}\partial_j\left(\frac{1}{2}h_{00}\right) + m\delta^{ik}(\partial_j h_{0k} - \partial_k h_{0j})\frac{\mathrm{d}z^j}{\mathrm{d}t} \,,$$

where m is the mass of the test particle. The second term on the right-hand side is velocity dependent and can be neglected on the grounds of the slow motion hypothesis (2.72) or attributed to a rotational effect and removed by declaring a non-rotating, quasi-canonical coordinate system to be one in which

$$\partial_j h_{0k} - \partial_k h_{0j} = 0 \,.$$

In this case,

$$m\frac{\mathrm{d}^2z^i}{\mathrm{d}t^2} \approx -m\delta^{ij}\partial_j\left(\frac{1}{2}h_{00}\right) \,.$$

We can now make the identification [see (2.50) on p. 17]

$$h_{00} = 2\Phi \,, \tag{2.74}$$

so that

$$\frac{\mathrm{d}^2z^i}{\mathrm{d}t^2} = -\frac{1}{2}h_{00,i} = -\Phi_{,i} \,, \tag{2.75}$$

where we make the interpretation that Φ is what would be called the Newtonian gravitational potential.

But who would call this a Newtonian potential? If one could find an observer who would set up these coordinates in her neighbourhood, and if this observer were to pretend that they were Minkowskian coordinates in a flat spacetime under the jurisdiction of special relativity, in which gravity really is a force field modelled by a potential, then observation of freely falling particles would lead this observer (under these illusions!) to attribute a Newtonian gravitational potential to this region of spacetime. This is much more sophisticated than the Newtonian idea that accelerating observers can pretend that they are not accelerating and attribute free particle motions to some gravitational potential. What makes the Newtonian version so simple is just the fact that time and distance are the same in the accelerating frame.

Note that, in the Newtonian spacetime, the accelerating observer who pretends she is not accelerating and attributes free particle motions to a gravitational field will be at pains to understand what sourced that field. We shall come in a moment to the question of the source. But let us just return to the observation on p. 18 that the metric components (2.48) on p. 17 would give us a quite different Newtonian gravitational potential. Of course, to get the metric components into the form (2.48), we have changed coordinates, and what has happened is that the quantity h_{00} in the above theory has not remained unchanged. This in turn is telling us that we have made a disallowed coordinate transformation for the above theory. This is not one of the coordinate changes that kept the coordinates in the quasi-stationary and quasi-canonical form. And yet the semi-Euclidean coordinates giving the intervals (2.45) or (2.48) are stationary, and they are quasi-canonical in DeWitt's sense, provided we restrict to coordinate values satisfying

$$\left|\frac{gy^1}{c^2}\right| \ll 1 \qquad \text{or} \qquad \left|\frac{g'y'^1}{c^2}\right| \ll 1 , \qquad \text{respectively} .$$

So DeWitt's allowed sets of coordinate transformations would appear to divide coordinate systems into disjoint classes, each of which can be considered as quasi-stationary and quasi-canonical. Note in passing that the above restriction on the coordinate values keeps us well away from the singularity in these metrics, which occurs when $y^1 = -c^2/g$.

But the key point to note here is that, in order to see a gravitational potential from an initial context, viz., general relativity, in which gravity is not modelled that way, we have to pretend that slightly non-inertial coordinates are in fact inertial, in the sense that they are things we can understand in that way, as though we were doing Newtonian gravitation. This point will be picked up in Sect. 15.7 when we consider Boulware's claim that there is an extremely strong gravitational field near the horizon of any uniformly accelerating semi-Euclidean observer.

Now if one knew the source of gravity, e.g., a star, one would expect this to help in establishing what part of a particle motion could be attributed to gravity and what part could be explained as an artefact of the coordinates. It would at least establish that there is some gravity. So let us return to the case of a massive point source described by (2.54) on p. 19, viz.,

$$T^{00} = M\delta(\mathbf{x}) , \qquad (2.76)$$

and see rather briefly how DeWitt's theory in [24] makes this second link between GR and observation. DeWitt's aim was to establish the value of the constant κ in Einstein's equations

$$G^{\mu\nu} := R^{\mu\nu} - \frac{1}{2}Rg^{\mu\nu} = \frac{1}{2\kappa}g^{-1/2}T^{\mu\nu} , \qquad (2.77)$$

where $G^{\mu\nu}$ is the Einstein tensor, $R^{\mu\nu}$ the Ricci tensor, g the modulus of the determinant of the covariant metric components $g_{\mu\nu}$, and $T^{\mu\nu}$ the energy–momentum tensor for the matter and energy distribution in the spacetime. Our aim will of course be to see whether this analysis throws any light on the SHGF.

In the quasi-canonical coordinate systems described above, it is first shown that the curvature tensor takes the approximate form

$$R_{\mu\nu\sigma\tau} = -\frac{1}{2}(h_{\mu\sigma,\nu\tau} + h_{\nu\tau,\mu\sigma} - h_{\mu\tau,\nu\sigma} - h_{\nu\sigma,\mu\tau}) , \qquad (2.78)$$

also called the linearised Riemann tensor, which turns out to be invariant under the approximate coordinate transformation law (2.62) on p. 25. This invariance is basically due to the fact that, unlike the metric tensor, the Riemann tensor is already first order in small quantities. DeWitt remarks that the physical significance of this invariance is that the presence or absence of a real gravitational field is characterised by the presence or absence of a nonzero Riemann tensor. Since this tensor represents the gravitational field, the weak field approximation is going to provide an invariant characterisation of the field, that is, an expression for the field that is independent of which quasi-canonical coordinate system we are using.

The problem for the SHGF is of course that the Riemann tensor is zero, so it would not count as a real gravitational field in DeWitt's remark. And as we just noted when examining the worldlines of free particles, different SE observers can attribute any Newtonian potential they like to the spacetime neighbourhood, precisely because changing from one uniformly accelerating SE coordinate system to another does not keep one inside the same class of quasi-canonical coordinate systems.

The linearised Ricci tensor and curvature scalar are

$$R_{\mu\nu} = \eta^{\sigma\tau}R_{\mu\sigma\nu\tau} = -\frac{1}{2}\eta^{\sigma\tau}(h_{\mu\nu,\sigma\tau} + h_{\sigma\tau,\mu\nu} - h_{\sigma\nu,\mu\tau} - h_{\mu\tau,\sigma\nu}) , \qquad (2.79)$$

$$R = \eta^{\mu\nu}R_{\mu\nu} = -h_{,\mu}{}^{\mu} + h^{\mu\nu}{}_{,\mu\nu} . \qquad (2.80)$$

The linearised Einstein equations are then

$$\frac{1}{2\kappa}T^{\mu\nu} = -\frac{1}{2}\left(l^{\mu\nu}{}_{,\sigma}{}^{\sigma} - l^{\mu\sigma}{}_{,}{}^{\nu}{}_{\sigma} - l^{\nu\sigma}{}_{,}{}^{\mu}{}_{\sigma} + \eta^{\mu\nu}l^{\sigma\tau}{}_{,\sigma\tau}\right) . \qquad (2.81)$$

This is where the supplementary condition (2.63), viz., $l^{\mu\nu}{}_{,\nu} = 0$, comes into its own. When coordinates are chosen so that this condition is satisfied, which is described as choosing the de Donder gauge, the linearised Einstein equations take the simple form

$$\Box^2 l^{\mu\nu} := l^{\mu\nu}{}_{,\sigma}{}^{\sigma} = -\frac{1}{\kappa} T^{\mu\nu} \; . \tag{2.82}$$

We can now tackle the point mass source at the origin of some coordinate system of the kind singled out above, for which we propose $T^{00} = M\delta(\mathbf{x})$, with all other components of the energy–momentum tensor being zero.

Of course there is problem of circularity here, because we do not know a priori whether the spacetime around such an object will allow us to find a stationary, quasi-canonical system. Since the point mass is at least stationary relative to the spatial coordinates $\mathbf{x}$, one might well expect a stationary system, but it is only with hindsight that we could justify this process. So we simply push ahead by requiring the supplementary condition $l^{\mu\nu}{}_{,\nu} = 0$ and observing that $l^{\mu\nu}$ can then have only one nonzero component, viz., l^{00}, because (2.82) implies $\Box^2 l^{\mu\nu} = 0$ whenever $(\mu, \nu) \neq (0,0)$. [Note that the supplementary condition can still be imposed in a quasi-stationary coordinate system, as evidenced by (2.71) on p. 26.] Equation (2.82) also gives

$$\nabla^2 l^{00} = \frac{1}{\kappa} T^{00} = \frac{1}{\kappa} M\delta(\mathbf{x}) \; . \tag{2.83}$$

The solution satisfying

$$\lim_{|\mathbf{x}|\to\infty} l^{00} = 0 \; , \tag{2.84}$$

and only taken seriously within the context of the weak field approximation when $|\mathbf{x}| \gg GM$, is

$$l^{00}(x) = -\frac{M}{4\pi\kappa} \frac{1}{|\mathbf{x}|} = \frac{1}{4\pi\kappa G}\Phi \; , \tag{2.85}$$

where we have introduced the usual definition for the Newtonian potential of a point particle:

$$\Phi(x) = -\frac{GM}{|\mathbf{x}|} \; . \tag{2.86}$$

The last manoeuvre is a sleight of hand! Although Φ may be the same function of x as the Newtonian potential, x itself is not a Cartesian coordinate for spacetime, except in an approximate sense. We now retrieve $h_{\mu\nu}$ from

$$l = l^{00} = l_{00} \; , \qquad h_{00} = l_{00} - \frac{1}{2}l = \frac{1}{2}l_{00} = \frac{1}{8\pi\kappa G}\Phi \; ,$$

$$h_{ij} = l_{ij} + \frac{1}{2}\delta_{ij}l = \frac{1}{2}\delta_{ij}l_{00} = \delta_{ij}\frac{1}{8\pi\kappa G}\Phi\,, \qquad h_{0i} = l_{0i} = 0\,.$$

As noted in (2.69) on p. 26, if we transform to another quasi-stationary coordinate system in the same class, h_{ij} and h_{0i} will take other values, but h_{00} will remain unchanged to this level of approximation. Note, however, that it is the boundary condition (2.84) which rules out the addition of some affine function of the space coordinates to the proposed solution for l^{00}, and such an affine function, e.g., gz, is precisely the kind of linear potential that leads to a static, homogeneous component to the gravitational field. If one switched to some other class of quasi-stationary and quasi-canonical coordinate system, one would then have to change the boundary condition (2.84) in order to obtain an approximation to Newtonian gravity.

In his presentation [24], DeWitt uses this theory to deduce the value of κ in Einstein's equations (2.77). For general relativity to agree with Newtonian theory about the motion of bodies under the action of gravity in this weak field approximation, we noted that $h_{00} = 2\Phi$ [see (2.74) on p. 28]. It follows that

$$\kappa = \frac{1}{16\pi G}\,, \tag{2.87}$$

whence the full Einstein equations take the form

$$G^{\mu\nu} := R^{\mu\nu} - \frac{1}{2}Rg^{\mu\nu} = 8\pi G g^{-1/2}T^{\mu\nu}\,. \tag{2.88}$$

Now we were looking at all this in the hope that it might throw light on the SHGF. What could source a static, homogeneous gravitational field in the Newtonian theory? The answer is of course an infinite plane of matter of infinitesimal thickness and unchanging, uniform surface density σ. Let us take this as the plane $z = 0$. The Newtonian gravitational potential at a height z above the plane is the sum of all the potentials due to surface elements on the plane. Since sets of points on the plane that are equidistant from the field point constitute rings, one carries out a simple integration over the whole plane by dividing it into thin rings centered on the point directly below the field point. The result is

$$\Phi(z) = 2\pi G\sigma z \qquad (z > 0)\,. \tag{2.89}$$

The force field per unit mass in this Newtonian gravitational (NG) theory is then

$$(-\Phi_{,1}, -\Phi_{,2}, -\Phi_{,3}) = (0, 0, -2\pi G\sigma)\,,$$

which is constant in space and time, and pointing downward for a field point above the plane. When $z < 0$, the sign of Φ is reversed, so that we have a spatiotemporally constant gravitational force field pulling up toward the plane $z = 0$. But note that for a given point $z > 0$, this same matter plane could be anywhere below it and the gravitational effect would be the same. Presumably it could even be moving back and forth up the z axis and the observer at some z that always remained above it would not notice, according to NG.

Let us thus consider the possibility

$$T^{00} = \sigma \delta(z) \,, \tag{2.90}$$

in the weak field approximation of general relativity, where σ is the constant mass per unit area of a plane through the spatial origin of some coordinate system, taking $T^{\mu\nu} = 0$ whenever $(\mu, \nu) \neq (0,0)$. If we can arrange with hindsight for the coordinates to be quasi-stationary and quasi-canonical, and satisfy the supplementary condition as in the analysis for a point particle source, then Einstein's equations take the simple form

$$\nabla^2 l^{00} = \frac{1}{\kappa} T^{00} = \frac{1}{\kappa} \sigma \delta(z) \,. \tag{2.91}$$

Of course, assuming that l^{00} is a function of z alone, this becomes

$$\frac{\mathrm{d}^2 l^{00}}{\mathrm{d}z^2} = \frac{\sigma}{\kappa} \delta(z) \,, \tag{2.92}$$

which is easy to solve because

$$\delta(z) = \frac{\mathrm{d}^2}{\mathrm{d}z^2} \left(\frac{|z|}{2} \right) \,.$$

We find

$$l^{00} = c_1 + c_2 z + \frac{\sigma}{2\kappa} |z| \,,$$

where c_1 and c_2 are constants. Since we expect reflection symmetry in the spatial plane $z = 0$, this implies $c_1 = 0 = c_2$, and the proposed solution is

$$l^{00} = \frac{\sigma}{2\kappa} |z| \,.$$

This time there is no solution with

$$\lim_{|\mathbf{x}| \to \infty} l^{00} = 0 \,,$$

but in fact we are expecting our approximations to be valid close to $z = 0$. We observe with hindsight that this solution does satisfy the supplementary condition $l^{\mu\nu}{}_{,\nu} = 0$. Once again, $l = l^{00} = l_{00}$, and inserting the value of κ,

$$h_{00} = l_{00} - \frac{1}{2} l = \frac{1}{2} l_{00} = 4\pi G \sigma |z| \,.$$

Since $h_{00} = 2\Phi$ is supposed to give the corresponding Newtonian potential, we then find

$$\Phi = 2\pi G \sigma |z| \,, \tag{2.93}$$

the same as in the purely Newtonian theory (NG) [see (2.89)].

This looks very nice, but there are two reasons why we have not really achieved our uniformly accelerating SE metric:

- The value of $g_{00} := \eta_{00} + h_{00}$ differs from the SE metric when $z < 0$, and this is obviously an essential part of the solution!
- The components h_{ij} of the above solution are not zero, as they would be for the SE metric.

Still, in general relativity, if we had an infinite plane of constant mass density σ, we might expect something like the uniformly accelerating SE metric, which is a flat space solution, on either side of the plane, just because in NG we find that this gives a solution with no tidal effects (implying zero curvature in GR). But if this SE metric does satisfy the Einstein equations for such a plane, where is it, and what is it doing? In the uniformly accelerating SE coordinates, it is amusing to imagine that the plane $y^1 = 0$ would do the trick, proposing

$$T^{00} = \sigma \delta(z) \,, \qquad T^{\mu\nu} = 0 \quad \forall\, (\mu,\nu) \neq (0,0) \,.$$

In fact, rather than trying to solve the full Einstein equations for this idealistic energy–momentum distribution, one can calculate the Einstein tensor for the metric

$$g_{00} = \left(1 + \frac{g|z|}{c^2}\right)^2 \,, \qquad g_{0i} = 0 = g_{i0} \,, \qquad g_{ij} = -\delta_{ij} \,, \quad i,j \in \{1,2,3\} \,. \tag{2.94}$$

It is interesting to note that this gives the following results. The connection coefficients are all zero except for

$$\Gamma^3_{00} = \frac{g}{c^2}\left(1 + \frac{g|z|}{c^2}\right)\theta(z) \,, \qquad \Gamma^0_{30} = \frac{g}{c^2}\left(1 + \frac{g|z|}{c^2}\right)^{-1}\theta(z) = \Gamma^0_{03} \,,$$

where

$$\theta(z) := \begin{cases} -1 & \text{for } z < 0 \,, \\ +1 & \text{for } z > 0 \,, \end{cases}$$

and the components of the Ricci tensor are all zero except for

$$R_{00} = \frac{g}{c^2}\delta(z) \,, \qquad R_{33} = -\frac{g}{c^2}\delta(z) \,.$$

The curvature scalar is then

$$R = \frac{2g}{c^2}\delta(z) \,,$$

and the Einstein tensor is

$$G_{00} = 0 = G_{33} \,, \qquad G_{11} = \frac{g}{c^2} \delta(z) = G_{22} \,, \qquad G_{\mu\nu} = 0 \quad \forall\, \mu \neq \nu \,.$$

Raising the indices, one finds that the energy–momentum distribution required to give this metric is specified by

$$T^{00} = 0 = T^{33} \,, \qquad T^{11} = \frac{g}{8\pi G c^2} \delta(z) = T^{22} \,, \qquad T^{\mu\nu} = 0 \quad \forall\, \mu \neq \nu \,.$$

The interpretation of this result is left to the reader. What is clear is that Einstein's equations were never intended to fulfill our Newtonian dreams.

So let us make a brief conclusion. What concerned us in the last section was how an observer at fixed SE spatial coordinate might try to understand the uniformly accelerating SE metric, or at least the way freely falling particles move around relative to her coordinate systems. We have seen that, on a local basis, any Newtonian gravitational potential whatever could be viewed as responsible for such motions. Furthermore, in a Newtonian world, where such effects could be generated by an infinite matter plane, one could measure accelerations of test particles relative to this source and distinguish them from accelerations of the observer, whereas in the world of general relativity one cannot make this kind of distinction so easily, because one cannot identify even an idealistic source so easily. On the other hand, one expects the SE metric to be a good approximation in certain spacetime regions, e.g., in a laboratory situated a long way from a star. In the latter case, which comes under the jurisdiction of the Schwarzschild solution to Einstein's equations, one can to a certain extent and with a certain proviso use the proper distance from the source to distinguish coordinate accelerations (of a test particle) due to the source and those due to choice of coordinates. The inevitable level of approximation is what determines the extent to which this will succeed, while the fact that proper distance depends on a subjective choice of spacelike hypersurface stands as a proviso.

In the highly idealistic, uniformly accelerating semi-Euclidean spacetime, one has to accept that one does not know where the gravitational source is located so that one cannot use it to make deductions about the motion of free particles. In that case, even though the curvature is zero, whence there are no tidal effects in this spacetime, one has to accept that different SE observers can choose to explain free particle motions by quite different Newtonian potentials in the weak field approximation. And we shall see in Chap. 15 that some authors interpret this gravitational field as having different strengths at different locations in spacetime. But it should be remembered that such an observer is not using locally inertial coordinates. In order to see a gravitational potential from an initial context, viz., general relativity, in which gravity is not modelled that way, we have to pretend that slightly non-inertial coordinates are in fact inertial, in the sense that they are things we can understand in that way, as though we were doing Newtonian gravitation.

To put it another way, in the weak field approximation to free fall, one can only get a Newtonian potential if the connection coefficients are not all zero at the event in question, whence the coordinate system is not locally inertial. Furthermore, this potential does not make a distinction between what one might call artefacts of the choice of coordinates, which might possibly be ascribed to the motion of an observer sitting at the spatial origin of that coordinate system, and gravitational effects that are due to the energy–momentum distribution. Even in the context of the weak field analysis for a point particle source described above, Einstein's equations allow the addition of any affine function of a space coordinate to the proposed Newtonian potential, and this affine function is only ruled out by boundary conditions imposed by the requirement that things look roughly as they would according to NG. And finally, that Newtonian potential may be the usual function of the space coordinates for such a source, but those coordinates have to be assimilated with the Cartesian coordinates of Newtonian physics to make this interpretation.

2.5 Geodesic Principle

This section can be considered as a slightly off-beat review of some basic relativity theory. We return to the interesting question of the geodesic principle brought up in Sect. 2.3. This states that, when point particles are not acted upon by forces (apart from gravitational effects), their trajectories take the form

$$\frac{d^2 x^i}{ds^2} + \Gamma^i_{jk} \frac{dx^j}{ds} \frac{dx^k}{ds} = 0 \,, \tag{2.95}$$

where $x^i(s)$ gives the worldline as a function of the proper time s of the particle and Γ^i_{jk} are the connection coefficients in the given coordinate system. In the literature, this is often derived from an action principle. One writes the worldline as a function $x^i(\lambda)$ of some arbitrary parameter λ, whence the appropriate action for the worldline between two points $P_1 = x(\lambda_1)$ and $P_2 = x(\lambda_2)$ of spacetime is

$$s(P_1, P_2) := \int_{\lambda_1}^{\lambda_2} \left(g_{ij} \frac{dx^i}{d\lambda} \frac{dx^j}{d\lambda} \right)^{1/2} d\lambda = \int_{\lambda_1}^{\lambda_2} L d\lambda = \int_{\lambda_1}^{\lambda_2} ds \,, \tag{2.96}$$

with Lagrangian

$$L := \left(g_{ij} \frac{dx^i}{d\lambda} \frac{dx^j}{d\lambda} \right)^{1/2} . \tag{2.97}$$

The Euler–Lagrange equations extremising the action under variation of the worldline are

$$\frac{d}{d\lambda} \left(\frac{\partial L}{\partial \dot{x}^i} \right) - \frac{\partial L}{\partial x^i} = 0 \,, \tag{2.98}$$

with $\dot{x}^i := dx^i/d\lambda$, and these lead to the above geodesic equation (2.95).

The action for some particles labelled by a is

$$\mathscr{A} = -\sum_a c m_a \int ds_a , \tag{2.99}$$

where m_a is the mass of particle a and s_a is its proper time. This is the action *because* variation of the worldline of particle a gives its equation of motion as

$$\frac{d^2 a^i}{ds_a^2} + \Gamma^i_{jk} \frac{da^j}{ds_a} \frac{da^k}{ds_a} = 0 . \tag{2.100}$$

Likewise, if some of the particles are charged with charge e_a for particle a, and there are some EM fields F_{ik}, one declares the action to be

$$\mathscr{A} = -\sum_a c m_a \int ds_a - \frac{1}{16\pi c} \int F_{ik} F^{ik} (-g)^{1/2} d^4 x - \sum_a \frac{e_a}{c} \int A_i da^i , \tag{2.101}$$

where A_i is a 4-vector potential from which F_{ij} derives, simply *because* variation of the worldline of particle a gives its equation of motion as

$$\frac{d^2 a^i}{ds_a^2} + \Gamma^i_{jk} \frac{da^j}{ds_a} \frac{da^k}{ds_a} = \frac{e_a}{m_a} F^i{}_j \frac{da^j}{ds_a} , \tag{2.102}$$

the minimal generalisation of the Lorentz force law to a curved manifold, while variation of A_i gives the EM field equations as the minimal extension of Maxwell's equations to the curved spacetime. Of course, these actions are designed to give appropriate field equations, and we are just decreeing here that the appropriate field equation for the particle labelled by a is a geodesic equation, or an equation like (2.102).

So what is the physical motivation? The appropriate physical argument supporting (2.95) is an application of the strong principle of equivalence, and it is here that we discover exactly how we are to link what happens in the curved manifold with measurements in our own world. We start from an action, but at some point we must say what the point of contact would be with physical reality. Let us suppose we impose a strong principle of equivalence, that is, we say roughly speaking that any physical interaction other than gravitation behaves in a locally inertial frame as though gravity were absent (but see Chap. 4 for more details). Relative to such a frame, any particle that is not subject to (non-gravitational) forces will then move in a straight line with uniform velocity, i.e., it will follow the trajectory described by

$$\frac{dv^i}{ds} = 0 , \tag{2.103}$$

where v^i is its 4-velocity. This is expressed covariantly through (2.95). If there are non-gravitational forces, we start with $F = ma$ in the locally inertial frame and we find (2.95) with a force term on the right-hand side. It is quite clear that we still have a version of Newton's second law $F = ma$, so we have certainly not explained inertia and inertial effects by this ploy, but merely extended this equation of motion to the new theory.

Let us look more closely at the claim that (2.103) is expressed covariantly through (2.95). A cheap way is to set the connection coefficients equal to zero in (2.95). This is basically the observation that the two equations are the same relative to Cartesian coordinates in a flat spacetime. This misses out some of the machinery of the connection construction that lies at the heart of non-Euclidean geometry, and this is not the place to expose all that. However, the following is an amusing way to look a little more deeply into this affair.

Let us set the scene by assigning a connection to the manifold in the following way: for each point and coordinate neighbourhood of that point, we have a function whose value at any point in the neighbourhood is a set of N^3 numbers Γ^a_{bc}, where N is the dimension of the manifold. Given another neighbourhood, the quantities transform by the rule

$$\Gamma^{a'}_{b'c'} = \Gamma^d_{ef} X^{a'}_d X^e_{b'} X^f_{c'} + X^{a'}_d X^d_{c'b'} \,,$$

where we have used the notation

$$X^{a'}_d := \frac{\partial x^{a'}}{\partial x^d} \,, \qquad X^d_{c'b'} := \frac{\partial^2 x^d}{\partial x^{c'} x^{b'}} \cdot$$

One can show with a little algebra that this transformation is transitive.

Now suppose U is a coordinate neighbourhood of M and $\lambda^a(u)$ is defined along a curve γ, given by $x^a(u)$. Then

$$\dot{\lambda}^a = \frac{\mathrm{d}\lambda^a}{\mathrm{d}u}$$

is not a vector, in general. We define the absolute derivative of λ^a to be

$$\frac{\mathrm{D}\lambda^a}{\mathrm{d}u} = \frac{\mathrm{d}\lambda^a}{\mathrm{d}u} + \Gamma^a_{bc}\lambda^b\frac{\mathrm{d}x^c}{\mathrm{d}u} \,,$$

for any N^3 numbers Γ^a_{bc} associated with each point and coordinate neighbourhood. If this is a vector field for all vector fields λ^a along γ, then it can be shown that Γ^a_{bc} must transform as stated above, using the quotient theorem for tensors. Conversely, if

$$\frac{\mathrm{D}\lambda^a}{\mathrm{d}u} = \frac{\mathrm{d}\lambda^a}{\mathrm{d}u} + \Gamma^a_{bc}\lambda^b\frac{\mathrm{d}x^c}{\mathrm{d}u} \,,$$

and Γ^a_{bc} transforms as above, then $\mathrm{D}\lambda^a/\mathrm{d}u$ is a vector.

We can now extend the Euclidean idea of parallelism. Suppose P has coordinates x^a and some neighbouring point Q has coordinates $x^a + \delta x^a$. We seek some kind of bijection from $T_P(M)$ to $T_Q(M)$, so that we can call corresponding vectors parallel. We demand that the correspondence be linear and reduce to the identity when Q approaches P. (The point about linearity is that, if v is parallel to w, then λv ought to be parallel to λw, for any $\lambda \in \mathbb{R}$, and so on.)

Let $\lambda^a + \delta^* \lambda^a$ be the vector at Q which is parallel to λ^a at P. Then there exists Y_b^a, dependent on P and Q, but not on λ^a, such that

$$\lambda^a + \delta^* \lambda^a = Y_b^a \lambda^b .$$

To first order in δx^a, we would like to obtain

$$Y_b^a = \delta_b^a - \Gamma_{bc}^a \delta x^c ,$$

noting that Γ_{bc}^a depends only on P, and δx^c depends on both P and Q. This will be possible provided that

$$\delta^* \lambda^a = -\Gamma_{bc}^a \lambda^b \delta x^c . \tag{2.104}$$

We see that

$$\frac{D\lambda^a}{du} = \lim_{\delta u \to 0} \frac{\delta \lambda^a - \delta^* \lambda^a}{\delta u} ,$$

where $\delta \lambda^a = \lambda^a (u + \delta u) - \lambda^a(u)$. Note that $\delta \lambda^a - \delta^* \lambda^a$ is a vector at Q.

For a flat space, parallelism is well-understood, and it is from this context that we wish to generalise. In Cartesian coordinates x^a, the connection components are zero, but in general coordinates $x^{a'}$, they are given by

$$\Gamma_{b'c'}^{a'} = \frac{\partial x^{a'}}{\partial x^d} \frac{\partial^2 x^d}{\partial x^{b'} \partial x^{c'}} ,$$

according to the transformation equation stipulated above.

Assuming this to be the general definition of connection coefficients in a flat space, relative to some arbitrary curvilinear coordinate system $\{x^{a'}\}$ expressed as functions of a Cartesian coordinate system $\{x^a\}$ (kept fixed throughout), it is instructive to derive the general transformation law for the connection in this flat space context. That is, we consider some other curvilinear coordinate system $x^{a''}$, in which the connection has coefficients

$$\Gamma_{f''g''}^{e''} = \frac{\partial x^{e''}}{\partial x^d} \frac{\partial^2 x^d}{\partial x^{f''} \partial x^{g''}} ,$$

and then check that

$$\Gamma_{f''g''}^{e''} = \Gamma_{b'c'}^{a'} \frac{\partial x^{e''}}{\partial x^{a'}} \frac{\partial x^{b'}}{\partial x^{f''}} \frac{\partial x^{c'}}{\partial x^{g''}} + \frac{\partial x^{e''}}{\partial x^{h'}} \frac{\partial^2 x^{h'}}{\partial x^{f''} \partial x^{g''}} .$$

This motivates the definition and transformation of connection coefficients in a general curved space, where the connection cannot be made zero on any neighbourhood.

Still working within our flat space, we can now see how these non-zero connection coefficients implement parallelism in general (non-Cartesian) coordinates.

In the Cartesian coordinates x^a, a vector v^a at x_0 is parallel to the vector with the same components v^a at $x_0 + \delta x$. Let us see what this looks like in the non-Cartesian primed coordinate system. Let us write the primed components of each of these two vectors. Firstly, at x_0,

$$v^{b'} = \left.\frac{\partial x^{b'}}{\partial x^a}\right|_{x_0} v^a \,,$$

and secondly, at $x_0 + \delta x$,

$$v^{b'} + \delta^* v^{b'} = \left.\frac{\partial x^{b'}}{\partial x^a}\right|_{x_0 + \delta x} v^a \,.$$

We can evaluate the last term to first order in δx, to obtain

$$v^{b'} + \delta^* v^{b'} = \left\{ \left.\frac{\partial x^{b'}}{\partial x^a}\right|_{x_0} + \left.\frac{\partial}{\partial x^c}\frac{\partial x^{b'}}{\partial x^a}\right|_{x_0} \delta x^c \right\} v^a \,.$$

Comparing with (2.104), we are expecting to find

$$-\left.\Gamma^{b'}_{d'e'}\right|_{x_0} v^{d'} \delta x^{e'} = \left.\frac{\partial^2 x^{b'}}{\partial x^c \partial x^a}\right|_{x_0} \delta x^c \, v^a \,.$$

Inserting

$$v^a = \frac{\partial x^a}{\partial x^{d'}} v^{d'} \qquad \text{and} \qquad \delta x^{c'} = \frac{\partial x^c}{\partial x^{e'}} \delta x^{e'} \,,$$

we are hoping to find that

$$-\left.\Gamma^{b'}_{d'e'}\right|_{x_0} = \left.\frac{\partial^2 x^{b'}}{\partial x^c \partial x^a}\right|_{x_0} \frac{\partial x^c}{\partial x^{e'}} \frac{\partial x^a}{\partial x^{d'}} \,.$$

The manipulation here is straightforward, first writing the right-hand side as

$$\frac{\partial}{\partial x^{e'}} \left\{ \frac{\partial x^{b'}}{\partial x^a} \right\} \frac{\partial x^a}{\partial x^{d'}} \,,$$

and then as

$$\frac{\partial}{\partial x^{e'}} \left\{ \frac{\partial x^{b'}}{\partial x^a} \frac{\partial x^a}{\partial x^{d'}} \right\} - \frac{\partial x^{b'}}{\partial x^a} \frac{\partial^2 x^a}{\partial x^{e'} \partial x^{d'}} \,.$$

The first term here is zero, and we are left with exactly what we defined to be $-\Gamma^{b'}_{d'e'}$.

The moral of this story is that, for general coordinates in a flat space we require non-zero connection coefficients, and we carry over both the transformation equation and the way they come into vector parallelism when we generalise to manifolds which do not have Cartesian coordinate systems. All this is supposed to comfort the reader that the covariant generalisation of

$$\frac{\mathrm{d}^2 x^i}{\mathrm{d}s^2} = 0 \tag{2.105}$$

is the geodesic equation

$$\frac{\mathrm{d}^2 x^i}{\mathrm{d}s^2} + \Gamma^i_{jk} \frac{\mathrm{d}x^j}{\mathrm{d}s} \frac{\mathrm{d}x^k}{\mathrm{d}s} = 0 \,. \tag{2.106}$$

This is then taken as the equation of motion of a point particle upon which no forces are acting, unless one counts gravity as a force. We observe that there is no mention of any parameters characterising the point particle. In particular there is no mention of its inertial mass. Of course there is no mention of parameters describing its inner make-up. After all, it is supposed to be a point particle. One has to wonder what would happen to a slightly spatially extended particle, i.e., with a world tube that intersects spatial hypersurfaces in a small region rather than a single mathematical point. This object might be spinning in some sense, or contain a charge distribution, for example. We shall return to this point in a moment, but let us begin with the disappearance of the inertial mass since this is directly relevant to the notion of equivalence principle.

Indeed, one of the great equivalence principles of physics, in fact an experimental result that already features in Newtonian physics, is the equivalence of passive gravitational mass, the measure of how much a particle is supposed to feel the Newtonian force of gravity, and the inertial mass, the measure of how much a particle is supposed to resist being accelerated. This meant that, in Newtonian gravitational theory, an observer could not tell whether the acceleration of a test particle relative to her Euclidean coordinate system was due to some gravitational field or due to her own acceleration and the consequent acceleration of her comoving Euclidean coordinate system. (The acceleration of that Euclidean system must not be rotational here.) In pre-relativistic theory, one had no difficulty setting up Euclidean coordinate systems in spatial hypersurfaces and time was the same for everyone.

So where did the inertial mass of the particle go? If we look back to (2.102), viz.,

$$m_a \left(\frac{\mathrm{d}^2 a^i}{\mathrm{d}s_a^2} + \Gamma^i_{jk} \frac{\mathrm{d}a^j}{\mathrm{d}s_a} \frac{\mathrm{d}a^k}{\mathrm{d}s_a} \right) = e_a F^i{}_j \frac{\mathrm{d}a^j}{\mathrm{d}s_a} \,, \tag{2.107}$$

we find the inertial mass m_a multiplying the acceleration term in the equation to give a force on the right that is determined by an external field F_{ij} and a coupling constant e_a characterising the particle. This is a typical equation of motion when there is some non-gravitational force (in this case electromagnetic) acting on the particle. Now in Newtonian gravitational theory, if the external field happens to be a gravitational potential Φ, one gets an equation of motion like this:

$$m_a \left(\frac{\mathrm{d}^2 a^i}{\mathrm{d}s_a^2} + \overline{\Gamma}^i_{jk} \frac{\mathrm{d}a^j}{\mathrm{d}s_a} \frac{\mathrm{d}a^k}{\mathrm{d}s_a} \right) = m_a h^{ij} \Phi_{,j} \, , \tag{2.108}$$

where $(h^{ij}) = \mathrm{diag}\,(0,1,1,1)$ and $\overline{\Gamma}^i_{jk}$ is the connection appropriate to Newtonian spacetime and relative to whatever coordinates we have chosen to describe it. The coupling factor on the right-hand side is just m_a, the same factor as we have on the left-hand side. The only relevant characteristic of our point particle thus cancels out.

This explains how the inertial mass disappears from the equation, but how do we get rid of the gravitational potential we have just introduced? Of course, we can absorb it into the connection, following the much more detailed account of all this in [11]. We now have a new connection

$$\Gamma^i_{jk} := \overline{\Gamma}^i_{jk} + h^{il} \Phi_{,l} t_j t_k \, , \tag{2.109}$$

where $(t_i) := (1,0,0,0)$, so that $t_i = \partial t / \partial x^i$. Equation (2.108) becomes

$$\frac{\mathrm{d}^2 a^i}{\mathrm{d}s_a^2} + \Gamma^i_{jk} \frac{\mathrm{d}a^j}{\mathrm{d}s_a} \frac{\mathrm{d}a^k}{\mathrm{d}s_a} - 0 \, , \tag{2.110}$$

still in this Newtonian context. So the equality of inertial and passive gravitational mass allows us to treat the trajectories of particles subjected only to gravitational effects as geodesics of a non-flat connection, because we do expect this new connection in (2.109) to be non-flat in general.

As Friedman says in [11], the equality of inertial and passive gravitational mass implies the existence of a connection Γ such that freely falling objects follow geodesics of Γ. This does not work for other types of interaction, where the ratio of m_a to the coupling factor, e.g., e_a for a charged particle, is not the same for all bodies. The worldlines of charged particles in an EM field cannot be construed as the geodesics of any single connection, because m_a/e_a in (2.102) varies from one particle to another. To quote from Friedman [11, p. 197]:

> The principle of equivalence (strict proportionality) of inertial and gravitational mass must therefore be true if any theory of gravitation like general relativity, in which gravitational interaction is explained by the dependence of a nonflat connection on the distribution of matter, is to be possible. But, of course, general relativity is not the only theory of this type. Classical gravitational theory can also be formulated in this way by taking advantage of the very same equivalence [of inertial and passive gravitational mass].

Friedman's book is recommended for anyone who thinks that Newtonian gravitational theory cannot be given a fully covariant and totally geometric treatment. The essential difference with general relativity is that, in this treatment of Newtonian gravity, there is a flat connection $\overline{\Gamma}$ living alongside the non-flat connection Γ of (2.109). The deep fact here is that, in general relativity, the non-flat connection is the *only* connection of the spacetime. We return to this point when we introduce the strong equivalence principle in Chap. 4.

Anyway, we now know what a point particle is supposed to do. But what about a spatially extended particle? Well, it turns out that the geodesic principle is not a principle at all, and neither is it likely to be better than an approximation for a real particle that cannot be treated as a mathematical point. This will be directly relevant in Chap. 12 when we discuss DeWitt and Brehme's extension of the Lorentz–Dirac equation to curved spacetimes [4], so it is worth spelling things out right away.

The point is that the geodesic principle follows from Einstein's equations in general relativity. We cannot go into all the details here. They can be found in the standard textbooks. However, it is worth seeing a proof of sorts. Recall that Einstein's equations can be written

$$G_{ij} = -\kappa T_{ij} , \qquad (2.111)$$

where

$$G_{ij} := R_{ij} - \frac{1}{2} g_{ij} R \qquad (2.112)$$

is the Einstein tensor expressed in terms of the Ricci tensor R_{ij} and curvature scalar R, κ is a constant that turns out to be expressible as

$$\kappa = \frac{8\pi G}{c^4} , \qquad (2.113)$$

and T_{ij} is the energy–momentum tensor expressing the distribution of mass and energy in the spacetime.

Now the covariant derivative of the Einstein tensor is zero in many circumstances, in particular when something called the torsion is zero. But the torsion is indeed often zero. In fact, it is sourced by the spin currents of matter in such a way that, in contrast to curvature, it does not propagate in spacetime, so it could only be nonzero in regions where there is matter or energy with some rotational property. A very clear, though somewhat sophisticated account of all this can be found in [20, Chap. 5]. Anyway, in a region where there is no spinning matter, Einstein's equation (2.111) implies that the covariant derivative of the energy–momentum tensor is zero. This is what we shall use to derive the geodesic 'principle'.

To do this, we shall consider an almost-pointlike particle. So when almost-point particles are not acted upon by forces (apart from gravitational effects), we would like to show that their trajectories take the form

$$\frac{\mathrm{d}^2 x^m}{\mathrm{d}s^2} + \Gamma^m_{kj} \frac{\mathrm{d}x^k}{\mathrm{d}s} \frac{\mathrm{d}x^j}{\mathrm{d}s} = 0 . \qquad (2.114)$$

We consider a small blob of dustlike (i.e., zero pressure) matter with density ρ and velocity field

$$v^i = \frac{\mathrm{d}x^i}{\mathrm{d}s} . \qquad (2.115)$$

This equation expresses the fact that we view each component dust particle as having its own path $x^i(s)$. The energy–momentum tensor for this matter is then

$$T^{ik} = \rho \frac{\mathrm{d}x^i}{\mathrm{d}s} \frac{\mathrm{d}x^k}{\mathrm{d}s} \tag{2.116}$$

and we are saying that Einstein's field equation (2.111) implies that

$$T^{ik}{}_{;k} = 0 . \tag{2.117}$$

We analyse (2.117) by inserting (2.116) and the result is the geodesic equation (2.95). For completeness, here is the argument. We have

$$\rho_{,k} \frac{\mathrm{d}x^i}{\mathrm{d}s} \frac{\mathrm{d}x^k}{\mathrm{d}s} + \rho \left(\frac{\partial}{\partial x^k} \frac{\mathrm{d}x^i}{\mathrm{d}s} + \Gamma^i_{km} \frac{\mathrm{d}x^m}{\mathrm{d}s} \right) \frac{\mathrm{d}x^k}{\mathrm{d}s} + \rho \frac{\mathrm{d}x^i}{\mathrm{d}s} \left(\frac{\partial}{\partial x^k} \frac{\mathrm{d}x^k}{\mathrm{d}s} + \Gamma^k_{km} \frac{\mathrm{d}x^m}{\mathrm{d}s} \right) = 0 . \tag{2.118}$$

If we did not have the idea of a velocity field v^i, it would be difficult to interpret partial derivatives of $\mathrm{d}x^i/\mathrm{d}s$ with respect to the coordinates. But as things are, we can say

$$\frac{\mathrm{d}x^k}{\mathrm{d}s} \frac{\partial}{\partial x^k} \frac{\mathrm{d}x^i}{\mathrm{d}s} = \frac{\mathrm{d}x^k}{\mathrm{d}s} \frac{\partial v^i}{\partial x^k} = \frac{\mathrm{d}v^i}{\mathrm{d}s} = \frac{\mathrm{d}^2 x^i}{\mathrm{d}s^2} . \tag{2.119}$$

The terms in the second bracket of (2.118) are

$$\frac{\partial v^k}{\partial x^k} + \Gamma^k_{km} v^m = \mathrm{div}\, v , \tag{2.120}$$

and the whole thing can now be expressed by

$$\mathrm{div}(\rho v) \frac{\mathrm{d}x^i}{\mathrm{d}s} + \rho \left(\frac{\mathrm{d}^2 x^i}{\mathrm{d}s^2} + \Gamma^i_{km} \frac{\mathrm{d}x^k}{\mathrm{d}s} \frac{\mathrm{d}x^m}{\mathrm{d}s} \right) = 0 . \tag{2.121}$$

By mass conservation,

$$\mathrm{div}(\rho v) = 0 , \tag{2.122}$$

and the result follows.

This proof purports to show that each constitutive particle of the blob follows a geodesic. But then we did not allow these particles to jostle one another. For example, we have zero pressure, as attested by the form of the energy–momentum tensor in (2.116). And we did not allow the particles to generate any torsion by revolving about the center of energy of the blob. And neither did we endow them with electric charge. It is in this sense that the geodesic 'principle' is in fact just an approximation, unless the test particle is not a blob, but a mathematical point.

Some argue that inertia is explained in general relativity, precisely because of the above proof (or better variants of it). This point of view is expressed in the philosophical study by Brown [21, p. 141]:

> GR is the first in the long line of dynamical theories, based on that profound Aristotelian distinction between natural and forced motions of bodies, that *explains* inertial motion.

This is not the view taken here, for reasons to be explained shortly. However, other issues discussed in Brown's book, in particular what he refers to as the dynamical approach to spacetime structure, should mark a turning point in our understanding of relativity theories that is exactly in line with the approach advocated in the present book. Another quote concerning the explanation of inertial motion is [21, Sect. 9.3]:

> Inertia, in GR, is just as much a consequence of the field equations as gravitational waves. For the first time since Aristotle introduced the fundamental distinction between natural and forced motions, inertial motion is part of the dynamics. It is no longer a miracle.

Here is an argument against that view. Recall the discussion just after (2.102) on p. 37. It was pointed out that actions like (2.99) and (2.101) are designed to give appropriate field equations, and that the appropriate field equation for the particle labelled by a is a geodesic equation, or an equation like (2.102). Now in GR, one adds a gravitational part to the action, viz.,

$$\mathscr{A}_{\mathrm{grav}} := \frac{c^3}{16\pi G} \int R(-g)^{1/2} \mathrm{d}^4 x . \qquad (2.123)$$

Some textbooks motivate this as follows. When the metric is varied in $\mathscr{A}_{\mathrm{grav}}$, a constant multiple of the Einstein tensor pops out. The point about this is the observation that, when the metric is varied in an action like (2.101), the energy–momentum tensor T_{ij} pops out. One gets a sum of contributions to this tensor from the matter as encapsulated in the action term

$$-\sum_a cm_a \int \mathrm{d}s_a ,$$

and from the EM fields as encapsulated in the action term

$$-\frac{1}{16\pi c} \int F_{ik} F^{ik} (-g)^{1/2} \mathrm{d}^4 x .$$

Setting the variation of the full action with respect to the metric equal to zero, one then obtains the Einstein equations, with the Einstein tensor on one side and the total energy–momentum on the other side.

Now the covariant derivative of the Einstein tensor is zero (assuming zero torsion) and this could in principle be worrying, because the Einstein equation then implies that the covariant derivative of the total energy–momentum is zero. However, there is a general result that the energy–momentum tensor derived from an action of the form

$$\int L(-g)^{1/2} \mathrm{d}^4 x \qquad\qquad (2.124)$$

by varying the metric always has zero covariant derivative when L is a scalar, and more sophisticated versions, e.g., invariance of the matter action under the group of diffeomorphisms is sufficient to guarantee zero covariant derivative of the corresponding energy–momentum tensor on shell if the torsion is zero [20, Sect. 6.5]. (As mentioned above, if the torsion is not zero, the covariant derivative of the Einstein tensor is not zero either. This case is not considered here.) Of course, the action $\mathscr{A}$ in (2.99) does not have the form (2.124) but one expects some general theorem to ensure that the resulting energy–momentum tensor will have zero covariant derivative on shell, i.e., when the field equations, that is, the geodesic equations, are satisfied.

So it looks as though the geodesic equations, and their variants with a force on one side, are built in by construction of the action. It is no surprise therefore that they should pop out again when we set the covariant derivative of the energy–momentum tensor equal to zero. Perhaps one should be more suspicious of arguments from actions. They are neat, and bring a level of unity in the sense that one can derive several dynamical equations from the same action by varying different items. On the other hand, we are only getting out what we put in somewhere else.

One way to make progress with explaining inertia might be to pay more attention to the fact that test particles are not likely to be well modelled by mathematical points. We have already mentioned the spinning particle and the effect of curvature on its motion. A possibly different effect (although possibly similar?) occurs when the particle is a source of some classical force field, typically electromagnetic. The spatially extended particle then exerts forces on itself and in simple cases it can be shown that these forces oppose acceleration in flat spacetime and explain why a force is needed to keep the particle off a geodesic in curved spacetime. If all inertia were due to these self-forces, the geodesic equation, or relevant extension of $F = ma$ would then be replaced by an equation of the form $\sum F = 0$, where the F summed over include self-forces.

Chapter 3
Gravity as a Force in Special Relativity

We said at the end of Chap. 1 that there had been a certain intellectual leap from the case of a charge uniformly accelerating in Minkowski spacetime to one that is statically supported in a gravitational field. So at the beginning of Chap. 2, we then sought to infer something about a statically supported particle in a gravitational field from the description, entirely within the special theory of relativity, of a uniformly accelerating particle in a spacetime with no gravitational fields. This is hopefully what was achieved, although we have yet to consider the electromagnetic fields that might be generated when this particle is charged.

But it seems worth pausing for a moment to consider what special relativity theory would have said about the gravitational field. The aim is to spell out what changes occur when one moves to general relativity. In special relativity, gravitational effects are just forces like any other. We can well imagine that the original uniformly accelerating body in Minkowski spacetime might just be undergoing this motion due to a uniform gravitational field. This is indeed possible as we shall see now.

Consider a particle with rest mass m_0 moving along the x axis in Minkowski spacetime with uniform acceleration. According to (2.9) on p. 8, its position is given by $x(t)$ (with $y = 0 = z$) at coordinate time t, where

$$x(t) = \frac{c^2}{g} \left[\left(1 + \frac{g^2 t^2}{c^2} \right)^{1/2} - 1 \right] . \tag{3.1}$$

In special relativistic dynamics, we have a four-force

$$F^\mu = m_0 a^\mu , \tag{3.2}$$

where

$$a^\mu = \frac{\mathrm{d}^2 x^\mu}{\mathrm{d}\tau^2} \tag{3.3}$$

is the four-acceleration and τ the proper time. As in (2.10) on p. 8, we have

$$v(t) := \frac{\mathrm{d}x}{\mathrm{d}t} = \frac{gt}{(1+g^2t^2/c^2)^{1/2}} \, , \qquad a(t) := \frac{\mathrm{d}^2x}{\mathrm{d}t^2} = \frac{g}{(1+g^2t^2/c^2)^{3/2}} \, . \qquad (3.4)$$

The first of these implies that

$$\gamma(t) := \left[1 - v(t)^2/c^2\right]^{-1/2} = (1+g^2t^2/c^2)^{1/2} \, , \qquad (3.5)$$

whence

$$\frac{\mathrm{d}t}{\mathrm{d}\tau} = \gamma = (1+g^2t^2/c^2)^{1/2} \, . \qquad (3.6)$$

Therefore,

$$a^0 = c\frac{\mathrm{d}^2t}{\mathrm{d}\tau^2} = \frac{g^2t}{c} \qquad (3.7)$$

and

$$a^1 = \frac{\mathrm{d}^2x}{\mathrm{d}\tau^2} = g(1+g^2t^2/c^2)^{1/2} \, , \qquad a^2 = 0 = a^3 \, . \qquad (3.8)$$

Finally, the four-force is

$$(F^\mu) = (m_0 a^\mu) = m_0 g\left(gt/c, (1+g^2t^2/c^2)^{1/2}, 0, 0\right) \, . \qquad (3.9)$$

We write

$$(F^\mu) = \gamma(v)\left(\dot{m}c, f\right) \, , \qquad (3.10)$$

where $\dot{m} := \mathrm{d}m/\mathrm{d}t$ and $m := m_0\gamma$ is the (speed-dependent) mass. Then

$$f := \frac{\mathrm{d}}{\mathrm{d}t}(mv) \, , \qquad (3.11)$$

is the 3-force, with v the coordinate velocity in the inertial frame. So we have

$$f = (m_0 g, 0, 0) \, . \qquad (3.12)$$

The fact that the 3-force is constant means that the coordinate acceleration in the inertial frame [given by the second equation of (3.4)] decreases. This is because the inertial mass is increasing with the speed.

This is presumably how one would model motion in an SHGF (with force g per unit mass in the x direction) in special relativity theory. We then have the equation of motion

$$mg = \frac{\mathrm{d}}{\mathrm{d}\tau}(mv) \, . \qquad (3.13)$$

It is a striking thing in a way that one should be able to obtain uniform acceleration by precisely this kind of force in the context of special relativity, because it is a very different scenario to the one proposed by general relativity.

In the present model, the object is freely falling under a gravitational force, whereas in the general relativistic view of a static homogeneous gravitational field that we outlined in the last chapter, this same object would be viewed as being held still against the effects of a gravitational source, and it is then the Minkowski frame that is freely falling. Of course both models consider the object to be accelerating, and both models consider the Minkowski frame to be inertial. But in special relativity, a freely falling object is not undergoing inertial motion, whereas it is in general relativity.

Furthermore, note in passing that the origin of the Minkowski coordinate frame that is taken to be freely falling in the general relativistic view of a static homogeneous gravitational field does not appear to the semi-Euclidean observer at the spatial origin of the SE coordinate frame to have uniform acceleration. This is seen from (2.12) and (2.13) on p. 9 by looking at the SE coordinate version of the worldline $(t = \sigma, 0, 0, 0)$ (for varying σ) in the Minkowski system:

$$
y^0(\sigma) = \frac{c^2}{g} \tanh^{-1} \frac{g\sigma}{c}, \qquad y^1(\sigma) = \frac{c^2}{g} \left(1 - \frac{g^2 \sigma^2}{c^2} \right)^{1/2} - \frac{c^2}{g}, \qquad y^2 = 0 = y^3.
$$

$$\tag{3.14}$$

Eliminating σ in order to write y^1 as a function of y^0 for this worldline (which is of course a geodesic), we find

$$
y^1(y^0) = \frac{c^2}{g} \left[\frac{1}{\cosh(gy^0/c^2)} - 1 \right] .
\tag{3.15}
$$

Naturally, this does not look like anything in particular to the person using these coordinates! There is an important message here: a freely falling object in the special relativistic version of the static homogeneous gravitational field may have uniform acceleration, but the freely falling object in the GR version of the SHGF has no acceleration, and although it does have a *coordinate* acceleration for the SE observer, this acceleration does not look anything like a uniform acceleration. Why should it?

Enough has been said about this for the moment and we must return to the question raised in Chap. 1. But as yet, we have not said anything about the theory of electromagnetism in these different models. The next chapter aims to do this.

Chapter 4
Applying the Strong Equivalence Principle

We are perhaps better armed to consider the problem raised by Bondi and Gold in the quote on p. 3. They make the claim that it is impossible to distinguish between the action on a particle of matter of a constant acceleration or of static support in a gravitational field. This looks like the weak equivalence principle WEP4 discussed on p. 23, which comes down to the idea labelled WEP2 on p. 18 that an observer in the spacetime with structure specified by the intervals (2.45) or (2.48) cannot say whether the fields she observes via the spacetime metric are due to her own motion or a (linear) gravitational potential.

But Bondi and Gold's paradox concerns a charged particle, and for the moment we have not considered the way electromagnetism is to be formulated in general relativity. In particular, we have to examine the fields of a charged particle when it is statically supported in a gravitational field. We can see what this means despite the vagueness of the statement: the charge is held still relative to the coordinates $\{y^\mu\}$ of Chap. 2 and we are interpreting the metric of (2.16) on p. 9 as due to an SHGF. We observe that, not only is the charge stationary relative to these coordinates, but the components of the metric are independent of the time coordinate and the mixed time–space components g_{0i} and g_{i0} for $i = 1, 2, 3$ are all zero.

This introduces the notion of a static metric: it is one for which there exist coordinates such that the latter conditions are satisfied (see also p. 127). We observe immediately that a static metric can always be made to look non-static by choice of coordinates. For example, for an arbitrary acceleration in (2.2) on p. 6, the metric of the adapted semi-Euclidean system generally has a time-dependent g_{00} component [given by (2.7) on p. 8]. The Minkowski metric is of course static, and looks static in any inertial coordinates. As shown in Sect. 2.2, when it comes to semi-Euclidean systems, only uniformly accelerating observers will end up with metrics that still look static in their corresponding semi-Euclidean systems. So what we are discussing here is a static metric and coordinates in which it looks static.

Now in special relativity, one can formulate Maxwell's equations for electromagnetism in the standard way relative to the inertial frame. We may suppose that this is what Bondi and Gold were doing in the first part of their paper. But in general relativity a further principle is required in order to extend Maxwell's theory.

This is often referred to as the strong principle of equivalence (SEP). It says that, in the locally inertial frames whose existence is guaranteed by the weak principle of equivalence, all physics looks roughly as it does in the context of special relativity. This is of course a qualitative statement that would not help us much with any real problems. In practice, one takes the special relativistic formulation of the physics, e.g., electromagnetism, in an inertial frame and replaces all coordinate derivatives by covariant derivatives.

The covariant derivatives have to refer to the Levi-Civita connection. One could envisage using different connections to extend different bits of physics, but the strong principle of equivalence tells us always to use the same one (the reader should refer back to the idea of a unique connection introduced on p. 42). Consequently, in a locally inertial coordinate system where the connection coefficients are almost zero and the metric has the Minkowski form, we are not actually changing anything by switching from coordinate derivatives to covariant derivatives. At least, we are not changing very much. The only changes will depend on the connection coefficients, which we have arranged to be very small.

Now in the scenario we are discussing here, one might say that we have the application *par excellence* of the strong principle of equivalence: for we really are not changing anything. The locally inertial frame promised by the weak equivalence principle is exactly and globally inertial! Electromagnetism is thus formulated exactly in the inertial system, and exactly as it was in the special relativistic formulation. But there should be no doubt that we have had to apply a new principle here, and that we have a new theory, viz., the minimal extension of Maxwell's equations (MEME) to general relativity theory. (This is minimal because one can introduce curvature-dependent terms into any extension and it will still reduce to Maxwell's equations when there are no gravitational effects.)

So what does this minimal extension predict for the statically supported charge in this static, homogeneous gravitational field? It says that, viewed from the freely falling Minkowski frame, it will produce exactly the same electromagnetic fields as the uniformly accelerating charge viewed from an inertial frame in a spacetime with no gravitational effects. If the latter radiates, then so will our statically supported charge, and vice versa.

Now although they do not say it explicitly, Bondi and Gold would appear to conclude that the uniformly accelerating charge in a Minkowski spacetime with no gravitational effect radiates electromagnetic energy, since they dispose of two arguments in favour of the non-radiation hypothesis, and they refer shortly afterwards to the travelling energy at $t = 0$. Indeed, this is how Fulton and Rohrlich later interpreted Bondi and Gold's conclusions [3]:

> [...] they claimed that the δ-function provides proof for the radiation from a uniformly accelerated charge.

(However, we argue in Chap. 9 that Fulton and Rohrlich did not understand what the problem was for Bondi and Gold.) This means that Bondi and Gold expect the statically supported charge in this static, homogeneous gravitational field to radiate as viewed from the freely falling Minkowski frame. (In Sect. 15.12.1 on p. 246 ff., we examine these same fields, in particular, the energy–momentum tensor derived from them, when viewed in the semi-Euclidean frame moving with the charge, since this too has led to much debate.) Now how could this raise a paradox? We are told that:

> [...] it might be thought that a radiation field is required to assure that no distinction can be
> made between the cases of gravitation and acceleration.

The statement is a little vague because considerably abbreviated, but we may suppose that we understand what they mean by the cases of gravitation and acceleration. The point is that we expect the same electromagnetic fields in the inertial frame for a uniformly accelerating charge in a spacetime with no gravitational effects, not even non-tidal ones, and for another charge held still relative to coordinates in which some (static, homogeneous) gravitational effects can be described by the metric (2.16). It is our strong principle of equivalence that is at stake here.

The paradox arises for Bondi and Gold because they do not expect a static charge in a static gravitational field to be able to radiate. The paraphrase here is taken from the paper by Mould [5], to be discussed later. Bondi and Gold themselves employ a less crisp formulation in this case:

> [...] if the charge had an associated radiation field, this would reach distant regions unaf-
> fected by the gravitating body, where Maxwell's equations apply without any modification
> due to gravitational effects. According to these equations there can be no static radiation
> field; and as the whole system is static the electromagnetic field cannot depend upon time.

Support for this presumptious claim would later be obtained by examination of the fields of a uniformly accelerating charge in a Minkowski spacetime with no gravitational effects, not even non-tidal ones, when viewed in the semi-Euclidean frame adapted to the charge (see Sect. 15.12.1 on p. 246). It turns out that the magnetic field components are identically zero and that the electric field components are constant in time. The above formulation quoted from Bondi and Gold is rather qualitative and it is not clear whether they themselves calculated the field components relative to the coordinates $\{y^\mu\}$ used to express the metric in (2.16).

However, there is an obvious problem with the idea they express here, just as there are problems with conclusions reached by considering electromagnetic field components relative to a semi-Euclidean frame (see Sect. 16.8). The most striking point is that the attribute of being static, either for the electromagnetic field or for the metric components, is clearly coordinate-dependent. In order to define a static metric (or spacetime) above, we had to say that there exist coordinates in which the metric appears to be static, noting that it could not have time-independent components relative to all coordinate systems. And relative to the inertial coordinate system, there is nothing static at all about the electromagnetic fields generated by a uniformly accelerating charge (although the metric itself is still static in that frame). The electromagnetic field components depend on Minkowski time, even if they do not turn out to depend on semi-Euclidean time.

A qualitative argument cannot stand alone here. Indeed, to push this problem into a corner, we should be asking what theory of electromagnetism is being used here in conjunction with general relativity in order to deduce that a static charge in a static gravitational field will not radiate. What theory predicts this, if it is not the minimal extension of Maxwell's equations (MEME) via the strong principle of equivalence? But if it is MEME, then we have the remarkable result that MEME is self-contradictory.

The suggestion in the present book is that static charges in static gravitational fields (to use the rather too brief paraphrase due to Mould) can in fact radiate, and that they do in fact radiate in the rather idealistic case considered here in which one can apply a kind of strong principle of equivalence *par excellence*. At least, the claim in the present book is that this is the prediction of general relativity in combination with the strong principle of equivalence as stated above (which we abbreviate to GR+SEP), and that there is no paradox because one must in fact accept that a static charge in a static gravitational field can radiate.

Naturally, all the claims here are open to experimental test, or they would be if one could detect this kind of radiation. It is interesting to note in passing the quite different prediction that would be made by the special relativistic theory of gravity outlined in Chap. 3. In that case, one has a global Minkowski frame and any gravitational effect is modelled by an acceleration, presumably toward some massive object. The freely falling charge is accelerating here. Now Maxwell's theory applies directly in inertial coordinates for the Minkowski spacetime. (Indeed, one could say with hindsight that the Minkowski spacetime was designed to house Maxwell's theory as-is.) If we can approximate the gravitational effects by a constant force per unit mass as we usually do in an Earth-based laboratory (it is a force in this special relativistic view), the charge will be uniformly accelerating according to the analysis in Chap. 3. Applying Bondi and Gold's conclusions, it will thus radiate.

In contrast, if the same charge were held still relative to the global inertial system and not allowed to fall, it would not radiate. The predictions are the exact opposite of those made by GR+SEP, precisely because the notion of (local) inertial frame is the exact opposite in the two theories. This is spelt out in more detail in Chap. 10 and should be borne in mind for the discussion in later sections.

Before going on to consider Fulton and Rohrlich's continuation of the debate, let us try to understand how Bondi and Gold proposed to resolve their paradox, because it will be important to see why it is not in fact a solution:

> There is [...] no paradox where the principle of equivalence is considered with its limitations. The surveying of a sufficiently large region can always reveal the presence of a gravitational field through the inhomogeneity that is associated with it, which is absent for an acceleration field. The presence or absence of radiation from the charge could only be established by surveying the space out to a distance $c^2 g^{-1}$ from the charge; but at that distance the presence of a gravitational field can in general be inferred from its inhomogeneity, and there is hence no requirement from the electromagnetic effects to conceal the distinction between the two cases. In the special case where a gravitational field is homogeneous over the entire distance $c^2 g^{-1}$, the potential difference is c^2 over that interval. That case is known to result in several anomalies and it appears from other considerations [...] that it must be excluded from any physical discussion.

The point in the second sentence is that one can only obtain intervals of the form (2.45) or (2.48) in the semi-Euclidean coordinate systems adapted to uniformly accelerating observers in Minkowski spacetimes with no gravitational effects, and these are perfectly homogeneous, whereas any real gravitational effect modelled in GR will not have a perfectly homogeneous metric of this form. The paraphrase by Fulton and Rohrlich [3] (to be discussed further in Chaps. 5, 7, and 9) is:

> They resolve this paradox by arguing that hyperbolic motion requires an infinite homogeneous gravitational field and that such a field does not exist in nature.

Put like this, it does not sound quite so convincing. After all, what we are trying to resolve is a purely theoretical conundrum, and one would like a purely theoretical solution. Note also that Bondi and Gold are raising the theoretical question as to whether the presence or absence of radiation from a charge can only be established by surveying the space out to *large* distances, which seems later to have caused problems for Fulton and Rohrlich, and even Boulware. Parrott argues forcibly against this claim in [7, Sect. 5], the gist being that the field components are analytic functions off the worldline of the charge generating them, and an analytic function is uniquely determined by its values on any open set, no matter how small. In principle, one can pick out terms going to zero as $1/r$ when $r \to \infty$ just by looking at a few points as close to the charge worldline as we wish.

The idea in the above quote is that, at any distance over which one could affirm the observation of EM radiation, the presence of a gravitational field would be revealed by its inhomogeneity, and the electromagnetic effects do not then have to be the same in the two cases. In other words, in a typical curved spacetime, the inertial frames are only inertial locally, so one expects the minimally extended Maxwell equations (MEME) to make at least a slightly different prediction to Maxwell's equations in a flat spacetime with no gravitational effects (not even non-tidal ones). This is supposed to save the charge from having to radiate in the case where it is stationary relative to coordinates in which the metric static.

This is not convincing at all, given that we do have a purely theoretical situation in which the locally inertial frame turns out to be globally inertial. The last two sentences of the above quote from [2] look like a last-ditch attempt to divert attention. After all, if it is true that a stationary charge in a coordinate system for which the metric has static form cannot radiate electromagnetic energy, what is in doubt is GR+SEP, which definitely predicts that the charge will radiate in the perfectly homogeneous system, where there is an application *par excellence* of the strong equivalence principle.

There is also every reason to expect GR+SEP to contravene this no-radiation requirement on stationary charges in other static spacetimes. In order to understand what a charge will do at a given event, one has to change to the coordinates of a freely falling frame with the charge at the origin at that event. This will clearly reveal the acceleration of the charge. Acceleration is effectively measured relative to freely falling frames not just arbitrary coordinate systems. Application of MEME will then predict radiation to a first approximation as viewed in the freely falling frame, and there is no reason to expect closer approximation to do any more than slightly adulterate this state of affairs, especially since we do have a theoretical case (the infinite, homogeneous case) where there is no closer approximation.

One might say that the problem raised by Bondi and Gold is more serious than they seem to realise. If they really intend to maintain the idea that a stationary charge in a coordinate system for which the metric has static form cannot radiate, then they do appear to be at loggerheads with GR+SEP. One has to ask why they do not notice this, or why they assume that the problem with the equivalence principle has gone away. In fact they are using the notion of locality in the equivalence principle in combination with a received idea that radiation fields can only be detected at some distance from their source to solve their problem. If one rejects this received idea, it becomes clear that there must be situations where one can detect radiation within distances over which the gravitational effects (the metric) still appear to be homogeneous. What they then have to sacrifice is the idea that a stationary charge in a coordinate system for which the metric has static form cannot radiate.

Put another way, if one wishes to maintain the latter idea, one is in big trouble, because GR+SEP predicts otherwise. But GR+SEP is the only way we have of extending electromagnetism (or any other theory) to curved spacetimes. So the real problem is here: how would we be able to say anything at all about electromagnetic effects in general relativity? In particular, how could we even formulate electromagnetism in general relativity in order to show that a static charge in a static gravitational field must not radiate? Perhaps it is no surprise that Bondi and Gold's argument is qualitative.

If Bondi and Gold really intend to uphold the idea that a charge that is stationary relative to coordinates in which the metric has the static form will not radiate, they really do need the mistaken notion that EM radiation can only be detected by observations made over some minimal region of spacetime within which gravitational effects will always have revealed their inhomogeneity. But the situation for them is worse than that, because even then it is still not at all clear that this inhomogeneity would be sufficient to mean that MEME would not predict radiation over that region. This would also be essential to save their point of view, because it is not sufficient just to be unable to detect radiation over a given region. They are saying that, in some circumstances, there actually is no radiation.

This is perhaps where one sees the trouble caused by versions of the equivalence principle that speak about whether two situations can be distinguished. One at once assumes that one is talking about whether some observer or other could measure any differences. But that is not at all the idea of SEP. How could it be a serious theoretical consideration? The accuracies with which we can measure EM fields may depend on techniques that are quite different from those used to measure gravitational effects, and in any case, they are both purely technical questions. The strong equivalence principle provides an exact means of transposing pre-GR physics to GR, although whether this leads to a good theory or not remains open, of course.

The problem with statements referring to the indistinguishability of two situations is that they are almost never strictly true, either in Newtonian gravitational theory or in GR. This is why they have to refer to local observations. Real gravitational fields are always inhomogeneous and if one has sufficiently accurate observations, they can be distinguished from effects due to the motion of an observer by observations made over any neighbourhood of the observer, no matter how small.

Furthermore, these observations may concern only the motion of neutral test particles and pay no attention whatever to other physical effects, such as electromagnetic effects. For effects due to the motion of an observer to be confused with gravitational effects in GR, the gravitational effects have to be non-tidal, i.e., with zero curvature, so the strict application of such a principle, insofar as there is one, concerns only this zero curvature case. Although WEP and SEP may be loosely stated with this kind of reference to local observations, these are only loose statements, because the mathematical formulation is exact, as befits fundamental theory, whether the theory successfully corresponds to reality or not.

To sum up, in order to save their idea that a charge that is stationary relative to coordinates in which the metric components are static cannot radiate, Bondi and Gold are having to assume that EM radiation can only be detected by observations made over some minimal region of spacetime within which gravitational effects will always have revealed their inhomogeneity; but even if this were true, it would only rescue a loose version of the strong equivalence principle, while the exact theory GR+SEP which pays no lip service to questions of local measurements or what some observer may or may not be able to distinguish quite clearly requires charges that are stationary relative to coordinates in which the metric components are static to radiate in some circumstances. The example of the SHGF provides as clear an example as one could hope to find, since it is an application of SEP *par excellence* likely to reveal the tendencies of the theory, and since it is in any case a good approximation in many situations where one could clearly determine the presence of an EM radiation field by sufficiently accurate measurement.

Chapter 5
The Debate Continues

In 1960, Fulton and Rohrlich [3] produced a simpler derivation of the electro-magnetic fields due to an eternally uniformly accelerating charge. These authors consider a uniformly accelerating charge in a Minkowski spacetime with no gravitational effects (not even non-tidal ones) [3]. The main achievement of their paper is to obtain the EM fields proposed by Bondi and Gold in a simpler and more convincing way, without assuming any specific physical process. Bondi and Gold consider a process in which a positive and negative charge are simultaneously created, which may of course be interesting for other reasons. Fulton and Rohrlich achieve the same end by mathematics alone.

The main claim is that such a charge will radiate. They devote some time to the question of how energy can then be conserved, given the fact that the radiation reaction force as it is usually calculated is zero in this case, and in a last section they too address the question of whether radiation in uniformly accelerated motion contradicts the principle of equivalence.

Let us pick up their discussion from the expressions for the EM fields and potentials, which agree with those of Bondi and Gold (except for a printing mistake in the latter). It is interesting to examine the way Fulton and Rohrlich view the inclusion of distributions on $x + t = 0$ (see Sect. 15.9 for the full mathematical analysis). When they give the Lienard–Wiechert potentials used to calculate the fields in the region $x + t > 0$, they state a causality condition

$$t - t_Q = R := |\mathbf{r} - \mathbf{r}_Q| > 0 \,, \tag{5.1}$$

where $(\mathbf{r}, t)$ is the field point and $(\mathbf{r}_Q, t_Q)$ is then the unique retarded point on the charge worldline (see Fig. 5.1). This just says that t_Q was prior to t. Forgetting two space dimensions, the condition (5.1) is equivalent to $x + t > 0$. Indeed, the union of all future light cones of points on the particle worldline is precisely the open set $x + t > 0$. Now Born overlooked the need to restrict to $x + t > 0$, and apparently it was Schott who first pointed out the need for the restriction here. However, Schott did not worry about the region $x + t = 0$, which is all-important in the debate. One

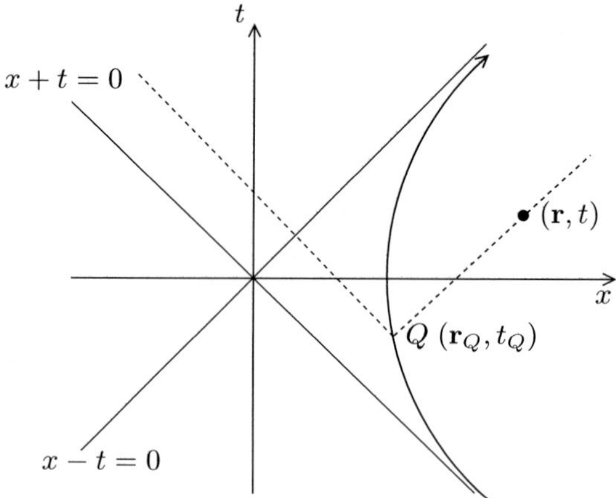

Fig. 5.1 The curve is the worldline of an eternally uniformly accelerating charge. Starting with any field point $(\mathbf{r}, t)$, there exists a unique retarded point Q on the charge worldline with the property that $(\mathbf{r}, t)$ lies on the future light cone of Q. *Dashed lines* indicate the future light cone of the retarded point

cannot just set the fields equal to zero here, or equal to the values suggested by continuously extending the Born solution, and hope to satisfy Maxwell's equations.

Fulton and Rohrlich's causality condition (5.1) restricts solutions to the open region $x + t > 0$, rather than the closed region $x + t \geq 0$. They argue here that fields on $x + t = 0$ would have had to have been emitted at a time $t = -\infty$, whereas the particle would then have been moving at the speed of light, which cannot happen because it has nonzero rest mass. This seems to be a confusion of physics and mathematics. The delta function arises as a limit of the Coulomb field of the charge at times when it was moving very fast. The whole idea of the limit is that we do not have to talk about a time when the charge was moving at the speed of light. We only have to have a situation in which, for any speed v, there was a time prior to which it was always moving faster. Just this situation, ideal as it is, is sufficient to generate a delta function from the Coulomb fields.

Indeed Maxwell's equations do require fields on $x + t = 0$. Fulton and Rohrlich note that Born's fields satisfy Maxwell's equations in $x + t > 0$, but not on $x + t = 0$ if one accepts that all fields are identically zero in $x + t < 0$. Because of their pseudophysical argument about the nonzero rest mass of the charge, they are not enthusiastic about the analysis by Bondi and Gold, preferring to merely state the result and show that it achieves its aim of satisfying Maxwell's equations everywhere. The real problem here is that the motion we are talking about is indeed an ideal motion, which could not happen. The question then is: do we not have a good approximation nevertheless when the uniformly accelerating motion started a long time ago?

Fulton and Rohrlich also point out that the Lienard–Wiechert potentials are not valid in the region $x + t = 0$ and explain this by saying that the derivation of these potentials assumes that the source is not at infinity. This is, of course, a paraphrase for a much longer statement concerning limits, because the eternally uniformly accelerating charge is never at infinity. Presumably if the potentials were derived directly from

$$A^\mu(x) = \frac{e}{2\pi} \int_{-\infty}^{\infty} d\lambda' \frac{dx^\mu}{d\lambda}(\lambda')\theta\left(t - x^0(\lambda')\right)\delta\left([x - x(\lambda')]^2\right), \qquad (5.2)$$

the general rule for a point particle of charge e moving on the worldline $x(\lambda)$, where λ is its proper time, then we would obtain the modified potentials due to Bondi and Gold.

It is then restated that the modified fields for $x + t = 0$ arise from the time $t = -\infty$ when the charge was at $x = +\infty$. Of course, to continue this reasoning, the charge was then unphysical because it was moving at the speed of light! But this is not a valid way of talking about the limiting process mentioned above, i.e., for any speed v, there was a time prior to which it was always moving faster. The charge was never actually moving at the speed of light, but there is no bound on the speeds it had, a kind of intermediate situation which is apparently quite sufficient to produce delta functions on the null hypersurface.

The authors now go on to say that a hyperbolic motion should be regarded as asymptotic in the sense that the times $t_Q = \pm\infty$ can only be approached but never reached and conclude that this corresponds to restricting the domain of validity of the fields to the region $x + t > 0$. Of course, the idea of an asymptote is precisely that we do not have to actually attain some value, such as $\pm\infty$, which would not be useful in a physical model anyway. The delta function terms arise without the need to talk about $t_Q = \pm\infty$. It is true that restricting to $x + t > 0$ avoids the unphysical event $x_Q = +\infty$, $t_Q = -\infty$ where a particle of finite mass has exactly the velocity of light, and is being decelerated, as they say. But the delta function terms on the null hypersurface do not require this unphysical event. Indeed, there is no need to refer to this unphysical event at all.

As a result of the long discussion attributed to this point, during which the authors reveal their uneasiness with regard to the ideality of the motion, it is claimed that the question of whether there is radiation during the hyperbolic motion has nothing to do with whether one adopts the original solution or the modified solution for the fields. According to the authors, the modification suggested by Bondi and Gold is not physically meaningful. One wonders, however, whether the delta function may not be an essential part of the approximation if the motion is a good approximation itself to hyperbolic motion over a long period. It may, as Boulware later suggested [6], be the only way to understand the flow of energy in spacetime, always a delicate problem in electromagnetism (see Sect. 15.12.2).

It turns out that Fulton and Rohrlich have a better reason to make this claim: they eventually show that one can calculate a radiation rate $\mathscr{R}$ for any event on the charge worldline. Since the fields on $x + t = 0$ are not required to find $\mathscr{R}$ for any such event, they are in fact irrelevant to the radiation question. It should be noted immediately that this puts paid to Bondi and Gold's solution to their paradox, wherein one can only check for radiation far from the charge. Here the radiation is associated with specific events on the charge worldline. Oddly enough, however,

Fulton and Rohrlich would appear to concord with Bondi and Gold's solution. Let us examine their arguments more closely.

First of all they try to establish the idea that the radiation should be invariant for all observers, using the following qualitative argument from quantum theory. When a photon is emitted, it can be registered by a counter located at a great distance from the source, and such an effect can be seen by every observer. Indeed this would appear to show that the existence of radiation should be independent of inertial observer, but it does not show that the rate of radiation should be independent of inertial observer, which is what they eventually conclude from their quantitative (pre-quantum) arguments. One would not a priori expect a rate of energy flow to be Lorentz independent, because one has to consider the energy of these photons and the time interval over which they are registered. Although the counting of photons is sufficient to decide whether or not there is radiation, the energy of the photons and the time scale used to count their rate of arrival both depend on the frame.

The authors also require the radiation to reduce to the conventional definition given by the Poynting vector in the instantaneous rest system of the source. Indeed they begin their calculation by constructing the Poynting vector $\mathbf{S} = (S_1, S_2, S_3)$ from the energy–momentum tensor

$$4\pi T^{\mu\nu} = -F^{\mu\lambda} F^{\nu}_{\lambda} + \frac{1}{4} g^{\mu\nu} F^{\lambda\sigma} F_{\lambda\sigma} ,$$ (5.3)

by the usual definition

$$T^{k0} = -S_k , \quad k = 1, 2, 3 .$$ (5.4)

The retarded four-velocity of the charge at time t_Q is v^{μ}_Q. Interestingly, they describe t_Q as the time when radiation was emitted, already affirming the idea that radiation is associated with specific events on the charge worldline. If n^{μ} is a spacelike unit vector orthogonal to v^{μ}_Q, we are told that the energy flux density emitted at event Q on the worldline in the direction specified by n^{μ} is then

$$I = T^{\mu\nu} v^Q_{\mu} n_{\nu} .$$ (5.5)

This is a scalar quantity, i.e., unchanged by change of Lorentz frame. In fact, it gives the energy flux density (per unit time and per unit area) with all quantities measured in the instantaneous rest system for the charge at event Q, denoted by S_Q. This works because in S_Q, the expression reduces to

$$I = \mathbf{S} \cdot \hat{\mathbf{n}} ,$$ (5.6)

where $\hat{\mathbf{n}}$ is a 3-vector defined by $n^{\mu} = (0, \hat{\mathbf{n}})$, bearing in mind Fulton and Rohrlich's convention for the metric, in which spacelike vectors have positive length.

Parrott gives a very detailed analysis of the energy–momentum tensor and how it can be interpreted [13, Sect. 3.8]. He discusses the derivation of the Lorentz–Dirac equation by the method first discussed by Dirac [14] and carries out the integration of energy–momentum flow through the walls and caps of a Bhabha tube, composed of retarded spheres centered on events on the worldline (see Chap. 6 for more details). Parrott's book [13] will be a good point of comparison for what is being discussed here, precisely because it is so detailed. In this respect, Fulton and Rohrlich's discussion is a little disappointing.

According to Parrott's rule on p. 111 of [13], with the opposite sign conventions for the Minkowski metric, and using the notation $\hat{T}$ for the energy–momentum when it is construed as a vector-valued 1-form: If n is spacelike, then $\hat{T}_x(n)$ is the energy–momentum per unit area and per unit time at x, flowing in the n direction relative to a frame in which n is purely spacelike, i.e., a frame in which $n = (0, \mathbf{n})$. The presence of v^Q in I of (5.5) just picks out the zero component in that same frame. What is important to note is that the rate we are talking about is with respect to the proper time of the charge at the retarded time, i.e., it is not a rate with respect to coordinate time in the inertial frame. The rate of a 4-vector with respect to coordinate time would not be a 4-vector.

Now at each time in this instantaneous rest frame, the space part of the light cone is a light sphere of radius R, where R will be large for large times. We take a 2D surface element $\mathrm{d}^2\sigma$ on its surface, with normal $\hat{\mathbf{n}}$. Then $I\mathrm{d}^2\sigma$ is the rate at which energy is radiating through this element at the appropriate time, i.e., the time $t = t_Q + R$ when the energy emitted at time t_Q reaches the distance R. The total rate of emission of radiated energy at time t_Q, which we denote by $\mathscr{R}$, can be found by integrating over the whole surface of the light sphere in the limit of infinite $R = t - t_Q$ for the fixed emission time t_Q. Hence,

$$\mathscr{R} = \int T^{\mu\nu} v^Q_\mu n_\nu \mathrm{d}^2\sigma \qquad (\text{limit } R \to \infty, \text{ fixed } t_Q) . \qquad (5.7)$$

We note that we have a rate here, because we would have to integrate over a 3D hypersurface with one dimension being time to obtain an energy. Why do we take the light sphere to infinity? The idea is that we might otherwise pick up quantities that are not radiation. For Fulton and Rohrlich, radiation has to be something that can get a long way from the source.

However, Parrott shows that one does not have to be infinitely far from the source at its emission time in order to pick up the radiated energy in an integration (see Sect. 5 of [7]). This is a key point, because Fulton and Rohrlich argue later that an observer who wants to detect radiation cannot do so in the neighbourhood of the particle's worldline and this is a major step in their argument concerning the equivalence principle. Furthermore, as mentioned above, Fulton and Rohrlich clearly assume that the radiation is associated with specific events on the charge worldline, which would also appear, at least intuitively, to contradict the idea that it cannot be detected in the neighbourhood of the worldline.

Anyway, in the inertial frame S_Q in which the charge was at rest at the retarded time, the formula for $\mathscr{R}$ becomes

$$\mathscr{R} = \int \mathbf{S} \cdot \hat{\mathbf{n}} R^2 \mathrm{d}\Omega \qquad (\text{limit } R \to \infty, \text{ fixed } t_Q) , \qquad (5.8)$$

where $\mathrm{d}\Omega = \sin\theta\,\mathrm{d}\theta\mathrm{d}\phi$ indicates the integration over angles. If $\mathscr{R} \to 0$ as $R \to \infty$, then we say that no radiation is emitted. If the limit is finite, then there is radiation. However, the reckoning at very large distances has become a fundamental feature of this definition.

There is one more tricky point to come here, which rather spoils the clarity of the approach. The authors note that (5.8) is valid in any inertial system, provided that a factor of $\mathrm{d}t/\mathrm{d}t_Q$ is inserted into the integral. This factor is to be computed from the causality condition (5.1) and it is claimed that this makes $\mathscr{R}$ into an energy emission rate $\mathrm{d}W/\mathrm{d}t_Q$ relative to the source time t_Q. From the causality condition (5.1), we have

$$\frac{\mathrm{d}t}{\mathrm{d}t_Q} = 1 + \frac{\mathrm{d}R}{\mathrm{d}t_Q}\,. \tag{5.9}$$

We are taking t and R to be functions of t_Q by simply fixing the space coordinates and asking how R would change if the source event were a little different. Our freedom to do this comes from the fact that we also adjust t so that the new $t_Q + \mathrm{d}t_Q$ is the retarded time for $(t + \mathrm{d}t, x, y, z)$. We then have

$$R^2 = (\mathbf{r} - \mathbf{r}_Q)\cdot(\mathbf{r} - \mathbf{r}_Q)\,, \tag{5.10}$$

whence

$$2R\frac{\mathrm{d}R}{\mathrm{d}t_Q} = -2(\mathbf{r} - \mathbf{r}_Q)\cdot\mathbf{v}_Q\,, \tag{5.11}$$

so that

$$\frac{\mathrm{d}R}{\mathrm{d}t_Q} = -\frac{\mathbf{R}}{R}\cdot\mathbf{v}_Q = -\hat{\mathbf{n}}\cdot\mathbf{v}_Q\,, \tag{5.12}$$

and finally

$$\frac{\mathrm{d}t}{\mathrm{d}t_Q} = 1 - \frac{\mathbf{R}}{R}\cdot\mathbf{v}_Q = 1 - \hat{\mathbf{n}}\cdot\mathbf{v}_Q\,. \tag{5.13}$$

It is not explained why this factor, when inserted into the integrand of (5.8), should deliver the rate of energy emission as measured relative to the arbitrary inertial frame in which the charge has coordinate 3-velocity $\mathbf{v}_Q$.

This is not an entirely trivial matter, because it has an angle dependence via the scalar product. The rest of the discussion here is confusing too, but fortunately, these are standard results and one can pick and choose one's preferred exposition. The much more detailed version in [13] is recommended here and will be cited. Oddly, Fulton and Rohrlich themselves cite a general result from another source, although they are still disappointingly vague about where the quantities mentioned should be measured. They quote the very general result that the radiation rate $\mathscr{R}$ for any given source point Q with given four-velocity $v^\mu(\tau)$ is found using the Lienard–Wiechert potentials to be

$$\mathscr{R} = \frac{2}{3}e^2 a_\mu a^\mu \,,\tag{5.14}$$

where

$$a^\mu_Q := \frac{\mathrm{d}v^\mu_Q}{\mathrm{d}\tau}$$

is the retarded four-acceleration of the source. They then triumphantly point out that this is an invariant, which is not helpful, because we do not know in what frame the energy is to be measured and relative to what time. And as mentioned above, one does not a priori expect an energy rate to be invariant because both energy and time depend on the frame. There are subtle points here that do not get a sufficiently careful treatment.

So before seeing how Fulton and Rohrlich go from here, let us take a look at Parrott's very detailed discussion in [13].

Chapter 6
A More Detailed Radiation Calculation

One of the main points in the present context is the calculation in Sect. 4.4 of [13] leading to the result

$$\int_{S_r(\tau_1,\tau_2)} {}_\perp T = \frac{q^2}{2r}\big[u(\tau_2) - u(\tau_1)\big] \quad \text{(mass renormalisation)}$$

$$-\frac{2}{3}q^2 \int_{\tau_1}^{\tau_2} u(\tau)\langle a(\tau), a(\tau)\rangle \, d\tau \quad \text{(radiation)}$$

$$-q \int_{\tau_1}^{\tau_2} \hat{F}_{\text{ext}}(u) \, d\tau \quad \text{(Lorentz force)} . \tag{6.1}$$

The notation and line-by-line comments require some explanation.

T is a vector-valued 1-form representing the energy–momentum of all the electromagnetic fields, i.e., fields generated by the charge itself and any external fields acting on it. The symbol $S := {}_\perp T$ denotes the Hodge dual of this object, a vector-valued 3-form which Parrott interprets in his Sect. 3.8.

The integration is carried out over the walls $S_r(\tau_1, \tau_2)$ of a Bhabha tube (see Fig. 6.1). We choose some fixed radius $r \in \mathbb{R}^+$. For a given $z(\tau)$ on the worldline, we choose a Lorentz frame in which the event $z(\tau)$ has coordinates $(\tau, 0)$ and in which the particle is at rest at time τ. If light pulses are emitted in all directions from this event, the light forms an expanding sphere with radius r at time $\tau + r$ in this frame. As τ varies from some τ_1 to some $\tau_2 > \tau_1$, these spheres of radius r form our hypersurface:

$$S_r(\tau_1, \tau_2) := \bigcup_{\tau_1 < \tau < \tau_2} S_r(\tau) , \tag{6.2}$$

where

$$S_r(\tau) := \Big\{ z(\tau) + r\big[u(\tau) + w\big] : \langle w, w\rangle = -1 , \quad \langle u(\tau), w\rangle = 0 \Big\} . \tag{6.3}$$

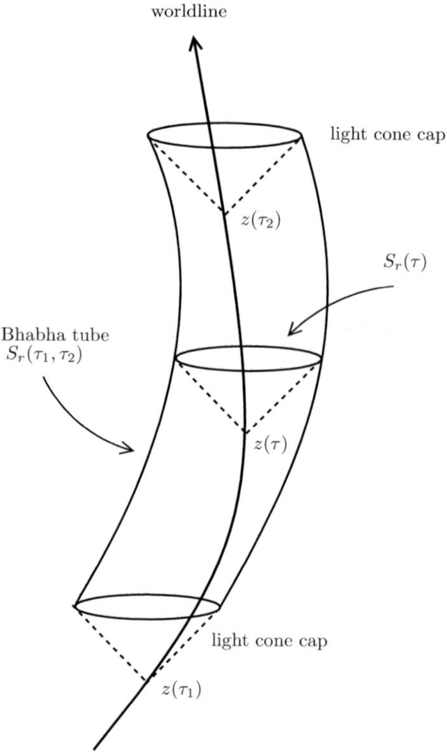

Fig. 6.1 Construction of the Bhabha tube showing light cone caps at the ends of the tube

The object $S_r(\tau_1,\tau_2)$ is the Bhabha tube. The spheres $S_r(\tau)$ are known as the retarded spheres for obvious reasons: τ is the retarded time for all points on $S_r(\tau)$. The $S_r(\tau)$ are genuine spheres within the hyperplane $t = \tau + r$ in the emission frame. They contain the particle, but it is not generally at the centre at this time. This is because the sphere is centered on the spatial origin in that frame, and there is no reason why the particle should remain at the spatial origin of this frame! It will only do so if it is moving inertially, i.e., without acceleration.

In the Bhabha tube, the Minkowski spacetime vectors connecting the emission point $z(\tau)$ with the corresponding sphere $S_r(\tau)$ are not orthogonal to the worldline, i.e., they are not purely spacelike in the rest frame of the particle at time τ, but are in fact null. What Parrott calls the Dirac tube, used by Dirac in his paper [14], is

$$\bar{S}_r(\tau_1,\tau_2) := \bigcup_{\tau_1 < \tau < \tau_2} \bar{S}_r(\tau) , \tag{6.4}$$

where

$$\bar{S}_r(\tau) := \left\{ z(\tau) + rw : \langle w,w \rangle = -1 , \quad \langle u(\tau),w \rangle = 0 \right\} . \tag{6.5}$$

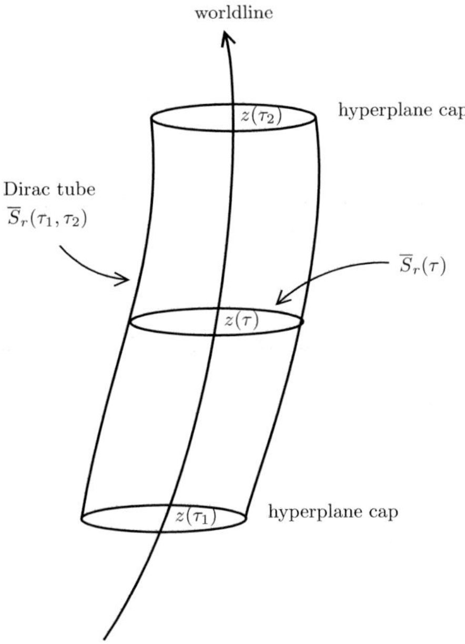

Fig. 6.2 Construction of the Dirac tube showing the hyperplane caps at the ends of the tube

For given τ and r, the associated sphere is now in the spacelike hypersurface of the particle rest frame at time τ (see Fig. 6.2). This explains why the Dirac and Bhabha tubes do not generally coincide anywhere. For given τ, the intersection of the Bhabha tube with the spacelike hypersurface of the particle rest frame at time τ is not generally going to be $\overline{S}_r(\tau)$.

The Bhabha tube is used because the integration works out more easily. The integration on the left-hand side of (6.1) gives the total energy–momentum flowing out of the retarded sphere that follows the charge between times τ_1 and τ_2. In order to derive an equation for the conservation of energy–momentum, which can in turn be used to derive an equation of motion for the charge, one must also integrate over the caps closing the two ends of the tube. This is discussed later (see Chap. 11). In fact, those integrals contribute only to mass renormalisation.

It is worth noting that the walls of these tubes (Bhabha or Dirac) are 3D hypersurfaces in spacetime, with two spatial dimensions and one temporal dimension. In general, coordinate-independent integration over such a submanifold can only act on an integrand that is a 3-form, in this case $S := {}_\perp T$. However, since S takes vector values, one must restrict to inertial coordinates, in which case one can obtain Lorentz covariant results precisely because Lorentz transformations are constant over spacetime. When attempting the same kind of investigation in general relativity, a very large part of the extra work can be put down to this problem. The paper on this subject by DeWitt and Brehme is discussed briefly in Chap. 12 [4]. This also explains the comment by Fulton and Rohrlich with regard to the fact that $\mathscr{R}$ as given by their definition (5.7) is an invariant quantity because the retarded sphere over which they integrate is invariant. In general, it is worth being very suspicious of this kind of apparently innocuous remark. Coordinate-independent integration can be a problem even in special relativity.

On the right-hand side of (6.1), one has other mass renormalisation terms, which we do not discuss here, terms which lead to the usual Lorentz force law for the external electromagnetic field $\hat{F}_{\text{ext}}$, and a radiation term

$$R(\tau_1, \tau_2) = -\frac{2}{3}q^2 \int_{\tau_1}^{\tau_2} u(\tau)\langle a(\tau), a(\tau)\rangle \, d\tau \,, \tag{6.6}$$

interpreted as the energy–momentum radiated by the charge over its proper time interval $[\tau_1, \tau_2]$. The energy–momentum is measured in an arbitrary inertial frame. In these formulas, $u(\tau)$ is the charge four-velocity and a the charge four-acceleration, both taken at its proper time τ, and the notation $\langle a, a\rangle$ indicates the usual pseudo-inner product defined by the Minkowski metric.

For the moment, let us concentrate on the term $R(\tau_1, \tau_2)$. We have the generalised Larmor radiation law, which says that the total energy–momentum radiated in any finite proper time interval $\tau_1 \leq \tau \leq \tau_2$, as seen from the laboratory frame, is

$$R(\tau_1, \tau_2) = -\frac{2}{3}q^2 \int_{\tau_1}^{\tau_2} \langle a, a\rangle u \, d\tau \,, \tag{6.7}$$

where the minus sign arises because $\langle a, a\rangle \leq 0$. Dividing by the proper time interval $d\tau := \tau_2 - \tau_1$ and taking the limit as it tends to zero, one has the formula

$$\begin{array}{c}\text{energy–momentum radiation rate}\\ \text{per unit proper time}\end{array} = -\frac{2}{3}q^2\langle a, a\rangle u \,. \tag{6.8}$$

The energy–momentum here is measured in an arbitrary inertial frame, in which the four-velocity of the charge at the emission time was u, but its rate of emission is with respect to the proper time of the charge. It says that, in the instantaneous rest frame of the charge at the relevant time, there is no radiation of pure momentum. The expressions (6.7) and (6.8) are covariant (because proper time is a scalar) and imply the usual Larmor formula

$$\frac{2}{3}q^2 a \cdot a \tag{6.9}$$

for the rate of outflow of field energy through a retarded sphere, as viewed in the instantaneous rest frame of the charge at the retarded time, with $(0, a)$ the four-acceleration in this instantaneous rest frame.

We shall return to these results and Parrott's derivation of the Lorentz–Dirac equation later (see Chap. 11), but for the moment let us continue the discussion of [3].

Chapter 7
Defining the Radiation
from a Uniformly Accelerating Charge

The result (5.14) cited in [3] looks very like (6.8), except that the zero component of the latter contains $u^0 = \gamma(u)$, where $u := v_Q$ is the instantaneous coordinate speed of the charge at the retarded event Q. We are about to find out exactly what $\mathscr{R}$ is in the expression of (5.14). We will then understand what is involved in the claim that it is invariant.

Of course, one can convert the quantity in (6.8) from a rate with respect to the proper time τ of the charge to a rate with respect to the time coordinate t in the arbitrary inertial frame, using the well known result

$$\frac{d\tau}{dt} = \gamma(u)^{-1} \ .$$

Since (6.8) says that

$$\frac{dE}{d\tau} = -\frac{2}{3}q^2 \langle a, a \rangle \gamma(u) \ , \tag{7.1}$$

we have the remarkable result that

$$\frac{dE}{dt} = \frac{dE}{d\tau}\frac{d\tau}{dt} = -\frac{2}{3}q^2 \langle a, a \rangle \ . \tag{7.2}$$

This shows precisely why the radiation rate $\mathscr{R}$ in (3.5) of [3] is an energy rate with both energy and time measured in an arbitrary inertial frame. One would not a priori expect such a quantity to work out the same in any inertial frame, but it does, because of the form of the energy–momentum radiation found in (6.8), and because $\langle a, a \rangle$ (or $a_\mu a^\mu$ in [3]) is an invariant quantity. We observe that this invariance holds for any charge motion.

Hopefully, this is clearer than what is said in [3]. The meaning of $\mathscr{R}$ has changed in going from (3.3) and (3.4) (where it is the same) to (3.5) of the cited paper. In the former, it was a rate of radiation of energy measured in the instantaneous rest frame of the charge with respect to the proper time of the charge; in the latter it is a rate of radiation of energy measured in an arbitrary inertial frame with respect to the time coordinate in that frame. Of course, this is what Fulton and Rohrlich are explaining between these equations, and it is achieved with the fudge factor they denote by dt/dt_Q of (5.13).

The interlude of Chap. 6 introducing the method used by Parrott in [13] will not be wasted because we shall use his account of the Lorentz–Dirac equation to see what is at stake in the next part of the great debate: how come the radiation reaction in the Lorentz–Dirac equation is zero for a uniformly accelerating charge if the charge manages to radiate electromagnetic energy?

But first, we need to review Fulton and Rohrlich's conclusions with regard to the new $\mathscr{R}$ of (5.14), because there is another way in which the exposition by Parrott can cast light upon the debate. They carry out the calculation in the specific case of the uniformly accelerating charge, using (5.8) but putting in the fudge factor of (5.13), with the result

$$\mathscr{R} = \frac{2}{3}\frac{e^2}{\alpha^2},\tag{7.3}$$

where α is their notation for the reciprocal of the proper acceleration. This clearly agrees with the general result (5.14). So the already remarkable result (5.14), which shows that this radiation rate will come out the same relative to any inertial frame, leads to the even more remarkable result (7.3) which shows that for a uniform acceleration this radiation rate will not change as time goes by. That is, it does not depend on the event Q on the charge worldline which we identify with the emission event.

It is worth stressing two things here. First, this radiation rate (7.3) is the rate with respect to the coordinate time in the chosen inertial frame, not the rate with respect to the proper time of the charge, as suggested by Fulton and Rorhlich in their conclusions after stating (7.3). Secondly, one does not need to integrate over a retarded sphere (the light sphere for Fulton and Rohrlich) at infinity in order to obtain the radiation rate. We need to be quite clear about this so let us devote several paragraphs to it.

Looking back at Parrott's result (6.1) on p. 67, we observe that it does not depend on the radius r of the retarded spheres making up his Bhabha tube. In the Dirac approach to obtaining an equation of motion by integrating the flow of energy–momentum through a world tube containing the charge worldline, one obtains terms that go as $1/r$ and terms that go as r, r^2, etc. But one has to let r tend to zero in the end to get the equation of motion (discussed later). The terms that go as $1/r$ can fortunately be absorbed into the particle mass in the classical version of the renormalisation process. The terms that go as powers of r greater than unity go to zero, and one is left with the radiation term.

This is also a well known feature of radiation reaction terms when one uses self-force calculations on extended charge distributions to calculate them. This is discussed in more depth later (see p. 110), but a simple exposition can be found in [15, Chap. 28]. Briefly, an extended charge distribution generates EM fields that act on other parts of the distribution. Some terms in the resulting self-force go as $1/r$, the radius of the distribution, while others are independent of it, and yet others go as powers of r greater than unity. Here one avoids mass renormalisation because r is fixed, but the terms of order $1/r$ contribute to the mass. The terms independent of r can be interpreted as the extra force against which one must do work in order to produce the required electromagnetic radiation when accelerating the charge. (One actually has an explanation for why a charged particle radiates, which is lost when the particle is treated as a mathematical point.)

In another sense, it is not really surprising that the radiation rate should be independent of the radius of the retarded sphere across which its passage is measured. As mentioned earlier, Fulton and Rohrlich clearly assume that the radiation is associated with specific events on the charge worldline, and this would also appear, at least intuitively, to contradict the idea that it cannot be detected in the neighbourhood of the worldline.

Returning to the conclusion that the radiation rate (7.3) is independent of the emission event Q, there is a reason why one might expect this, already mentioned at the beginning of Chap. 1: the charge worldline looks exactly the same to any inertial observer moving along the axis of acceleration, up to change in coordinate origin. Let us examine this idea. If in the inertial frame S the charge decelerates down the x axis, stopping at $x = \alpha$ at time $t = 0$ and accelerating back up the x axis, then in an inertial frame S' found by Lorentz boost along the x axis, the charge decelerates down the x' axis, stopping at $x' = \alpha$ at time $t' = 0$ and accelerating back up the x' axis.

It is worth proving this simple claim. If the charge worldline is given by $x^2 - t^2 = \alpha^2$ in S, and if the coordinates x', t' are given as usual by

$$x' = \gamma(x - vt) , \qquad t' = \gamma(t - vx) ,$$

in S', then

$$x'^2 - t'^2 = \alpha^2 ,$$

by straight calculation. This is related to the fact that the worldlines of uniformly accelerating particles are the flow lines of a Killing vector field associated with the Lorentz boosts in a way we shall make explicit later (see Sect. 16.2).

One obvious consequence of this remarkable symmetry, which largely characterises the Minkowski spacetime, is that the retarded fields of the uniformly accelerating charge will also look the same to any inertial observer moving along the x axis. As pointed out by Fulton and Rohrlich, one can write

$$F'_{\mu\nu}(x',y',z',t') = F_{\mu\nu}(x',y',z',t') , \tag{7.4}$$

where the left-hand side is the EM field tensor component in the primed frame at the point with primed coordinates (x',y',z',t'), and the right-hand side is the EM field tensor component in the unprimed frame at the point with unprimed coordinates $(x,y,z,t) = (x',y',z',t')$. Note that we are not saying that the primed and unprimed field components are the same at some given event in spacetime. If we set $(x,y,z,t) = (x',y',z',t')$, then (x,y,z,t) does not label the same event in S as (x',y',z',t') labels in S' (see also Sect. 15.9.4).

It should be stressed here that one has the same energy radiation rate per unit coordinate time in S' as in S for any charge worldline, according to our analysis. In other words, the invariance of $\mathscr{R}$ in (5.14) is not a special consequence of the symmetry described here, but a general result. However, we can obtain the constancy of our radiation rate from the above symmetry for uniform acceleration, because it shows that the radiation rate (7.3) is independent of the emission event Q.

In any case, we do have the result that $\mathscr{R}$ in (7.3) is constant, and it is a consequence of the fact that $a_\mu a^\mu$ is constant, i.e., the acceleration is uniform. Furthermore, we can agree with Fulton and Rohrlich that every inertial observer will measure the same rate of energy radiation with respect to their own coordinate time, and also that every inertial observer will measure the same intensity pattern. One could say, as they do, that radiation cannot be transformed away by a Lorentz transformation. Likewise, radiation cannot be transformed into existence by a Lorentz transformation. It is not an artifact of one's inertial motion. (Neither, as it turns out, is it an artifact of one's accelerating motion. At least, this is what we shall argue on pp. 153 ff. in Chap. 14.)

Fulton and Rohrlich reiterate the idea that the radiation rate $\mathscr{R}$ can be computed for an arbitrary event on the hyperbolic worldline, which as we have said, suggests that it does not have to be detected at infinity. They use this observation to conclude something rather different, namely that the fields on $x+t=0$ are irrelevant to the question of whether the charge radiates. This is certainly what one would conclude from the analysis by Parrott too. So the value of $\mathscr{R}$ follows from the Born solution alone, and in the region $x+t>0$.

With regard to Bondi and Gold's argument, they interpret it as meaning that radiation is emitted at $t=-\infty$, producing the fields on $x+t=0$, and claim that nothing is proved concerning radiation emission at other times. Here we come back to the problem with the limit. Of course, the mathematical model does not contain a point with $t=-\infty$, so it is not strictly true that nothing is proved with regard to the charge on its worldline at other times. For it is the fact that the uniform acceleration occurs over an unbounded set of times prior to any given time that appears to produce the fields on $x+t=0$, and so one is indeed saying something about the charge on its worldline at past times. However, in another sense, it is true that Bondi and Gold prove nothing about the radiation rate, and it is not even explicitly stated in [2] that there is radiation!

Although Fulton and Rohrlich discard the fields on $x+t=0$ as physically meaningless, as they do the idea of hyperbolic motion at $t=-\infty$, this is clearly a misunderstanding of the mathematical notion of limit, and we shall see in Sect. 15.12 what Boulware has to say about it in his celebrated solution to the radiation question for uniformly accelerating charges [6]. Boulware devotes a lot of attention to the flow of energy in the region where the Born solution is valid. He suggests that a delta function field is produced on the null hypersurface $z+t=0$ by Lorentz transformation of a Coulomb field produced in early times of the motion and that there is a flow of radiation from the null hypersurface which focuses onto the charge and then radiates out from it again. In a sense, he is trying to justify the Bondi/Gold picture. He is also offering an explanation for the fact that the radiation reaction prescribed by Dirac is zero, and yet we still have radiation (see Chap. 8). However, all these things will turn out to be quite irrelevant to the problem of the equivalence principle.

There is one more point which is effectively dealt with in the discussion of radiation from a uniformly accelerating charge given by Fulton and Rohrlich, and this is Pauli's argument against there being any radiation. Indeed, it appears that Pauli takes a limit to large distances R at a fixed coordinate time $t = 0$. This means that he integrates over a Dirac sphere that is always simultaneous with the proposed emission point on the worldline, rather than the retarded sphere considered in [3]. This trivially gives zero for the radiation, because, for the chosen coordinate time $t = 0$, the Poynting vector is identically zero everywhere on the corresponding spacelike hypersurface. Pauli's definition of radiation is clearly not a good one.

Fulton and Rohrlich conclude that the fact that the magnetic field $\mathbf{H} = 0$ at $t = 0$ is unusual for uniformly accelerated motion and of some interest, but that it has nothing to do with the presence or absence of radiation. The interesting point is that the magnetic field is zero in the instantaneous rest frame of the charge. In fact, as mentioned earlier, in the semi-Euclidean frame adapted to a hyperbolic motion, which borrows the instantaneous rest frames of the charge to provide spatial coordinates, the magnetic field turns out to be identically zero at all times. Furthermore, the electric field is also static relative to the proper time of an observer moving with the charge when the semi-Euclidean coordinate system is used. But we have to agree with Fulton and Rohrlich, and this is the import of Parrott's paper [7] on the equivalence principle, that this has nothing to do with the presence or absence of radiation. This is what Parrott calls Boulware's error. The Poynting vector in the co-accelerating frame is zero, but this does not mean that there is no radiation. On the other hand, we shall argue here that this is still not relevant to the question of the equivalence principle.

Chapter 8
Energy Conservation
for a Uniformly Accelerated Charge

We have shown that energy is radiated indefinitely far from the charge and now we have the problem of showing where that energy came from. Fulton and Rohrlich examine the equation of motion for the charge because they are concerned with its kinetic energy. For this reason they discuss the Lorentz–Dirac equation, but without much enthusiasm. Indeed this equation of motion is not very well established by observation, or even by theory, predicting some rather odd behaviour at times [13]. It does have the merit that one can derive it in two apparently different ways:

- Consider the self-force exerted on a small charge distribution, sum of forces exerted by one part on another, when it is accelerated.
- Consider the energy–momentum tensor and obtain an equation of motion by requiring conservation of energy–momentum.

The two methods do lead to the same equation of motion, up to their respective limitations. Fulton and Rohrlich describe the second method as an application of Dirac's classical electrodynamics, first published in 1938 [8]. If they had worded it as we have here, their discussion might have seemed somewhat circular.

For the record, let us just repeat the presentation of this equation as given in [3], with roughly the same notation. The Lorentz–Dirac equation is

$$
m\frac{\mathrm{d}v^{\mu}}{\mathrm{d}\tau} = F^{\mu}_{\mathrm{ext}} + \Gamma^{\mu}\,,
\tag{8.1}
$$

where

$$
\Gamma^{\mu} := \frac{2}{3}e^2\left(\frac{\mathrm{d}^2v^{\mu}}{\mathrm{d}\tau^2} - v^{\mu}\frac{\mathrm{d}v_{\lambda}}{\mathrm{d}\tau}\frac{\mathrm{d}v^{\lambda}}{\mathrm{d}\tau}\right)\,.
\tag{8.2}
$$

If the externally applied force is electromagnetic, then

$$
F^{\mu}_{\mathrm{ext}} = eF^{\mu\nu}_{\mathrm{ext}}v_{\nu}
$$

is the Lorentz force.

Now one of the strange things when the motion is a uniform acceleration is that the so-called radiation reaction term Γ^μ is zero in the above equation (see Sect. 17.1). The equation of motion becomes in this case

$$m\frac{\mathrm{d}v^\mu}{\mathrm{d}\tau} = F^\mu_{\mathrm{ext}} \,, \tag{8.3}$$

where F^μ_{ext} is just whatever external force we require to keep the charge on its hyperbolic course. The authors now show that the Newtonian force required to produce uniform acceleration is in fact a constant force, already discussed in Chap. 3 for a slightly different reason. Their proof goes as follows. The covariant force F^μ is related to the Newtonian three-vector force $\mathbf{F}$ by

$$F^\mu = \gamma(\mathbf{F}\cdot\mathbf{v},\mathbf{F}) \,, \tag{8.4}$$

which produces the identity $F^\mu v_\mu = 0$. The latter is a necessary part of relativistic dynamics, because the 4-force is declared proportional to the 4-acceleration in a generalisation of Newton's law, and the 4-acceleration is always pseudo-orthogonal to the 4-velocity (in the Minkowski geometry) because the 4-velocity has constant (unit) pseudolength.

Now the equation of motion (8.3) proposed here gives

$$m\frac{\mathrm{d}(\gamma\mathbf{v})}{\mathrm{d}t} = \mathbf{F}_{\mathrm{ext}} \,. \tag{8.5}$$

If we feed in the details of the hyperbolic motion on the left-hand side using the notation of [3], viz.,

$$v = \frac{t}{(\alpha^2+t^2)^{1/2}} \,, \qquad \gamma(v) = \frac{(\alpha^2+t^2)^{1/2}}{\alpha} \,, \tag{8.6}$$

where $1/\alpha$ is the proper acceleration of the charge, we soon find that

$$\mathbf{F}_{\mathrm{ext}} = m\frac{1}{\alpha}\hat{\mathbf{z}} \,, \tag{8.7}$$

with acceleration along the z axis in [3]. This proves that the Newtonian 3-force producing uniform acceleration is a constant force, and conversely, that a constant Newtonian 3-force will produce a uniform acceleration. In this case, the 3-acceleration is

$$\mathbf{a} = \frac{\mathrm{d}(\gamma\mathbf{v})}{\mathrm{d}t} = \frac{1}{\alpha}\hat{\mathbf{z}} =: \mathbf{a}' \,, \tag{8.8}$$

where $\mathbf{a}'$ is the space part of the 4-acceleration in the instantaneous rest system.

As noted in [3], one can define one-dimensional uniform acceleration in an inertial system S to be that motion characterised by a constant Newtonian force

$$\mathbf{F}_{\text{ext}} := m\frac{\mathrm{d}(\gamma\mathbf{v})}{\mathrm{d}t} \, .$$

There is no mention of the connection with free fall in a static homogeneous gravitational field (SHGF) as it might be modelled by a constant Newtonian force in special relativity, thus leading to a prediction of hyperbolic motion in such a case (see Chap. 3).

Fulton and Rohrlich now set out to establish that zero radiation reaction does not necessarily imply that there is no radiation. Defining first the kinetic energy of the particle as

$$T := m(\gamma - 1) \, , \tag{8.9}$$

we examine the zero component of (8.3):

$$\frac{\mathrm{d}T}{\mathrm{d}\tau} = F_{\text{ext}}^0 = \frac{\mathrm{d}W_{\text{ext}}}{\mathrm{d}\tau} \, . \tag{8.10}$$

Naturally, we have equated the zero component of the external force with the rate of work it does, as dictated by (8.4). So we have here that the rate of work done by the external force is equal to the increase in kinetic energy of the particle. Interestingly, Fulton and Rohrlich now add that we know this to be the case for a neutral particle in a static gravitational field. Of course, there is doubt about whether this is the case for a charged particle, even if the gravitational field is also homogeneous, in which case the Lorentz–Dirac equation allows the charged particle to follow a hyperbolic motion, precisely because we are not sure that the Lorentz–Dirac equation is telling us the truth!

So it looks as though the work done by the external force is all used up changing the kinetic energy of the charge, if this analysis is right, and we would not appear to have any source for the radiated energy [3]:

> This conclusion seems to make the original question even more demanding: What supplies the energy which the particle radiates? Does (8.10) not prove that there cannot be radiation and that the usual notion is valid, viz., that no radiation reaction means no radiation?

Later we shall review the idea put forward by Boulware [6] that the energy ultimately comes from the early days of the motion, when it was generating the delta function field on the null hypersurface which now causes a flow of energy in towards the charge (see Sect. 15.12). This explanation, if it is one, is closely linked to the ideal nature of this motion, which has been going on forever. The fact that the explanation is linked to the ideal nature of the motion is an argument in its favour, because the whole situation here is indeed ideal.

On the other hand, Parrott shows very clearly that the derivation of the Lorentz–Dirac equation cannot be justified on the basis of energy–momentum conservation unless the motion starts and finishes with the same acceleration (see below) [13, p. 141]. He advocates the case of asymptotically free motion, where the particle starts and finishes with zero acceleration. We are not in that case here, and nor do we have $a(\infty) = a(-\infty)$ (see p. 294). Parrott's analysis shows that what is at fault here is the very derivation of the Lorentz–Dirac equation [13]:

> All derivations of the Lorentz–Dirac equation are motivated by conservation of energy–momentum: the change of energy–momentum of the particle over a given proper time interval should equal the energy–momentum furnished by the external forces driving the particle minus the radiated energy–momentum, assuming that it is legitimate to absorb infinite terms of a certain structure into a mass renormalization. Although this principle motivates the derivation, the final equation unfortunately does not guarantee such conservation of energy–momentum in general, but only in certain special cases. One such special case is when the particle is asymptotically free, meaning that its proper acceleration vanishes asymptotically in the infinite past and future.
>
> Thus it is not clear that the equation should apply to a particle which is not asymptotically free, such as a particle which is uniformly accelerated for all time. Since the equation does not guarantee conservation of energy–momentum for uniform acceleration for all time, the fact that the radiation reaction term is zero implies nothing about the additional force which must be furnished for perpetually uniformly accelerated motion.

Since we do not have $a(\infty) = a(-\infty)$ in the uniform acceleration case, it may be that Boulware's idealistic argument has to be the final explanation of how the energy moves around to guarantee a radiated component. But later we shall also consider Parrott's analysis of a charged rocket (see Chap. 17). There we shall obtain the most explicit possible explanation for why Dirac's radiation reaction does not tell the whole story. With this in mind, let us turn to Fulton and Rohrlich's solution.

We return to the situation for a general, not necessarily uniform acceleration, in which case

$$\Gamma^0 = \frac{2}{3}e^2 \left(\frac{\mathrm{d}a^0}{\mathrm{d}\tau} - \gamma a_\mu a^\mu \right) = \frac{2}{3}e^2 \frac{\mathrm{d}a^0}{\mathrm{d}\tau} - \gamma \mathscr{R} \,, \tag{8.11}$$

bringing in the radiation rate $\mathscr{R}$ for a general acceleration. With the convention for the Minkowski metric in [3], $\mathscr{R}$ is positive because a^μ is spacelike. Returning to the Lorentz–Dirac equation, dividing it by γ and recalling that $\mathrm{d}t = \gamma \mathrm{d}\tau$, we obtain

$$\frac{\mathrm{d}T}{\mathrm{d}t} - \left(\frac{\mathrm{d}Q}{\mathrm{d}t} - \mathscr{R} \right) = \frac{\mathrm{d}W_{\text{ext}}}{\mathrm{d}t} \,, \tag{8.12}$$

where

$$Q := \frac{2}{3}e^2 a^0 \,. \tag{8.13}$$

This is an energy conservation equation. What we have done is perhaps rather circular, because Dirac obtained his equation from energy–momentum conservation considerations.

We read the equation as follows: the rate of work done by the external force equals the rate of increase in kinetic energy minus the rate of work done by the radiation reaction. But we now have the latter analysed into two parts. One is a reversible rate $\mathrm{d}Q/\mathrm{d}t$, i.e., it can be positive or negative, and the other is an irreversible rate $-\mathscr{R}$ which is never positive. The sum of these two parts is not usually zero, although it is for the uniform acceleration. In this case then, the energy lost in the form of radiation is entirely accounted for by part of the work done by the radiation reaction. In the general situation, the remaining part of this work supplies an additional energy Q which may be positive or negative.

But what is this energy Q? Fulton and Rohrlich interpret it as part of the internal energy of the charged particle. Like the kinetic energy of the particle, it can be increased or decreased. Schott called this the acceleration energy, since Q increases when the velocity increases and decreases when the velocity decreases. (This is because a^0 is the rate of change of γ with respect to τ, and γ increases as v increases and decreases as v decreases.) All this is in grave danger of being a circular argument. The Lorentz–Dirac equation arose by considering energy–momentum conservation and now we are just checking that it does dictate energy–momentum conservation. However, we are going a step further in terms of physical interpretation, because we are trying to understand what this acceleration energy could be.

For example, if the motion is periodic or bounded and we average over sufficiently long times, the term $\mathrm{d}Q/\mathrm{d}t$ will average out. Under these conditions, Q represents a fluctuating term and over long time intervals all the work of the radiation reaction goes into radiation. The same is true if one considers the energy balance between an initial and a final state of motion which are equal. This is precisely Parrott's point about the condition $a(-\infty) = a(\infty)$. In these cases, no radiation reaction would indeed imply something like no radiation. Most practical problems are like this, e.g., charged harmonic oscillator, scattering of charged particles with asymptotic freedom. In general, however, there is always a nonzero instantaneous acceleration energy.

In the case of uniform acceleration, we have $\Gamma^0 = 0$ and the total work done by the radiation reaction force is zero, whence

$$\mathscr{R} = \frac{\mathrm{d}Q}{\mathrm{d}t} > 0 \,. \tag{8.14}$$

It is suggested that the internal energy of the particle must be considered as the difference $m - Q$, which is decreasing while energy is being radiated away. Fulton and Rohrlich consider this to be a rather unphysical picture: the accelerated electron decreases its internal energy, transforming it into radiation. They note, however, that this would not affect its rest mass, because $Q = 0$ in a frame in which it is instantaneously at rest.

One wonders therefore in what sense this energy could be internal, particularly as it is now concluded that the radiation energy is compensated by a decrease in that part of the field surrounding the charge which does not escape to infinity (in the form of radiation), and which does not contribute to the electromagnetic mass of the particle. Is this a faltering step towards Boulware's energy flow explanation?

In any case, they conclude that, no matter how (8.14) is interpreted, the principle of energy conservation is not contradicted. One could be a little more precise: if the Lorentz–Dirac equation is right in this context, then since it derives from the principle of energy conservation, this principle cannot be contradicted; but if this equation is wrong, then we have to go back to its derivation and try to put things right, and this means reinstating a correct conclusion from the principle of energy conservation, whence this principle will once again be upheld.

A final proposal is that we simply exclude all cases where the average

$$\left\langle \frac{\mathrm{d}Q}{\mathrm{d}t} \right\rangle_{\mathrm{av}} \neq 0 \, .$$

This is indeed an ad hoc restriction as it stands. On the other hand, one ought to bear in mind that a highly idealistic motion like perpetual uniform acceleration is likely to produce highly idealistic results, e.g., delta function terms, which can only be viewed as approximations to anything that might occur in reality.

Chapter 9
The Threat to the Equivalence Principle According to Fulton and Rohrlich

To review what has been said, if we assume that the Lorentz–Dirac equation is applicable, this equation of motion for a charged particle is satisfied by hyperbolic motion and differs in no way from the equation of motion of a neutral particle, assuming of course that the operative external forces compelling them to that motion are non-electromagnetic, since otherwise those forces could not affect a neutral particle. Of course, we have to be careful not to state a truism: the particle forced to follow hyperbolic motion will follow hyperbolic motion! The point is that the same external force can get both the neutral and the charged particle undergoing hyperbolic motion. This happens because the extra term in the Lorentz–Dirac equation for the charged particle happens to be zero whenever we arrange for the motion to be a uniform acceleration.

As a consequence, Fulton and Rohrlich conclude that, in a static homogeneous gravitational field (SHGF), a neutral and a charged particle will follow the same trajectory with the same time dependence. If Galileo had dropped a neutron and a proton from the leaning tower of Pisa, they would have fallen equally fast! A charged and a neutral particle in an SHGF behave exactly alike, except for the emission of radiation from the charged particle. At least, this is the conclusion from the Lorentz–Dirac equation, with the proviso that we remain within the jurisdiction of special relativity, i.e., they are referring to a special relativistic theory of gravity, as proposed in Chap. 3. The latter is an important proviso here. We apply the dynamical equation of special relativity when we analyse the 4-force due to an SHGF, and we conclude that this SHGF will accelerate the particles uniformly. Then the charged and neutral particle behave exactly alike in this case, except for the emission of radiation from the charged particle.

It should perhaps be noted in passing, that we are not in the conditions of *eternal* uniform acceleration here, if the two particles are dropped from the top of a tower. In the special relativistic view, they begin with zero acceleration and, as Galileo loosens his grip, their acceleration increases to the gravitational value. This kind of situation is discussed more carefully in Sect. 15.10, when we consider how the delta function terms in the EM fields of an eternally uniformly accelerating charge may build up on the null surfaces, and again in much more detail in Chap. 17, when we examine Parrott's analysis of the energy required to move a charged rocket. For the present discussion, the difference seems to be irrelevant, because we do not need to assume that we are in the situation where the charges are dropped from zero acceleration.

In any case, this is the point where Fulton and Rohrlich consider the principle of equivalence to enter the scene:

> A particle which is falling freely in a homogeneous gravitational field should appear to an observer who is falling with it like a particle at rest in an inertial frame (field-free surroundings). If we consider a neutral particle falling in a homogeneous gravitational field, this is indeed what happens. But when the particle is charged, the observer can establish the presence of a gravitational field by looking for radiation. If he observes radiation from the charge, he knows that he and the charge are falling in a gravitational field; if he observes no radiation, he knows that he and the particle are in a force-free region of space.

The first sentence in this quote is a loose statement of an equivalence principle for general relativity. If the word 'appear' includes being able to observe electromagnetic fields here, we may say that it is a loose statement of the strong principle of equivalence. The minimal extension of Maxwell's equations (MEME) to the general relativistic model of this SHGF tells us that, in the locally inertial frame (which happens to be globally inertial in this case) of the observer falling freely with the charged particle, the EM fields it generates will look locally (and in fact globally in this case) like those of a stationary charged particle in a flat spacetime, i.e., they will be Coulomb fields. (Note that the ambiguous term 'field-free surroundings' in the quote is interpreted to mean a spacetime with no gravity.)

We may deduce that, although they state (a loose version of) the strong equivalence principle (SEP) for general relativity, Fulton and Rohrlich are considering a special relativistic theory of gravity here, because they expect the freely falling charge to radiate. It is difficult to see how a prediction from special relativity could affect one of the principles of general relativity. In fact, general relativity and special relativity make different, even diametrically opposite, predictions with regard to this radiation question, as mentioned in Chap. 4 and discussed further in Chap. 10. General relativity with SEP (and hence MEME) tell us that neither a neutral nor a charged particle will radiate when in free fall, and each particle will indeed appear to an observer who is falling with it like the corresponding particle at rest in an inertial frame in a spacetime with no gravity.

For whatever problem it is they think they are dealing with, Fulton and Rohrlich appear to adopt the same solution as Bondi and Gold (compare with the quote from [2] on p. 54):

> The solution to this apparent difficulty is to be found by considering an actual measurement of radiation, using our definition [...]. Radiation is defined by the behavior of the fields in the limit of large distance from the source. Correspondingly, an observer who wants to detect radiation cannot do so in the neighborhood of the particle's geodesic. Rather, he must be at a large distance from it, where gravitational fields have different values. The principle of equivalence, however, is a locally valid principle, referring to the geodesic of the particle, whereas the discussion above shows that any observation of radiation is not a local observation. This point is implied in the related argument given by Bondi and Gold who first discussed this difficulty.

One understands now the earlier emphasis on the limit as the retarded sphere expands toward infinity. As mentioned, however, one does not need to go to infinity to detect the radiation (at least, in principle, because in practice it would not be easy to detect it anywhere). It is interesting to see that the gravitational fields are expected to have 'different values' at large distances from the particle worldline, even though we have been considering a homogeneous field. One gets the idea, however: we are talking about a realistic gravitational field. On the other hand, that would leave us in the lurch for the idealistic case that was the subject of [3], whatever its merits.

The claim that the principle of equivalence is a locally valid principle is unfair, although once again, one gets the point in the context of the loose statement made in the above quote: A particle which is falling freely in a homogeneous gravitational field should appear at least approximately (over some suitably restricted region of spacetime) to an observer who is falling with it like a particle at rest in an inertial frame. But we could say once again that this leaves us in the lurch for the idealistic case we have here, where the locally inertial frame turns out to be globally inertial.

One gets a clue as to what might be going wrong in the following extract:

> Since all our considerations here refer to the framework of the special theory of relativity and, in particular, require the existence of inertial coordinate systems, we are forced to consider the principle of equivalence from a similar point of view. Whatever gravitational field we introduce for the purpose of comparing it with an inertial field, we must be sure to have a distribution of distant stars which define our inertial systems. This means in particular that any homogeneous gravitational field is necessarily of finite extent, imbedded as it were, in an inertial coordinate system. We remark parenthetically that an infinite homogeneous gravitational field does not exist within the framework of general relativity either. The nonexistence of infinite homogeneous gravitational fields assures that the observation of radiation (observer at large distance from the source) takes place outside the homogeneous part of the gravitational field.

The first sentence makes it clear that the earlier statement interpreted as a statement of SEP in general relativity, i.e., that a particle falling freely in an SHGF should appear to an observer who is falling with it like a particle at rest in an inertial frame in field-free surroundings, was intended as an equivalence principle for special relativity. Perhaps the intention was something like this in the context of special relativity: as an observer, one cannot tell whether one is at rest and objects are moving under a gravitational field, or one is accelerating and objects are at rest. If this is their meaning, then one has to accept that Maxwell's theory in special relativity predicts that this is incorrect, precisely because it predicts that the observer only has to see whether the charged objects she observes are radiating.

This still requires us to check that a charge with inertial motion cannot appear to radiate to an accelerating observer, e.g., relative to the semi-Euclidean coordinates of Chap. 2. Let us make three points immediately:

- If one naively interprets SE coordinates as though they are inertial, one might find what 'looks like' radiation here, but it would not save the equivalence principle just proposed for SR, because, not surprisingly, the electromagnetic fields expressed relative to these coordinates do not 'look like' the EM fields of a uniformly accelerating charge when they are expressed relative to inertial coordinates.

- We do not need the proposed equivalence principle because it is not actually part of SR, or GR.
- One should take more care over naively interpreting SE coordinates as though they were inertial.

This is discussed further on pp. 153 ff. in Chap. 14, and again in Sect. 16.10, where it transpires that Rohrlich does indeed consider that a charge with inertial motion can appear to radiate to an accelerating observer.

The rest of the last quote concords once again with Bondi and Gold for the solution to the paradox and is not very satisfying if one considers that the very idealistic scenario of eternal uniform acceleration ought to be allowed to fit in at least theoretically (but see the end of Chap. 4 for a more detailed discussion of the other problems with this proposed solution).

All things considered, it is not easy to identify the standpoint from which Fulton and Rohrlich are viewing this problem and it is probably unproductive to go into more detail. However, to end this discussion of their key paper, it is worth asking whether the root of the problem might not be that Fulton and Rohrlich have not understood what the difficulty was for Bondi and Gold. The latter applied SEP in general relativity, and were clearly referring to general relativity rather than special relativity. But the import of [3], and it was also the conclusion in [2] although less explicitly stated, is that a charge that is stationary relative to coordinates in which the metric is static will radiate if we accept GR+SEP. It was the latter situation that Bondi and Gold found unphysical. But if this conclusion (that a charge that is stationary relative to coordinates in which the metric is static will radiate) were false, as Bondi and Gold suggest, then there must be something wrong with GR+SEP. Unfortunately, Bondi and Gold's proposed solution would not save GR+SEP (see the end of Chap. 4).

To sum up, there are two points to be retained from the above analysis of [3]:

- Fulton and Rohrlich are talking within the confines of special relativity. The freely falling particle is accelerating and should radiate if charged. This radiation will be detectable by the freely falling observer. Of course, a neutral particle in this context will not radiate. If we consider the same two particles at rest in a context where there is no gravitational effect, neither will radiate. The two situations are distinguishable in special relativity so their proposed equivalence principle, if indeed it is one, and if we have really understood what they intend, will not hold.
- Fulton and Rohrlich are not talking within the confines of general relativity. In that context, the freely falling particle is not accelerating and will not radiate, even if charged, provided we accept GR+SEP. Likewise, if we consider the same two particles at rest in a context where there is no gravitational effect, neither will radiate, because this is just a flat spacetime. One normally assumes that this strong equivalence principle holds, otherwise one would have no way of even formulating electromagnetism in general relativity.

Note that the question of whether the freely falling charge radiates or not is almost a matter of life and death for the whole theory of general relativity. This is because general relativity, insofar as it can say anything about freely falling charges, i.e., in conjunction with SEP, says that they will not radiate, whereas special relativity says that they will. At least this is the situation to a first approximation. In Chap. 12, we shall discuss the paper [4] by DeWitt and Brehme which shows that, still assuming SEP, one would not expect free fall to be a real option for charged particles in general relativity, if free fall is defined as geodesic motion, because of the self-force they exert on themselves. DeWitt and Brehme suggest an extension of the Lorentz–Dirac equation which gives some idea of how the electromagnetic fields of the charge will act on it to perturb it from a geodesic.

Anyway, one should not forget the other formulation of an equivalence principle in special relativity which may be the one Fulton and Rohrlich are actually discussing: an inertial observer watching objects fall freely in a gravitational field might think she was accelerating, that the objects were stationary, and that there was no gravitational field. This is a rather simplistic application of the idea that the inertial and passive gravitational masses are identical. Put another way, we are saying that one cannot distinguish a gravitational acceleration of a test particle from an acceleration of the observer, at least not by local experiments. It is important to understand that, in the second interpretation, wherein the observer is accelerating, the observer is in a non-inertial frame. (As an aside, note that the observer would have to be a point observer, not to feel the acceleration relative to the inertial frame.)

What does the case of the charged particle tell us here? The freely falling charge will radiate according to special relativity, because gravity is a force and the charge is considered to be accelerating. If it is the observer that is accelerating and the charge is inertial, the charge will not radiate. Nor will it appear to radiate to our non-inertial observer (at least such apparent radiation relative to SE coordinates will not 'look like' the radiation one would observe from a uniformly accelerating charge). So this equivalence principle, if indeed it is one, breaks down in special relativity, which is where we may presume that the authors would like it to work. However, they try to save it by a rather unsatisfactory argument wherein the equivalence principle is local, but radiation can only be observed far away.

Chapter 10
Different Predictions of Special Relativity and General Relativity

Given the risk of confusion here, it is perhaps useful to spell out the different predictions made by special relativistic and general relativistic theories of gravitation when we consider electromagnetic theories. We have to adjoin something to general relativity in order to see how to handle EM fields in curved spacetime, so we shall assume the strong principle of equivalence (SEP). We consider four cases for each theory and show that SR and GR make diametrically opposed predictions for EM radiation from a single charge source.

The four cases here are simply generated by the charge being in free fall or being supported in the gravitational field, and the observer being in free fall or being supported in the gravitational field. For each case, we consider a situation with a gravitational field, first in a special relativistic context and then in a general relativistic context.

10.1 Four Cases for Special Relativity

A charge in free fall is non-inertial, i.e., accelerating, in special relativity, and of course likewise for the observer. Cases are labelled 'free fall' and 'supported' precisely because the terms 'non-inertial' and 'inertial' are theory-dependent.

- Both charge and observer supported. In SR, both would be considered to have inertial motion. The observer would use Minkowski coordinates and the charge would not move relative to those coordinates. Therefore it would not radiate.
- Charge supported and observer in free fall. Now the charge is inertial, stationary in some Minkowski coordinates, and so would not radiate in that frame. The fact that the observer is accelerating and might adopt semi-Euclidean coordinates could of course mean that she would observe radiation in some sense. This is something that has to be carefully considered (see Sect. 14.3). It is a problem in itself. In fact we come up against another difficulty here: there are no privileged frames for accelerating observers. The best we can do is to find suitably adapted normal coordinates such as SE coordinates. This is to be contrasted

with the remarkable case of inertial motion, for which there are privileged frames making the whole of physics look exactly the same (we are talking about flat spacetimes here). The best option for an accelerating observer in flat spacetime may be to adopt inertial coordinates and use her knowledge of relativity! She would once again say that there was no radiation. But one should also consider what an accelerating detector would register and in Chap. 14 we shall discuss this issue as presented in [5]. We shall also consider in Sect. 16.10 the view expressed by Rohrlich [22].

- Charge in free fall and observer supported. The observer is in an inertial frame and can adopt Minkowski coordinates, relative to which the charge is now accelerating. The observer will observe radiation.

- Both charge and observer in free fall. The charge is accelerating and will radiate for an inertial, i.e., supported observer. If the observer falls with the charge and adopts semi-Euclidean coordinates adapted to her worldline, the present view is that one should still consider there to be radiation. This is a part of the Boulware/Parrott debate discussed in [7] and Chap. 16. Most of this debate concerns the rather idealistic special case of uniform acceleration in Minkowski spacetime, where the first attempts to actually write down the generated EM fields were incorrect (Pauli, Born), and where Dirac's radiation reaction term turns out to be identically zero.

So far we have only used ordinary Maxwell electromagnetism in Minkowski spacetime. To sum up, in the view advocated here, there is radiation when the charge is in free fall and no radiation when it is supported, no matter what the observer is doing. When the observer is accelerating, the question does remain debatable. One could define radiation in such a way that there is radiation for an observer accelerating through a Coulomb field, or no radiation for an observer accelerating with the source, but this is certainly not useful for the debate about the equivalence principle. In that context, and probably in others too, it is better to define radiation relative to inertial coordinates. There is no reason a priori why an accelerating observer should conveniently observe the opposite to what an inertial observer would observe. WEP and SEP certainly do not require it.

10.2 Four Cases for General Relativity

A charge in free fall is inertial, i.e., non-accelerating, in general relativity, whereas a supported charge is non-inertial, i.e., accelerating. The same goes for the observer. Comparing with the situation in special relativity, it is clear that the terms 'non-inertial' and 'inertial' are theory-dependent.

- Both charge and observer in free fall. By the weak equivalence principle (WEP), the observer can adopt locally inertial coordinates adapted to her worldline and, by SEP, the charge at rest in this frame will not radiate for an observer falling with it.

- Charge in free fall and observer supported. The charge is stationary in its locally inertial frame and will not radiate for an observer moving with it, according to SEP. The supported observer is accelerating relative to that locally inertial frame, but as explained above, the present view is that one should not define there to be radiation by naive calculations relative to the rather arbitrary coordinate system such an observer might adopt. In the case of an SHGF, where the locally inertial frame is globally Minkowskian, the field is Coulomb in the Minkowski system, and transforms to a rather complicated field in the SE system [see (14.15) and (14.16) on p. 154]. However, one of the themes of the present account is that there is no reason to say that this is a radiating field, or will 'look like' a radiating field to the supported observer, because these are after all only coordinates (but see the opposite view expressed by Rohrlich in Sect. 16.10 and the discussion of Mould's radiation detector in Chap. 14).
- Charge supported and observer in free fall. The charge is non-inertial, accelerating for the freely falling locally inertial observer. The latter therefore observes radiation, according to SEP.
- Both charge and observer supported. The charge is non-inertial, accelerating for a freely falling locally inertial observer. In the present view, the fact that the observer is accelerating with the charge should not be considered to transform away the radiation. This is a part of the Boulware/Parrott debate discussed later (see Chap. 16).

Here we have used SEP to extend Maxwell's equations to curved spacetime. To sum up, in this view, there is radiation when the charge is supported and no radiation when it is in free fall, no matter what the observer is doing, with the provisos introduced above with regard to accelerating observers.

10.3 Conclusion

When the gravitational field in question is static and homogeneous so that the freely falling frame is globally Minkowskian, the calculations in the two sequences SR and GR are identical, in the order they have been presented. This is why the sequence of conclusions NR, NR, R, R is the same (NR not radiate, R radiate). However, the premises have been reversed in the two sequences.

What we are discussing here is not the possible contradiction of the strong equivalence principle by the radiation phenomenon, but a different prediction by a different theory. SR predicts that the freely falling charge will radiate, whereas GR predicts that it will not. This difference is not totally surprising, because GR introduces an interaction between gravity and electromagnetism, and this interaction is formulated by the minimal extension of Maxwell's equations, replacing ordinary spacetime derivatives by covariant derivatives, a direct application of SEP. We note also that a fundamental aspect of the observers is quite simply reversed between GR and SR: freely falling observers are inertial in GR and accelerating in SR, whilst supported observers are non-inertial in GR and inertial in SR.

There remains one doubt here for the GR case: in a general spacetime, the metric will not be exactly Minkowskian over any region and the connection will not be exactly zero over any region, these effects being due to curvature. In Chap. 12 we examine the paper [4] by DeWitt and Brehme to see whether this makes a significant difference to the claims here, i.e., is it possible that curvature effects could dominate the first-order expectation implied by GR+SEP? For example, could the radiation effect expected for a supported charge in a real (non-homogeneous) gravitational field be somehow cancelled by curvature effects? One can already guess the answer: there is no reason why it should, because EM effects depend on the strength of their sources, while gravitational effects like curvature depend on the strength of their sources, and these strengths are physically independent. This is basically the conclusion at the end of Chap. 13, which discusses the whole question of a static charge in a static spacetime.

The above doubt is related to a pseudo-problem that has been rejected: radiation is often thought of as a long-distance phenomenon, whereas the freely falling frame moving with the charge is only locally inertial in GR (for a realistic gravitational field). The argument here is that one would expect to be able to compare length scales and show that such a frame would be big enough to observe that there is no radiation. In the case of an SHGF, the result is clear, but for a curved spacetime, it might seem more subtle. However, as discussed in Chap. 4 (see p. 55), the world tube used to calculate energy–momentum flow can be shrunk to any small size in the kind of analysis used to derive an energy–momentum conservation equation and the radiation terms can still be identified.

Chapter 11
Derivation of the Lorentz–Dirac Equation

Two problems are woven together in this debate: one is the question of whether there should be radiation from uniformly accelerating charges in flat spacetime with no gravitational effects (not even non-tidal ones) and the other is the question of whether there should be radiation from charges that are stationary relative to coordinates in curved spacetimes (or even flat spacetimes in which we consider that there is an SHGF to be treated by general relativity) in which the metric is static. The first question spawns others, as we have seen. The aim of this section is to consider one of these in more detail, viz., if uniformly accelerating charges in flat spacetime (without gravitational effects) do radiate, how is it that the radiation reaction derived from energy–momentum conservation turns out to be zero?

One is inclined to look at the derivation of the Lorentz–Dirac equation, because the radiation reaction terms turn up there. We begin with Parrott's derivation in [13] and compare it with Dirac's original 1938 derivation in [14].

11.1 Parrott's Derivation

The present continues from Chap. 6. We start with the law of conservation of energy–momentum in the form

$$\delta \left(T_{\mathrm{em}} + \frac{m}{q} \rho \, u \otimes u^* \right) = 0 \,, \tag{11.1}$$

which refers to a charged dust with charge density ρ made up of particles of mass m and charge q. The symbol δ denotes the covariant divergence and it applies here to a vector-valued 1-form. This law of conservation of energy–momentum is equivalent to the Lorentz equation

$$m \frac{\mathrm{d}u}{\mathrm{d}\tau} = q \hat{F}(u) \,, \tag{11.2}$$

remembering that, when we model a charged dust, u is a velocity vector field. The starred quantity u^* is the one-form associated with u by the metric.

We now try to model a point charge. One might envisage a distribution approach, replacing ρ in the above by some kind of delta function. This can indeed be viewed as the way one obtains the retarded field solutions for Maxwell's equations for a point charge in arbitrary motion. However, rather than implementing this idea in a complete and rigorous way, we adopt a hybrid argument as follows.

We consider a single charged particle with worldline $\tau \mapsto z(\tau)$, parametrised by proper time. F_{ret} is the retarded field it produces and F_{ext} the field due to any other sources. The total field is $F = F_{\text{ret}} + F_{\text{ext}}$. How will the charge move when subject to this total field? We surround the particle by a closed 2D spatial surface, such as a surface of fixed radius r_0 relative to some well-defined Lorentz frame. This 2D surface then moves through time with the particle in some well-defined way, thereby generating a 3D hypersurface in Minkowski spacetime $\mathbb{M}$. There are two possible tubes using the sphere idea:

- the Bhabha tube defined by (6.2) and (6.3) on p. 67 (see Fig. 6.1),
- the Dirac tube defined by (6.4) and (6.5) on p. 68 (see Fig. 6.2).

One has to consider how to close the ends of the tube. It is easy to see that the Dirac tube is closed by solid spheres of radius r in the spacelike hypersurfaces associated with times τ_1 and τ_2. However, such regions would not close the Bhabha tube very effectively. Should one close the earlier end of the tube by a spacelike hyperplane at time τ_1? The problem with this is that this hyperplane does not even intersect the tube walls. So what about the spacelike hyperplane at $\tau_1 + r$? The problem with this is that it excludes the point $z(\tau_1)$. The same remarks hold for the later end of the tube.

The answer is that one must close the Bhabha tube with portions of the light cones at τ_1 and τ_2. Physically, it is worth noting that, if the particle has constant velocity between times τ and $\tau + r$, then $\bar{S}_r(\tau + r) = S_r(\tau)$. If we are dealing with a scattered particle which, by the usual definition, has asymptotically constant velocity in the infinite past and the infinite future, then a Bhabha tube will approximate a Dirac tube.

One now considers the integral of $\perp T$ over the hypersurface represented by the tube, where T is now the energy–momentum tensor for the electromagnetic fields, not including any mechanical term. In other words, T is what we called T_{em} in (11.1). Recall also that $S := \perp T$ is the Hodge dual of T. One problem here is that the integral of $\perp T$ over a portion of a light cone has no obvious physical interpretation [7, p. 107]. However, it can still be evaluated in detail, and we shall also discuss the relevance of the result.

In contrast, the integral of $\perp T$ over a solid spherical cap in a spacelike hyperplane has an immediate interpretation [7, pp. 105, 111]. It turns out that the integral of S over a hypersurface spanned by infinitesimal triples of orthogonal spacelike vectors e_1, e_2, e_3 is a sum of things that are all proportional to $S(e_1, e_2, e_3) = T(v)$, where v is the (timelike) tangent to the particle worldline at the relevant value of τ. We interpret $T(v)$ as the energy–momentum per unit volume at the relevant point relative to the

Lorentz frame in which $v = (1,0,0,0)$. This is a sophisticated description, but it basically comes straight from the definition of S and T.

So the integral of $_{\perp}T$ over the spherical cap at time τ_2 is the total energy–momentum due to the total electromagnetic field F inside the sphere at proper time τ_2. What we should stress here is that this is not the interior of the retarded sphere, but the interior of the Dirac sphere. The integral over a similar cap at time τ_1 will give minus the energy–momentum of the field in the Dirac sphere associated with τ_1. The sign is reversed because the caps have opposite orientations.

What about the integral of $_{\perp}T$ over the sides of the tube, where we now consider the Bhabha tube? In fact, for the Dirac tube, we obtain the total energy–momentum flowing out of the sphere between times τ_1 and τ_2, and for the Bhabha tube, the same but between times $\tau_1 + r$ and $\tau_2 + r$. One might consider the latter to be the energy–momentum that flowed out of the particle itself between times τ_1 and τ_2.

Now a naive principle of energy–momentum conservation would tell us that the integral of $_{\perp}T$ over the entire tube, including sides and caps, plus the change in energy–momentum of the particle between proper times τ_1 and τ_2 must be zero. This seems clear enough when we have a Dirac tube closed by spacelike hyperplane caps. However, Parrott carries out the integration over the sides of a Bhabha tube, because it is easier.

It is interesting to recall Dirac's argument briefly [14]. In order to integrate over his tube, he had to produce a Taylor expansion of the fields and their energy–momentum tensor, which was tedious. Parrott escapes this with the Bhabha tube. Dirac then used a neat but (possibly) weak argument to avoid calculating the integrals over the caps. Indeed, Parrott suggests that the integrals over the Dirac caps have never been computed exactly [13, p. 161]. It is a salutary exercise to carry out the integrals over the Bhabha caps in full detail using the methods described in [13] and the results will be mentioned below.

There seems to be no problem interpreting the integral over the sides of the Bhabha tube as the total energy–momentum flowing out of the sphere between times $\tau_1 + r$ and $\tau_2 + r$. If we integrate over spacelike hyperplane caps at these times, we presumably get a good approximation to the change in the field energy–momentum over the proper time interval $[\tau_1 + r, \tau_2 + r]$. Presumably this could be said to give the decrease in mechanical energy–momentum of the particle over the period $[\tau_1, \tau_2]$. In fact it looks as though we get something very close to a Dirac tube, shifted a time r up the worldline from the region $[\tau_1, \tau_2]$. In the limit as $r \to 0$ one expects the results to coincide.

There is another question here: if we integrate over a Bhabha tube, complete with Bhabha caps, how can we interpret the result? Do we get the change in mechanical energy–momentum of the particle? One problem is the difficulty interpreting the integrals over light cone caps, mentioned above. Another is that we are not getting the mechanical energy–momentum from the formalism of the energy–momentum tensor, as we do for a charged dust. One would presumably need to represent the point charge by a delta function charge density in an expression of type (11.1). We shall skate over these technical details here.

The programme is as follows. Given the worldline, we compute the retarded fields F_{ret} generated by the charge, then the energy–momentum tensor T of the total field, including any external field F_{ext}, assumed given. We then integrate $_\perp T$ over the whole tube and apply the requirement of energy–momentum conservation in the form mentioned above, which is a hybrid form, involving our intuition for one part and the covariant divergence for the other: the integral of $_\perp T$ over the entire tube, including sides and caps, plus the change in energy–momentum of the particle between proper times τ_1 and τ_2 must be zero. This gives a relation between the change in energy–momentum of the particle and F_{ext}, in which we take the limit $r_0 \to 0$. We can also divide through by $\Delta\tau := \tau_2 - \tau_1$ and take the limit as $\Delta\tau \to 0$.

The first problem with our programme is that the integral of $_\perp T$ over (Dirac) solid sphere caps diverges (more about the Bhabha caps in a moment). The energy–momentum inside such a sphere is infinite. This is a standard calculation when the charge actually has constant velocity, in a Lorentz frame in which it is stationary and we assume F_{ext} is zero. The frame is chosen with the charge at the origin. The expression for F_{ret} reduces in this case to

$$\mathbf{E} = q\frac{\mathbf{w}}{r^2}\,, \qquad \mathbf{B} = 0\,. \tag{11.3}$$

where $\mathbf{w}$ is the unit vector toward the field point. The component T^0_0 of T is

$$\frac{\mathbf{E}^2 + \mathbf{B}^2}{8\pi} = \frac{q^2}{8\pi}\frac{1}{r^4}\,, \tag{11.4}$$

whence the integral of $_\perp T$ over either cap has energy component

$$\lim_{\varepsilon \to 0}\int_\varepsilon^{r_0} \frac{q^2}{8\pi}\frac{1}{s^4} 4\pi s^2\, \mathrm{d}s = \lim_{\varepsilon \to 0}\frac{q^2}{2}\left(\frac{1}{\varepsilon} - \frac{1}{r_0}\right)\,. \tag{11.5}$$

Other components of the energy–momentum in the caps are zero, more or less by good fortune, since the magnetic field is zero, and so therefore is the Poynting vector.

It is noted that the integral over the cap never changes as the particle moves along its worldline, in this special case. An unchanging infinity is no problem for the dynamics here. This is what allows mass renormalisation. However, one is beginning to change the equation of motion rather radically from the Lorentz equation to the Lorentz–Dirac equation.

Note in passing that what makes renormalisation possible here is precisely what makes the self-forces contribute to an inertial effect in the case of a finite charged particle. That is, renormalisation is possible because of the structure of the infinities which arise, allowing them to be absorbed into mechanical energy–momentum terms in the equation of motion; and this is precisely what allows some self-force terms to be interpreted as inertial forces. We shall say more about this later (see p. 110).

Now suppose the world tube is a Bhabha tube capped by suitable portions of the light cones at events on the worldline associated with proper times τ_1 and τ_2, denoted by $\mathscr{C}(\tau_1,\varepsilon)$ and $\mathscr{C}(\tau_2,\varepsilon)$, respectively, where

$$\mathscr{C}(\tau_i,\varepsilon) = \left\{(\tau_2+\rho,x,y,z):0<\rho<\varepsilon, \quad x^2+y^2+z^2=\rho^2\right\}, \qquad (11.6)$$

with $i=1,2$. Each cap is a union of spherical surfaces, one for each time in the interval $[\tau_i,\tau_i+\varepsilon]$, where the radius of the sphere at time $\tau_2+\rho$ is ρ. One finds

$$\lim_{\varepsilon\to0}\left[\int_{\mathscr{C}(\tau_1,\varepsilon)}{}_{\perp}T-\frac{1}{2}q^2u(\tau_1)\left(\frac{1}{\varepsilon}-\frac{1}{r}\right)\right]=rq\hat{F}_{\mathrm{ext}}(u(\tau_1))+O(r^2), \qquad (11.7)$$

and likewise for τ_2. In the limit as $r\to0$, the external fields contribute nothing. What can we conclude from this?

- The light cone cap is mathematically more convenient than the solid sphere cap, in that one can rather easily work out the integral of an object like ${}_{\perp}T$ over this region, even in a completely general case.
- Physically, the result is more difficult to interpret.
- Despite the lack of a simple physical interpretation, the result is suggestive. Indeed, it suggests that the caps do merely introduce a mass renormalisation, as proposed when we examined the Coulomb field in a Dirac (solid spherical) cap.

In the case of a scattered particle, i.e., one which comes in from spatial infinity on a straight line as a free particle, with constant velocity, interacts for a finite time with a spatially localised electromagnetic field, and then becomes free again, leaving the scene with constant velocity, one can cap the tube with Dirac spheres at times in the remote past and the remote future. Let $u(\tau)$ be the particle 4-velocity at proper time τ, and

$$u_{\mathrm{in}}:=\lim_{\tau\to-\infty}u(\tau), \qquad u_{\mathrm{out}}:=\lim_{\tau\to\infty}u(\tau). \qquad (11.8)$$

In a Lorentz frame in which the particle is at rest in the remote past, u_{in} has components $(1,0,0,0)$. Then the integral of ${}_{\perp}T$ over the past cap is

$$\int_{\mathrm{past\ cap}}{}_{\perp}T=\left(\lim_{\varepsilon\to0}\frac{q^2}{2}\int_{\varepsilon}^{r_0}\frac{\mathrm{d}s}{s^2}\right)u_{\mathrm{in}}. \qquad (11.9)$$

By the same reasoning,

$$\int_{\mathrm{future\ cap}}{}_{\perp}T=\left(\lim_{\varepsilon\to0}\frac{q^2}{2}\int_{\varepsilon}^{r_0}\frac{\mathrm{d}s}{s^2}\right)u_{\mathrm{out}}. \qquad (11.10)$$

As usual, for these Dirac caps, one interprets the difference between the left-hand sides of (11.10) and (11.9) as the difference between the energy–momentum of the electromagnetic field in the future cap and that in the past cap. Formally, the difference between the right-hand sides looks like

$$m_{\mathrm{f}}u_{\mathrm{out}}-m_{\mathrm{f}}u_{\mathrm{in}}, \qquad (11.11)$$

where m_{f} can be called the field mass, which we have found to be infinite.

The equation of energy–momentum conservation in our hybrid reasoning is then

$$\int_{\text{side}} {}^{\perp}T + \int_{\text{caps}} {}^{\perp}T + mu_{\text{out}} - mu_{\text{in}} = 0 \,, \qquad (11.12)$$

where m is the rest mass of the scattered particle. We now rewrite this as

$$m_{\text{ren}} u_{\text{out}} - m_{\text{ren}} u_{\text{in}} = -\int_{\text{side}} {}^{\perp}T \,, \qquad (11.13)$$

where

$$m_{\text{ren}} := m_{\text{f}} + m \,, \qquad m_{\text{f}} = \lim_{\varepsilon \to 0} \frac{q^2}{2} \int_{\varepsilon}^{r_0} \frac{\mathrm{d}s}{s^2} \,. \qquad (11.14)$$

We now seek an equation of motion that guarantees (11.13). The differential version
of this is

$$m_{\text{ren}} \frac{\mathrm{d}u}{\mathrm{d}\tau} = -\lim_{\Delta\tau \to 0} \frac{1}{\Delta\tau} \lim_{r_0 \to 0} \int_{S_{r_0}(\tau, \tau+\Delta\tau)} {}^{\perp}T \,. \qquad (11.15)$$

The tube on the right-hand side is the Bhabha tube, despite the fact that we have
integrated over Dirac caps. We shall gloss over the fact that we have taken a limit
as $\tau_2 - \tau_1$ tends to zero, which does not quite fit in with the idea of the scattered
particle. Note that the left-hand side is independent of r_0, and this is why one finds
the limit as $r_0 \to 0$ on the right-hand side. Everything now hinges on evaluating the
integral on the right-hand side. The result is given in (6.1) on p. 67, repeated here
for convenience:

$$\int_{S_r(\tau_1, \tau_2)} {}^{\perp}T = \frac{q^2}{2r} \big[u(\tau_2) - u(\tau_1) \big] \quad \text{(mass renormalisation)}$$

$$\qquad - \frac{2}{3} q^2 \int_{\tau_1}^{\tau_2} u(\tau) \langle a(\tau), a(\tau) \rangle \, \mathrm{d}\tau \quad \text{(radiation)}$$

$$\qquad - q \int_{\tau_1}^{\tau_2} \hat{F}_{\text{ext}}(u) \, \mathrm{d}\tau \quad \text{(Lorentz force)} \,. \qquad (11.16)$$

Once again it is a salutary exercise to carry out these integrals in the way shown by
Parrott in [13].

As one would expect, the limits on the right-hand side of (11.15) are divergent.
Our task is to write

$$\lim_{\Delta\tau \to 0} \frac{1}{\Delta\tau} \int_{S_{r_0}(\tau, \tau+\Delta\tau)} {}^{\perp}T \qquad (11.17)$$

in the form

$$h(r_0) + f(r_0) \frac{\mathrm{d}u}{\mathrm{d}\tau} \,, \qquad (11.18)$$

where h and f are functions with

$$\lim_{r_0 \to 0} h(r_0) \text{ finite}, \qquad \lim_{r_0 \to 0} f(r_0) = \infty . \tag{11.19}$$

By a second mass renormalisation, terms proportional to $du/d\tau$ are absorbed into the mechanical energy–momentum. The final result is an equation of the form

$$\bar{m}\frac{du}{d\tau} = q\hat{F}_{\text{ext}}(u) + \frac{2}{3}q^2\langle a,a\rangle u , \tag{11.20}$$

where $\bar{m}$ is the new (doubly) renormalised mass and $a := du/d\tau$ is the proper acceleration. Basically, the right-hand side of (11.20) is the right-hand side of (11.15) with terms proportional to $du/d\tau$ discarded. This is still not the Lorentz–Dirac equation. Indeed, it is actually an equation that is impossible to satisfy.

The reason is simply that a and $\hat{F}_{\text{ext}}(u)$ are orthogonal to u, the former because $u^2 = 1$, the latter because $\hat{F}_{\text{ext}}$ is antisymmetric. We may deduce from (11.20) that $\langle a,a\rangle = 0$. In physical terms, the equation tells us that the change in energy–momentum of the particle per unit proper time differs from what is expected via the Lorentz equation by the term proportional to $\langle a,a\rangle u$. Now in a Lorentz frame in which the particle is instantaneously at rest, this last term represents a pure mass-energy loss per unit time proportional to the square of the proper acceleration. Unfortunately, there is nothing in the equation to account for such a loss.

Now we originally sought to obtain (11.13). This is the proposed equation of energy–momentum conservation. Instead we have obtained a differential version of it, viz., (11.15). Indeed, (11.15) implies (11.13), but it is not a necessary consequence. We shall now find another relation that implies (11.13). The integral over all proper time of our impossible equation (11.20) is

$$\bar{m}u_{\text{out}} - \bar{m}u_{\text{in}} = q\int_{-\infty}^{\infty} \hat{F}_{\text{ext}}\big(u(\tau)\big)\, d\tau + \frac{2}{3}q^2\int_{-\infty}^{\infty}\langle a(\tau),a(\tau)\rangle u(\tau)\, d\tau . \tag{11.21}$$

It is the last term that we shall rewrite now.

We observe that

$$\langle a(\tau),a(\tau)\rangle = \left\langle \frac{du}{d\tau},\frac{du}{d\tau}\right\rangle = \frac{d}{d\tau}\left\langle \frac{du}{d\tau},u\right\rangle - \left\langle \frac{d^2u}{d\tau^2},u\right\rangle .$$

This can be written

$$\langle a,a\rangle u = -\left\langle \frac{da}{d\tau},u\right\rangle u . \tag{11.22}$$

We now rewrite the second integral on the right-hand side of (11.21):

$$\int_{-\infty}^{\infty} \langle a(\tau), a(\tau)\rangle u(\tau)\, \mathrm{d}\tau = -\int_{-\infty}^{\infty} \left\langle \frac{\mathrm{d}a}{\mathrm{d}\tau}, u \right\rangle u\, \mathrm{d}\tau$$

$$= -\int_{-\infty}^{\infty} \frac{\mathrm{d}a}{\mathrm{d}\tau}\, \mathrm{d}\tau + \int_{-\infty}^{\infty} \left[\frac{\mathrm{d}a}{\mathrm{d}\tau} - \left\langle \frac{\mathrm{d}a}{\mathrm{d}\tau}, u \right\rangle u \right]\, \mathrm{d}\tau$$

$$= -a(\infty) + a(-\infty) + \int_{-\infty}^{\infty} \left[\frac{\mathrm{d}a}{\mathrm{d}\tau} - \left\langle \frac{\mathrm{d}a}{\mathrm{d}\tau}, u \right\rangle u \right]\, \mathrm{d}\tau \ .$$

$$(11.23)$$

Recalling the scattered particle ideal mentioned earlier, we now propose to put both $a(\infty)$ and $a(-\infty)$ equal to zero. More generally, one can assume that $a(\infty) = a(-\infty)$ for what it is worth. In this case,

$$\int_{-\infty}^{\infty} \langle a, a\rangle u\, \mathrm{d}\tau = \int_{-\infty}^{\infty} \left[\frac{\mathrm{d}a}{\mathrm{d}\tau} - \left\langle \frac{\mathrm{d}a}{\mathrm{d}\tau}, u \right\rangle u \right]\, \mathrm{d}\tau \ . \qquad (11.24)$$

So under this assumption about the acceleration at remote times, the proposed relation (11.21), which would appear to imply (11.13), itself follows from the alternative differential relation

$$\bar{m}\frac{\mathrm{d}u}{\mathrm{d}\tau} = q\hat{F}_{\text{ext}}(u) + \frac{2}{3}q^2 \left[\frac{\mathrm{d}a}{\mathrm{d}\tau} - \left\langle \frac{\mathrm{d}a}{\mathrm{d}\tau}, u \right\rangle u \right] \ . \qquad (11.25)$$

What we are saying then is that this new differential equation (11.25) implies the proposed equation of energy–momentum conservation (11.13). The great thing about the second term on the right-hand side of (11.25) is that it is orthogonal to u. So the whole equation refers to what is happening orthogonal to u. Note also that, physically, the second term on the right-hand side of (11.25) is (apart from the multiplicative factor) the component of $\mathrm{d}a/\mathrm{d}\tau$ orthogonal to u.

Since we have (11.22), the new equation (11.25) can also be written

$$\bar{m}\frac{\mathrm{d}u}{\mathrm{d}\tau} = q\hat{F}_{\text{ext}}(u) + \frac{2}{3}q^2 \left[\frac{\mathrm{d}^2 u}{\mathrm{d}\tau^2} + \left\langle \frac{\mathrm{d}u}{\mathrm{d}\tau}, \frac{\mathrm{d}u}{\mathrm{d}\tau} \right\rangle u \right] \ , \qquad (11.26)$$

or again,

$$\bar{m}\frac{\mathrm{d}u}{\mathrm{d}\tau} = q\hat{F}_{\text{ext}}(u) + \frac{2}{3}q^2 \left[\frac{\mathrm{d}a}{\mathrm{d}\tau} + \langle a, a\rangle u \right] \ . \qquad (11.27)$$

Naturally, $\bar{m}$ is usually replaced by m.

The chain of reasoning lacks only one step, namely, the one which establishes (11.17)–(11.19). This involves calculating the energy–momentum tensor and integrating it over the walls of the Bhabha tube. That part of the argument uses only the expression

$$F_{\text{ret}} = q \left[\frac{1}{r^2} w^* \wedge u^* + \frac{1}{r}(u^* + w^*) \wedge a_{\perp}^* \right] \qquad (11.28)$$

for the retarded field, where $a_\perp := a + \langle a, w \rangle w$ is orthogonal to both w and u and the stars denote one-forms associated with vectors, and the expression

$$T_{\text{em}}^{ij} = \frac{1}{4\pi} \left(F^i{}_k F^{kj} + \frac{1}{4} g^{ij} F^{lm} F_{lm} \right) \tag{11.29}$$

for the energy–momentum tensor of an electromagnetic field. The reader is referred to [13] for the details (highly recommended).

11.2 Dirac's Derivation

As mentioned earlier, Dirac avoided working out the integral over the caps by a characteristically clever mathematical argument [14]. In his notation, his integral over the tube walls gave

$$\int \int -\varepsilon^{-1} T_{\mu\rho} \gamma^\rho \, dS |d_2 x| = \int \left(\frac{1}{2} e^2 \varepsilon^{-1} \dot{v}_\mu - e v^\nu f_{\mu\nu} \right) ds . \tag{11.30}$$

To clarify the parallel here, note the following:

- ε is the sphere radius r,
- γ is similar to εw, being the vector from the field point on the Dirac sphere to the simultaneous point on the worldline in the rest frame of the charge for that point, with $\gamma \cdot \gamma = -\varepsilon^2$,
- dS is the measure on the standard 2-sphere in 3D Euclidean space,
- $|d_2 x|$ is a measure similar to our present $d\tau$, but with a multiplicative factor depending on the charge velocity,
- ds is just $d\tau$, since s was Dirac's proper time for the electron,
- e is the electron charge, represented now by q,
- v^μ is the 4-velocity of the charge,
- $f_{\mu\nu}$ is the actual field minus the mean of the advanced and retarded fields, which Dirac shows to be given by

$$f_\mu{}^\nu = F_{\mu \, \text{in}}^{\ \ \nu} + \frac{2}{3} e \left(\ddot{v}_\mu v^\nu - \ddot{v}^\nu v_\mu \right) ,$$

where F_{in} is our present F_{ext}.

Note that the full details do not matter here. The word 'similar' in the description of γ reminds us that it concerns the Dirac sphere, whereas Parrott uses the retarded sphere. In any case, we are intent on showing only the broad idea of Dirac's approach.

The quantity on the right-hand side of (11.30) is the (electromagnetic) energy and momentum that has escaped from the tube between the proper times that limit the integral $\int ds$. What can we do with this quantity? If we integrated the flow of electromagnetic energy and momentum out of the whole tube, including not only the integrals over the partly timelike walls, but also the integrals over the solid 3D spatial spheres at each end, we would not find zero because of the source within the tube. We could not deduce that the electromagnetic energy and momentum in the final sphere (containing the electron) was equal to the electromagnetic energy and momentum in the initial sphere minus the electromagnetic energy and momentum that had flowed out during the intervening period. Electromagnetic energy and momentum are not conserved. The energy and momentum of the electron also change over this period.

So what can we say on the basis of (11.30)? What we know is that, if we take the electromagnetic energy and momentum plus the (mechanical) energy and momentum of the electron in the initial sphere and we take away the right-hand side of (11.30), we end up with the electromagnetic energy and momentum plus the (mechanical) energy and momentum of the electron in the final sphere. But what we are subtracting is expressed as an integral over s from the initial to the final proper time, and we see that this integral has to be the difference of the total energy and momentum within the final and initial spheres, in other words, it has to depend only on the values of things at the initial and final proper times. This is why the integrand on the right-hand side of (11.30) has to be a perfect differential, i.e., there exists some vector-valued function of s which we call B_μ with the property that

$$\frac{1}{2}e^2\varepsilon^{-1}\dot{v}_\mu - ev^\nu f_{\mu\nu} = \dot{B}_\mu \ . \tag{11.31}$$

Up to a constant vector, $-B_\mu(s)$ is the total energy–momentum contained in the sphere at s.

We observe that

$$v \cdot \dot{B} = \frac{1}{2}e^2\varepsilon^{-1}v \cdot \dot{v} - ev^\mu v^\nu f_{\mu\nu} = 0 \ , \tag{11.32}$$

because $v \cdot \dot{v} = 0$ and $f_{\mu\nu}$ is antisymmetric in its indices. We now seek a simple solution to this relation. Dirac finds two: the first is the simplest, viz.,

$$B_\mu = kv_\mu \ , \tag{11.33}$$

for some constant k, and the second is intended to show that things get rapidly much more complicated, viz.,

$$B_\mu = \kappa\left[\dot{v}^4 v_\mu + 4(\dot{v} \cdot \ddot{v})\dot{v}_\mu\right] \ . \tag{11.34}$$

We now bring to bear the argument that the electron is simple and opt for the first choice. If (11.34) is the second simplest choice possible, we hardly need to invoke the simplicity of the electron. It would nevertheless be interesting to investigate it.

With this rather loosely motivated step behind us, we now have

$$\left(\frac{1}{2}e^2\varepsilon^{-1} - k\right)\dot{v}_\mu = ev^\nu f_{\mu\nu} \ . \tag{11.35}$$

Earlier we ignored terms from the integral that were linear or higher order in ε on the grounds that the cylinder radius could be made as small as we liked. We now have to face the fact that we have a term going as ε^{-1} in the equation of motion (11.35). We define a new constant m by

$$m := \frac{1}{2}e^2\varepsilon^{-1} - k\,. \tag{11.36}$$

Note that, if $B_\mu(s)$ is indeed the total energy and momentum in the sphere at proper time s, it is likely to be infinite on the usual reckoning of these things around a point particle. However, we choose not to think about such matters, and merely claim, not only that m is finite in the limit as $\varepsilon \to 0$, but that it is the inertial mass of the electron.

This is the essence of renormalisation. Two infinities cancel to leave what is actually observed. It is interesting to compare this approach with the one in which the electron itself is extended spatially. In that case, the tube considered is the worldtube of the electron and there is nothing to tend to zero. This is mathematically a much more satisfying procedure. However, Dirac obtains the same equation of motion for the electron (often called the Lorentz–Dirac equation). This is his justification for the above analysis at the end of the day.

We thus take it that

$$m\dot{v}_\mu = ev_v f_\mu{}^v\,. \tag{11.37}$$

This equation has the same form as the usual equations of motion (the Lorentz force law) for an electron in an external electromagnetic field, with m playing the part of the rest mass of the electron and $f_\mu{}^v$, the actual field minus the mean of the advanced and retarded fields, playing the part of the external field. However, we are not usually given $f_\mu{}^v$ so this equation is not in a form suitable for application to practical problems. What we are given is the incident field $F_{\mu\,\text{in}}^{\;\;v}$.

We can now substitute in the expression for $f_\mu{}^v$ and we have the Lorentz–Dirac equation. Put another way (see [13, p. 160]), we can say that Dirac has shown

$$\int_{\bar{S}_r(\tau_1,\tau_2)}{}^{\perp}T = -q\int_{\tau_1}^{\tau_2}\hat{F}_{\text{ext}}\big(u(\tau)\big)\,\mathrm{d}\tau - \frac{2}{3}q^2\int_{\tau_1}^{\tau_2}\left[\frac{\mathrm{d}a}{\mathrm{d}\tau} + \langle a(\tau),a(\tau)\rangle u(\tau)\right]\mathrm{d}\tau$$

$$+\frac{q^2}{2r}\big[u(\tau_2) - u(\tau_1)\big] + O(r)\,. \tag{11.38}$$

This is another version of (11.30), using Parrott's notation. One gets basically the same thing as Parrott's result (6.1) for the Bhabha tube, except that one immediately obtains the proper time derivative of the acceleration which caused so much trouble above, between (11.20) and (11.26). This is perhaps an argument in favour of Dirac's approach.

The crux of the latter is undoubtedly what comes just after (11.36). One does not explicitly calculate over the caps, because with this reasoning there is no need. Indeed, there is not even any interest in doing so. The total energy–momentum in the cap is $-B_\mu(s)$, up to a constant 4-vector. We propose $B_\mu = kv_\mu$, for some k which must actually be infinite.

What we see in either Dirac's or Parrott's calculation is that there is a mass renormalisation from the integral over the tube walls. This is the ε^{-1} in (11.30), which is not yet infinite, but will go to infinity as $\varepsilon \to 0$. But the mass renormalisations from the caps are hidden within B_μ in Dirac's argument, whereas Parrott brings them out very clearly. These are present before one shrinks the tube down to the worldline, i.e., no matter how thin the tube is, and in that sense are more serious.

As Dirac points out, B_μ could have the more complicated form (11.34), or some other form. So perhaps one ought to work out the integral over the Dirac caps! On the other hand, one does obtain a plausible looking theory (on the face of it). It is worth noting here that DeWitt and Brehme miss the renormalisation terms from the caps of their tube in the curved spacetime [4], as does Hobbs [8]. In fact, the infinities from the caps are renormalisable in that context. Presumably they missed these infinities (they do not even consider them) because they were following Dirac's line of argument which conceals them.

What we have here really is a point particle theory. One might argue that Dirac's argument is a convincing one, should one opt for a point particle. Indeed, in the limit as the tube shrinks down to the worldline ($\varepsilon \to 0$), it does seem reasonable to insist on a relation of type (11.31). This seems almost to formulate the idea that the particle is a mere mathematical point. But then the strange behaviour predicted for these particles in certain simple situations could be taken as a convincing argument against the point particle. So one of the aims of any theory proposing an extended particle would then be to get reasonable behaviour in those situations where the Lorentz–Dirac equation gives strange predictions.

11.3 Conclusion

Hopefully, this comparison between two ways of deriving the Lorentz–Dirac equation will illustrate just how unreliable the equation is. Parrott's point in his derivation was to show that the boundary conditions on the charge motion do make a difference. There is hope for the equation when the acceleration is zero outside some bounded proper time range (scattered particle), but it becomes much less obvious in the highly idealized case of hyperbolic motion (eternal uniform acceleration). Note also that one does not have $a(-\infty) = a(\infty)$ in the latter case, so one cannot save the derivation in this way (see p. 294 in Chap. 17).

In any case, one could simply argue that, since these derivations purport to start from energy–momentum conservation, if one then finds that there is no radiation reaction force in a case when there is radiation, either there is some other subtle explanation for this (perhaps Fulton and Rohrlich's acceleration energy mentioned earlier or Boulware's energy flow from the null cone [6]) or the derivation was unjustified. Indeed it is well known that there are peculiar solutions for the Lorentz–Dirac equation, as Parrott discusses at length [13, Sect. 5.5]. One of these is the so-called runaway solution. In a 2D flat spacetime, a charged particle at rest in the absence of external fields accelerates exponentially and reaches a speed arbitrarily close to the speed of light. However, Parrott notes that energy–momentum is not conserved here. The Lorentz–Dirac equation only guarantees conservation of energy–momentum under additional hypotheses that are not satisfied here. So this solution is inadmissible for this reason and the example reminds us that we must take care to know and impose appropriate auxiliary conditions with the equations of motion we have devised.

We return to the question of the validity of the Lorentz–Dirac equation in Chap. 17, which presents Parrott's analysis of the motion of a charged rocket. That example does indeed bring out all the strangeness involved in a literal acceptation of this equation, and shows explicitly how it arises in the context of Parrott's derivation.

It should be remembered that none of these issues are relevant to the doubt raised by Bondi and Gold concerning the strong equivalence principle. This concerns the question of whether a charged particle sitting still in a curved spacetime relative to coordinates for which the metric has static form should radiate. It is generally accepted that the uniformly accelerating charge in Minkowski spacetime without gravitational effects does radiate (this was the conclusion reached by Bondi and Gold, and also by Fulton and Rohrlich). But if SEP is valid, this means that a charge sitting still relative to semi-Euclidean coordinates for a uniformly accelerating observer who attributes her acceleration to being held stationary in the presence of an SHGF will radiate. (At least, it will radiate as observed by a freely falling observer, and the present view is that this is the only observer qualified to assess whether there is EM energy radiation or not.) The problem we are discussing here is how to reconcile the fact that such a charge radiates when the radiation reaction in its supposed equation of motion is zero. So in a sense the problem we are discussing here is a final doubt over whether such a charge can really be radiating.

It should further be noted that neither Dirac's nor Parrott's derivation involves examination of the fields remote from the charge, since integration takes place over a worldtube whose radius is made to tend to zero. This throws doubt on Fulton and Rohrlich's idea that radiation can only be determined far from the charge.

11.4 Self-Force Calculation

As mentioned on several occasions in the above, there is another way of obtaining the radiation reaction terms in the equation of motion, namely a self-force calculation. Indeed this was how the radiation reaction term was first obtained by Abraham and Lorentz. One considers the particle to have some finite extent over which charge is distributed. One has to make some hypothesis about this distribution. Yaghjian assumes that the charge is distributed uniformly over a spherical shell in an inertial frame of reference (the proper frame) moving instantaneously with the sphere center [16]. The shell is assumed to be relativistically rigid, meaning that it always looks exactly spherical in the proper frame of its center, no matter how it is accelerating.

This is precisely the notion of rigidity discussed further on p. 273, in the context of the so-called Rindler elevator.

In a remarkable calculation [16, Appendix B], he obtains the leading terms in the self-force directly from the relation

$$\mathbf{F}_{\mathrm{em}}(t) = \int_{\mathrm{charge}} \rho(\mathbf{r},t)\Big[\mathbf{E}(\mathbf{r},t) + \mathbf{u}(\mathbf{r},t)\times\mathbf{B}(\mathbf{r},t)\Big]\mathrm{d}V , \qquad (11.39)$$

where $\rho(\mathbf{r},t)$ specifies the charge distribution and $\mathbf{u}(\mathbf{r},t)$ its coordinate velocity distribution determined from the arbitrary coordinate velocity of the sphere center under the rigidity assumption. $\mathbf{E}(\mathbf{r},t)$ and $\mathbf{B}(\mathbf{r},t)$ are the electric and magnetic fields generated by this charge distribution itself. Expansions are made in terms of the radius a of the shell in its instantaneous rest frame and Yaghjian obtains the usual formula for the radiation reaction as the term of order a^0, i.e., with no dependence on a.

No particular assumption is made about the acceleration of the sphere center, so presumably this corroborates the usual formula for the radiation reaction, even in the dubious case of eternal uniform acceleration. Furthermore, one does not need to consider the EM fields in regions remote from the charged sphere. Note, however, that there does not appear to be any particular reason for accepting the rigidity assumption mentioned above, except that without it one would be at pains to carry out any kind of calculation at all, i.e., it is difficult to see what else one could assume about the charge distribution.

Chapter 12
Extending the Lorentz–Dirac Equation to Curved Spacetime

At the same time as Fulton and Rohrlich were writing [3], DeWitt and Brehme were transposing Dirac's 1938 derivation of the Lorentz–Dirac equation to the context of an arbitrary spacetime under the aegis of general relativity [4]. DeWitt and Brehme's paper is highly recommended for anyone interested in techniques for general relativity. However, we shall only be concerned with the final result and its interpretation, especially with regard to the role of the equivalence principle.

12.1 Equation of Motion of a Charged Particle

Apart perhaps from the opening sentence, the first paragraph of the abstract to [4] is a model of clarity:

The validity of the principle of equivalence is examined from the point of view of a charged mass point moving in an externally given gravitational field. The procedure is a covariant generalisation of Dirac's work on the classical radiating electron. Just as Dirac's calculation was kept Lorentz invariant throughout, so the present calculation is maintained generally covariant throughout. With the aid of two-point tensors, which are nonlocal generalisations of ordinary local tensors, the manifest general covariance of each step is achieved in an elegant and useful way. The Green's functions for the scalar and vector wave equations on a curved manifold are obtained and applied to the derivation of the covariant Lienard–Wiechert potentials. The computation of energy–momentum balance across a world tube of infinitesimal radius surrounding the particle worldline then leads to the equations of motion including radiation damping.

There is not much to add here. The gravitational field is described as externally given because they are assuming that the antics of the charged particle do not affect it. The result of the calculation is

$$
m\ddot{z}^{\alpha} = \frac{e}{c} F^{\alpha}_{\text{in } \beta} \dot{z}^{\beta} + \frac{2}{3} e^2 c^{-3} \left(\dddot{z}^{\alpha} - \frac{1}{c^2} \dot{z}^{\alpha} \ddot{z}^2 \right) + \frac{e^2}{c} \dot{z}^{\beta} \int_{-\infty}^{\tau} f^{\alpha}_{\beta \gamma'} \dot{z}^{\gamma'} (\tau') \mathrm{d}\tau'
$$

$$
- \frac{1}{3c} e^2 (R_{\gamma}{}^{\alpha} \dot{z}^{\gamma} + c^{-2} R_{\gamma \beta} \dot{z}^{\gamma} \dot{z}^{\beta} \dot{z}^{\alpha}) .
\tag{12.1}
$$

We need to explain the notation:

- $z^\alpha(\tau)$ is the charge worldline as a function of the charge proper time τ.
- $F_{\text{in}\ \beta}^{\alpha}$ is an external field in the form of electromagnetic waves that may affect the charge. When added to the retarded field $F_{\mu\nu}^{\text{ret}}$ due to the charge, we obtain the total field $F_{\mu\nu}$, i.e.,

$$F_{\mu\nu} = F_{\mu\nu}^{\text{in}} + F_{\mu\nu}^{\text{ret}}\,. \tag{12.2}$$

One can understand this physically, because the retarded field satisfies our preference for causal laws.

- m on the left is a renormalised mass. As in the case of flat spacetime, certain terms in the integral of the energy–momentum over the world tube are divergent, but fortunately they are proportional to the four-velocity $\dot{z}^\alpha$ and can be absorbed into this term of the equation of motion.
- The quantities $f_{\beta\gamma'}^{\alpha}$ are rather technical objects defined by

$$f_{\mu\nu\alpha} := v_{\mu\alpha\cdot\nu} - v_{\nu\alpha\cdot\mu}\,, \tag{12.3}$$

where the two-point tensor $v_{\mu\alpha}$ arises in the Green function for the vector wave equation. The dot denotes covariant derivative, while letters in the first half of the Greek alphabet correspond to one of the spacetime points with which the object is associated and letters in the second half correspond to the other spacetime point. A large part of the paper [4] is concerned with such two-point tensors. For example, one requires a two-point tensor of geodesic parallel displacement denoted by $\bar{g}_{\mu\alpha}(x,z)$ in order to define a coordinate-invariant integral of the energy–momentum tensor density, something that is not required in a flat spacetime. We shall not need to go into details here.

- $R_{\alpha\beta}$ is the Ricci tensor.

We need now to explain how the different terms in (12.1) can be interpreted:

- The first term on the right-hand side is obviously a Lorentz force term.
- The second term on the right-hand side corresponds exactly to what we have been calling the radiation reaction term in the flat spacetime case.
- The third term is what DeWitt and Brehme refer to as the tail, depending on the whole past history of the charge motion. This is something that does not arise in flat spacetime. Since $f_{\beta\gamma'}^{\alpha}$ derives ultimately from the Green function for the vector wave equation in the given spacetime, it depends on the metric, and in fact the curvature tensor for that spacetime. Indeed, part of the paper [4] is concerned with approximating $v_{\mu\alpha}$ in terms of the curvature tensor. The thing to bear in mind is that it is a curvature-dependent term.
- The last term on the right-hand side was missed by DeWitt and Brehme and is now known as the Hobbs term [12]. Presumably everyone who read [4] carefully will have spotted it between 1960 and the publication of Hobbs' paper in 1968, and one does not need to read [12] in order to see where it comes from. In fact

they truncate their expression for the quantity they denote by $P_{\mu\nu}$ a little too soon. The terms in what they call $\delta_{\pm}$ require one to add the further term

$$F_{\mu\nu}^{\pm}(\text{extra}) := \pm\frac{1}{3}e(\bar{g}_{\mu\alpha}\bar{g}_{\nu\beta} - \bar{g}_{\nu\alpha}\bar{g}_{\mu\beta})\kappa^{-2}R_{\gamma}^{\beta}\dot{z}^{\alpha}\dot{z}^{\gamma} \qquad (12.4)$$

to their quantity $F_{\mu\nu}^{\pm}$. This leads to a term

$$F_{\text{rad}}^{\alpha\beta}(\text{extra}) := \frac{2}{3}ec^{-2}(R_{\gamma}^{\alpha}\dot{z}^{\beta} - R_{\gamma}^{\beta}\dot{z}^{\alpha})\dot{z}^{\gamma} \qquad (12.5)$$

in the quantity $F_{\text{rad}}^{\alpha\beta}$ discussed further below.

The last three terms on the right-hand side of (12.1) are all considered as radiation reaction terms, arising from a quantity $F_{\text{rad}}^{\alpha\beta}$. Following Dirac [14], it is usual nowadays to make the following definitions for the electromagnetic fields. As mentioned above, the total field is

$$F_{\mu\nu} = F_{\mu\nu}^{\text{in}} + F_{\mu\nu}^{\text{ret}}\,. \qquad (12.6)$$

Not wishing to forget the advanced field defined by the charge motion, one also defines a field $F_{\mu\nu}^{\text{out}}$ by

$$F_{\mu\nu} = F_{\mu\nu}^{\text{out}} + F_{\mu\nu}^{\text{adv}}\,. \qquad (12.7)$$

One now defines the radiated field to be

$$F_{\mu\nu}^{\text{rad}} := F_{\mu\nu}^{\text{ret}} - F_{\mu\nu}^{\text{adv}} = F_{\mu\nu}^{\text{out}} - F_{\mu\nu}^{\text{in}}\,, \qquad (12.8)$$

which is clearly a free field, i.e.,

$$F_{\text{rad}\cdot\nu}^{\mu\nu} = 0 \qquad (\text{free field})\,. \qquad (12.9)$$

Concerning $F_{\mu\nu}^{\text{rad}}$, note the following points:

- In the regions of spacetime where one generally wishes to measure radiated fields, it barely differs from the retarded field.
- Dirac shows that it is finite on the charge worldline in the flat spacetime he considers in [14], unlike either $F_{\mu\nu}^{\text{ret}}$ or $F_{\mu\nu}^{\text{adv}}$.

One naturally expects the radiation field to depend on the curvature tensor and DeWitt and Brehme were indeed surprised to find that the curvature only appeared to arise through their tail term. Of course, in a spacetime with little curvature, these contributions to the radiated field are going to be very small.

Interestingly, DeWitt and Brehme also miss the fact that integration over the caps introduces a further mass renormalisation, presumably because they were following Dirac's example so closely, and he obscured this point (see p. 95). In fact they state [4, p. 253] that the integrals over the caps will retain contributions only from the particle stress density $T_{\text{P}}^{\mu\nu}$. However, if ε is the radius of the world tube, the retarded field goes as ε^{-2}, so that $T_{\mathrm{F}}^{\mu\nu}$ goes as ε^{-4}. On the other hand, the measures on the caps only go as ε^2.

Furthermore, there is an integral over V (the 4-volume enclosed within the world tube) which involves $T_{\mathrm{F}}^{\mu\nu}$ and may lead to the same kind of thing. This is also discarded in [4]. The measure is no more helpful here with regard to the problem of divergent integrations over ε, since it only contains $\varepsilon^2 \mathrm{d}\varepsilon$. The solid angle integration might remove a few terms due to symmetries. This integral is a delicate problem in the context of DeWitt and Brehme's presentation, since it does not arise in flat spacetimes. So one may refer to the discussion by Parrott to confirm the claim in the last paragraph (see p. 95 of the present book), but this integral over V is a new problem. Presumably, if it gives divergent terms, they do have the right form to be absorbed into the left-hand side of (12.1) by mass renormalisation.

It is striking that Hobbs did not pick this up in his 1968 paper [12]. Apart from being the first person to publish the missing term in (12.1), this author introduces a tetrad formalism for deriving the same results more simply. Parts of the explanatory text are copied verbatim from DeWitt and Brehme's paper and it is not at all clear how the tetrad formalism actually works in this context.

Although this is perhaps not really the place, one could say a lot more about the mass renormalisation. DeWitt and Brehme give the subject short shrift, probably because they only found one contribution to mass renormalisation. If they had noticed the other from the caps (and possibly another from the volume integral mentioned above), they might have been struck by the regularity with which these divergent terms turn up having exactly the right form to be absorbed into the mass term of the equation. This is presumably telling us something.

Although the idea is currently unpopular, it may be telling us that particles are not mathematical points, but extended in the spacelike hypersurfaces they intersect in spacetime. This view is supported by several observations:

- One can then understand why charged particles should radiate, and indeed why there should be a radiation damping term in the equations resulting from this type of energy–momentum integration. By accelerating it, one deforms the particle, and the various parts of the charge distribution within it exert forces on the other parts of the charge distribution, rather as happens in an emitting antenna.
- The miraculous possibility of renormalising the mass to get rid of divergent terms in the point-particle model simply becomes the widely accepted idea that binding energy in an extended particle should contribute to its inertial mass according to the usual Einstein relation between mass and energy.
- One can understand why such binding energy contributions to inertial mass should exactly equal their contribution to passive gravitational mass, via an application of the strong equivalence principle.

This is a big subject in itself, with many technical problems that need to be treated carefully.

One may have good fortune with the removal of divergent terms, but one is of course unlucky too, because the resulting equation (12.1), which generalises the

Lorentz–Dirac equation, surely suffers from at least as many vicissitudes as its forebear in flat spacetime.

However, when spacetime is flat and $F^{\text{in}} = 0$, a solution is $\ddot{z}^\alpha = 0$, i.e., geodesic motion, whereas when spacetime is curved, both the tail and the Hobbs term generally prevent this from being a solution, even when $F^{\text{in}} = 0$. One just has to require $\dot{z}^\alpha \neq 0$ at some point and then these terms deriving from the curvature get an acceleration going. But if $\ddot{z}^\alpha \neq 0$, then the other radiation damping terms in the equation of motion will not generally be zero either. This means that we expect radiation, even for free fall. But note that the notion of free fall has been subtly subverted here, because we have in fact demonstrated that a charged particle cannot fall freely in the sense of following a geodesic, simply due to the inevitable effects of its own electromagnetic fields on itself.

This is not really surprising. The curvature of spacetime means that, due to self-interaction of the charge with its own field, even in the absence of any external source of fields, the motion will not be geodesic. (One would also expect this situation if the particle were not a point particle, although in this case, the interaction of the field with its source will also explain to some extent the inertia of the particle.) It is therefore liable to involve radiation damping terms. More will be said about this below, in the context of the equivalence principle and the introduction to the DeWitt and Brehme paper.

Also mentioned in [4] are the so-called runaway solutions, which the authors describe as inadmissible. These are discussed in [13, Sect. 5.5]. In a 2D flat spacetime, there are solutions in which a charged particle at rest in the absence of external fields accelerates exponentially and reaches a speed arbitrarily close to the speed of light. Parrott notes that energy–momentum is not conserved here. The Lorentz–Dirac equation only guarantees conservation of energy–momentum under additional hypotheses that are not satisfied here. So such a solution is inadmissible for this reason and the example reminds us that we must take care to know and impose appropriate auxiliary conditions with the equations of motion we have devised.

Parrott also exposes one of the two-particle situations, first discussed by Eliezer, in which the solution turns out to be too strange to be taken as a serious physical possibility (although one has to remain open if no basic principle like energy–momentum conservation is flouted). One can consider this as a case of a charge in a certain type of external field. If two opposite charges of equal mass are moving along a line towards each other in a proper 4D spacetime, they will eventually begin to move apart with ever-increasing speed along that line, approaching the speed of light asymptotically!

As DeWitt and Brehme say, it would be a very interesting although probably difficult problem to see if this situation persists for the equations of a charged particle which moves in the curved spacetime produced according to Einstein's equations by charged mass points. This suggests the importance, in the n-particle problem, of taking into account the dynamical properties of the gravitational field and the fact that the metric is actually singular at the location of each particle.

12.2 The Equivalence Principle in All This

The issues discussed above show that the resulting equation of motion is somewhat pathological. But the most interesting part of the discussion in [4] undoubtedly concerns the threat to the equivalence principle. Before analysing the comments in [4], one should make two points right away. The first is that none of the analysis in any of these papers is easy to interpret because there is a certain vagueness in their discussion, in particular regarding the precise form of the equivalence principle that is being examined. This may in part be put down to subtle changes in the vocabulary used over the past fifty years. One is therefore obliged to impose one's own interpretation in a way that may not correspond to that of the original authors, rather as one sometimes does science by fitting a theory to the facts. (This is not suggested as a definition of science, only as a simplistic analogy.)

The second point is that DeWitt and Brehme use both the weak and the strong principles of equivalence, as they are now called, in order to derive their equation (12.1). They use WEP as soon as they accept the manifold formulation of spacetime with the Levi-Civita connection, and they assume SEP when they formulate the minimal extension of Maxwell's equations (MEME) by replacing coordinate derivatives in Maxwell's equations by covariant derivatives for this connection. If one now finds something wrong with the consequences, it may be that one should put it down to one or both of these equivalence principles.

Of course, one would prefer to sacrifice some other assumption if possible, given that one does not even have the usual theory of general relativity without WEP, and one cannot apply it to any other bit of physics without SEP. At least, with regard to SEP, one would have to find some other principle for extending flat spacetime physics to curved spacetimes. But this was the problem for Bondi and Gold [2]. Let us recapitulate for immediate comparison with the comments of [4] discussed below.

Bondi and Gold consider a flat spacetime but one in which semi-Euclidean coordinates are used and geodesic motion relative to these coordinates is imputed to the presence of an SHGF. A charge that is stationary relative to the coordinates is then found to radiate, provided that one can use SEP to extend Maxwell's theory to this model. However, the metric is static in this coordinate system, because the SE coordinates are those for a uniform acceleration, and Bondi and Gold find it impossible that such a charge in such a spacetime could radiate. This looks bad for SEP because it is an application *par excellence* of the principle. Their solution to the paradox is based on the idea that an SHGF is physically impossible. Another (proposed in the present book) is that one must just accept that a charge that is stationary relative to coordinates in which the metric is static can radiate.

Let us return to DeWitt and Brehme. The second paragraph of the abstract begins:

> Because of the nonlocal electromagnetic field which a charged particle carries with itself, its use as a device to distinguish locally between gravitational and inertial fields is really not allowable.

Perhaps this should be read as:

> Because the electromagnetic field which a charged particle carries with itself is nonlocal, its use as a device to distinguish locally between gravitational and inertial fields is really not allowable.

By implication, we are referring here to some version of the equivalence principle: one cannot distinguish locally between a gravitational field and an inertial field. It really is a pity that the principle in question and the context have not been specified more precisely, since a lot of the import of the paper is supposed to be concerned with this issue.

On the face of it, we could be talking about special relativity here, and the theory of gravity associated with it, since DeWitt and Brehme refer to acceleration by a gravitational field shortly after the above quote. However, gravitational fields do not accelerate anything in general relativity. In special relativity, the above statement might be construed as a statement of the equality of inertial mass and passive gravitational mass with the consequent idea that one cannot tell whether the motion of an object relative to one's coordinates is due to gravity or due to one's own acceleration. This principle is expected to break down for charged particles because, in this theory, gravitational accelerations are accelerations, so falling charges radiate when they are falling, according to Maxwell's theory; on the other hand, if it is the observer who has for some reason adopted non-inertial coordinates, e.g., semi-Euclidean coordinates adapted to an accelerating worldline, an inertially moving charge will no more appear to radiate than it would to an observer moving with it. At least, even if the accelerating observer in the second case did naively interpret quantities expressed relative to the SE coordinates as saying that there was EM radiation, she would not find the same value for the 'radiation' rate as an inertial observer measuring the radiation from a freely falling charge.

Of course, one expects the discussion here to be about general relativity, given that the whole paper applies to this theory with its in-built weak principle of equivalence. Here the idea that one cannot distinguish locally between a gravitational field and an inertial field looks very like the problem discussed in Chap. 2: in a Minkowski spacetime, one could not decide what interval to use, between (2.45) on p. 15 or (2.48) on p. 17, or indeed the usual Minkowski interval. This interpretation is perhaps corroborated by the fact that the above quote refers to the use of a charged particle to distinguish the two scenarios: we saw in Chaps. 2 and 4 that general relativity really cannot make any distinction between use of (2.45) and use of (2.48) because these are essentially different manifestations of the same metric, the Minkowski metric.

Let us just recycle some of the discussion in Chap. 2. In general relativity, one has to understand gravitational acceleration to mean free fall, even though this involves zero four-acceleration according to this theory. An inertial acceleration of a test particle refers to the fact that it appears to accelerate only because the observer is herself accelerating, or has chosen coordinates that give this appearance. For example, in a flat spacetime (but considered in general relativity), an accelerating observer may choose semi-Euclidean coordinates adapted to her worldline. The connection coefficients are no longer zero, even though there are no tidal effects. The worldline of a freely falling test particle, which is a geodesic, will not look like a straight line relative to such coordinates; it will look like an accelerating trajectory, with the connection coefficients playing the role of the acceleration [11, p. 135], as discussed in Chap. 2.

In a non-flat spacetime under the aegis of general relativity, one generally has to contend with a connection that cannot be made to go away over any neighbourhood owing to tidal gravitational effects. Some aspects of the connection come from gravity and some come from the coordinates. Might it be that the above-quoted principle, if it is a principle, is telling us that we cannot make the difference, at least on the basis of local observations? This idea is clearly related to the idea labelled WEP on p. 15, which says that one can get rid of the connection at any spacetime point by playing off the choice of coordinates with the gravitational effects, although generally, this can only be done at one point. Of course, one needs to see how test particles, either charged or neutral, could be used to make such a distinction.

If we imagine the observer with some arbitrary timelike worldline, then there exist coordinates adapted to that worldline, satisfying similar properties to those listed in the flat spacetime case on p. 5 [11, p. 181]. This is discussed on p. 20. Once again, the worldline of a freely falling test particle, i.e., a geodesic, will look like an accelerating trajectory in the adapted coordinates, with the connection coefficients playing the role of the acceleration [11, p. 182], just as displayed in (2.55) on p. 20. In this case, the constraint of adapting the coordinates to the observer's worldline in the sense of [11], in particular, the condition that the observer should remain at the origin of the spatial hypersurfaces of her coordinate system, usually precludes any possibility of making the connection components zero on her worldline. The exception arises when the observer is herself in free fall, in which case one can make the connection coefficients equal to zero right along the observer worldline (which is then a geodesic). But in a given coordinate system, can these considerations help to distinguish what aspects of the connection come from gravity and what aspects come from the coordinates?

One has to take into account the other crucial feature of the proposed principle, that one is only allowed local observations. The question of what is meant by local and nonlocal was also discussed in Chap. 2. If local means 'only at the relevant point', i.e., the event where the particle happens to be, then we could not use any detection of electromagnetic radiation from it, or detection of any other electromagnetic field it had caused, to decide whether the charge was being accelerated, or whether it was our own frame that was accelerating and it was ourselves who attributed the acceleration to it. The reason is just that we need to look at the electromagnetic fields in the spacetime neighbourhood of the charge at this event. But if local means 'in the neighbourhood of the relevant point', i.e., in the immediate spacetime neighbourhood of the particle, but leaving the meaning of 'immediate' vague for the moment, then we could at least try to use the nearby fields to distinguish whether the charge was being accelerated (in the sense of GR), or whether it was our own frame that was accelerating.

As noted with regard to the earlier papers, there seems to be an idea that electromagnetic radiation is a far-field effect and therefore does not count in the kind of local determination of whether a charge is being accelerated, or whether it is just being viewed from an accelerating frame. But the term 'far field' is relative. The far field may be much nearer than any significant effects due to curvature. In any case, any EM field emanating in some sense from the charge will already have set the scene for its future, and hence further, development at distances as close as we like to the charge. This is borne out by the fact that, in world tube calculations of energy–momentum conservation, we may shrink the world tube down to the limit of zero radius and still obtain the energy–momentum carried away by the field.

Here is another quote referring to this very issue, taken from the introduction to [4]:

> Before describing the procedure to be used, we should point out at once that the idea of using a charged particle to distinguish locally between gravitational and inertial fields is, of course, cheating. A charged particle carries with it an electromagnetic field, which is by no means local. A gravitational field can be readily distinguished from an inertial field by experiments carried out over an extended region, that is, by experiments which measure field gradients. The gradient of a field is a second derivative of a potential. In the general theory of relativity the potential is the spacetime metric, and second derivatives of the metric are expressed uniquely in a covariant manner by the components of the Riemann tensor, which describes the intrinsic spacetime curvature or, alternatively, the 'true' gravitational field. We should therefore not be surprised if, when the radiation reaction is included, we find the Riemann tensor entering explicitly into the dynamical equations of a charged particle moving in a gravitational field.

The word 'true' appears in inverted commas in [4]. Presumably, it stresses the fact that the field they are talking about really is gravitational, rather than an inertial one that might be confused with it. What is difficult to understand in the above quote is the idea that one might be cheating in some way by looking at electromagnetic fields. Of course, it depends on what one hopes to do by looking at them, and this is not very clear either. However, it obviously has something to do with checking some version of the equivalence principle.

We may wonder once again exactly what form the equivalence principle has assumed here. Following the suggestion in the quote, what does it mean to say that one cannot distinguish locally between gravitational and inertial fields? The problem is just that, if we allow the word 'local' to include some neighbourhood, then this version of the equivalence principle, if it is supposed to be one, is in danger of collapsing to nothing. As the authors themselves explain here, we do not need more than to get our hands on a small region in order to determine the metric, connection, and curvature in that region, the size of the region being determined by our measurement accuracy. If by gravitational field one is to understand tidal effects, i.e., nonzero curvature, then one can indeed distinguish it from nonzero connection due to coordinate effects, provided one is allowed to examine a big enough neighbourhood. The key here is presumably the question of measurement accuracy. For whatever error one is forced to tolerate for technical reasons, there will be some region small enough to ensure that one cannot distinguish a change in the connection from the zero value one may have arranged at some particular event by choice of coordinates.

There is a case, the one we have been specifically concerned with, where one would like to impute an acceleration relative to inertial coordinates in a completely flat spacetime to a force that is preventing a test particle from free fall in that spacetime. In that case, there is zero curvature, but one is considering observed effects as being due to a static, homogeneous gravitational field. Although it is a highly idealistic scenario, one would expect it to be a good approximation to certain regions of spacetime, such as a small region of an Earth-based laboratory. In this case, the nonzero connection coefficients relative to the semi-Euclidean coordinates one might adopt are coordinate-generated, and hence do not reflect a nonzero curvature. The idea that this might just be an approximation in some region of spacetime is close to the idea that the relevant neighbourhood is too small to be able to detect tidal effects for one's available measurement accuracy.

Since one is supposed to use test particles to understand one's coordinates, it should be noted that an equivalence principle that says (as suggested later in [4]), not that one cannot distinguish locally between gravitational and inertial fields, but that one cannot distinguish locally between gravitational and inertial accelerations of a test particle, is simply tautological in GR. (This too was also discussed in Chap. 2 and it reveals the risks of not making precise statements.) Presumably gravitational acceleration refers to geodesic motion, called (locally) inertial motion in GR. Of course, local experiments cannot distinguish between such a gravitational acceleration of a test particle, meaning geodesic motion, and an acceleration of the observer which makes the test particle appear to accelerate, because it is understood here that the test particle that appears to accelerate is actually undergoing (locally) inertial motion, so we are talking about the same particle and there is nothing to be distinguished.

More generally, one could say that any principle which proclaims that one cannot distinguish something locally has to be either tautological, because one is talking about distinguishing things that are actually the same, or simply wrong, because once one has determined the metric (and hence the connection and curvature) locally, one can distinguish geodesic motion in the region from any other kind of motion. This is the problem with the word 'locally' in this kind of statement of the equivalence principle.

But we were talking about charged particles here. The idea is that a charged particle will radiate when it is accelerating but not when it is just ourselves who are accelerating. Of course, we need to examine the electromagnetic fields in some neighbourhood to get any clue about this radiation, but then we could just look at the Riemann tensor in that neighbourhood. When we know the Riemann tensor, we know everything we need to know about the spacetime.

So why the fuss about radiation from charged particles? Well the theory in [4] is telling us that charged particles left to their own devices in the spacetime will follow worldlines with radiation damping terms that are nonzero, i.e., they will not generally follow a geodesic with $\ddot{z}^\alpha = 0$, basically because they generate EM fields that then act back on them in some way. (The way they act back is not explained in the point particle model, but it is in the extended particle model.) Here we have a case of a particle that only approximately satisfies the geodesic principle discussed in Sect. 2.5. Of course, it could be that the radiation damping terms do just happen to cancel out for certain motions in certain spacetimes. We do know that there are pathological cases. The problem with uniform acceleration in flat spacetime, i.e., the strange fact that the radiation damping terms can be zero in a case where there is acceleration, is relevant here, because we can even have a non-geodesic worldline with zero radiation damping terms. However, as we have seen, even in this case, the theory tells us that there is still radiation.

Can we force the charge to follow a geodesic and then look for radiation? If we do this, the DeWitt–Brehme equation of motion throws no light whatever on the question of whether the charge will radiate, because we have artificially introduced a new term in the equation of motion which cancels all the others to give $\ddot{z}^\alpha = 0$. However, we could feed this worldline into the Lienard–Wiechert formulas for the fields, work out the consequent energy–momentum of those fields and examine the flow of that energy–momentum to see whether there is a flow of energy away from the charge. The simplistic argument that the well-known radiation terms are zero in this case is not sufficient to be able to say that there is no radiation, as we know from the uniform acceleration case in flat spacetime.

In any case we need to reassess the question: What is free fall? For a neutral particle it is geodesic motion. But for a charged particle which, left to its own devices in the spacetime, will not generally follow a geodesic, we have to decide whether free fall should be: (A) geodesic motion, (B) motion according to (12.1), i.e., perturbed from a geodesic by self-forces exerted by the EM fields it itself generates. It seems reasonable to prefer (B). But then, having changed the meaning of a term, it is only to be expected that one must reword one's principles.

If we are talking about distinguishing different scenarios, one thing that we should be able to distinguish is the motion of a charged particle from the motion of a neutral particle. The first is not generally geodesic while the second is. Furthermore, the first is likely to fill the surrounding spacetime with EM fields, while the second is not.

However, we have wandered away from the problem at hand here, which is to ask whether there is any hope for an equivalence principle that talks about whether one can distinguish between a gravitational acceleration and an inertial acceleration, the latter being due to our own acceleration, merely by a local experiment. Perhaps the best approach here is to avoid such formulations, especially since one has built WEP and SEP into the theory mathematically and there is no doubt about how they should be employed. One could say that these formulations of the equivalence principle which refer to one's not being able to distinguish two scenarios on the basis of local observations are a kind of last step before leaving SR and replacing it by GR, or before leaving Newtonian gravitation (NG) and replacing it by a version that is both relativistic and geometrised. The equivalence principle then gets a new formulation in GR.

Indeed, in SR or NG, the equivalence principle in this version is not a principle. It is just something that one observes, at least to within experimental accuracy. The idea it refers to only becomes a principle, in the sense of an underlying hypothesis of the theory, in the formulation of GR. For example, it is absolutely clear what 'local' means: the connection coefficients can be made to equal zero at any point by suitable choice of coordinates, and then by continuity, the connection will be zero to all intents and purposes, i.e., within measurement accuracy, on some small enough spacetime region about that point. So we have flat spacetime physics to within a given accuracy on some small enough neighbourhood. The size of the neighbourhood depends on the required accuracy.

But a subtle change occurs in moving from SR to GR. In SR one can speak of gravitational accelerations, whereas in GR these are replaced by the notion of free fall, which is geodesic motion in a curved spacetime context. So there is no gravitational acceleration now. And what of inertial acceleration? This refers to artifacts in the formulas for the motion due to choice of coordinates. In GR, one can have nonzero connection coefficients by poor choice of coordinates and by intrinsic geometric effects, but they all look the same: they all just come out as nonzero connection coefficients. And they can all be made to go away at a point, or in some small enough spacetime neighbourhood for given measurement accuracy, by suitable change of coordinates.

One might say that, in GR, this whole formulation of an equivalence principle in terms of indistinguishable situations has become obsolete. Because it is replaced by a genuine principle which underlies the theory, any prediction of the theory can be said to test that principle, to some extent. The very formulation of the DeWitt–Brehme theory assumes a strong version of the equivalence principle applied to electromagnetic theory. DeWitt and Brehme wrote down an action and varied it. They found field equations that were just like the ones in flat spacetime, except that ordinary coordinate derivatives were replaced by covariant derivatives involving the connection. This directly implies that the physics of this extension of electromagnetic theory will look exactly like the ordinary Maxwell theory in flat spacetime whenever we can get the connection to zero. In other words, in any coordinates which make the connection zero at some point and hence, by continuity, arbitrarily small for some small enough neighbourhood, the extended theory of electromagnetism they are proposing will look like Maxwell's theory on that neighbourhood, to within the predetermined experimental accuracy.

Hence, to look at the results of this theory and ask whether, on theoretical grounds, the equivalence principle is valid as a consequence would be somewhat foolish. What we can do is to compare the DeWitt–Brehme equation of motion with observation of freely falling charges, where 'freely falling' means that the charges are influenced only by gravity and their own fields, and see whether this equation is a good prediction. If it is not, we may be led to suspect the equivalence principle, or we may be able to impute the problem to some other ingredient of the DeWitt–Brehme theory. But dredging up a version of the equivalence principle that still refers to gravitational accelerations, where we have replaced these by geodesic motion, and inertial accelerations, where we have replaced these by non-geodesic motion of the observer and correspondingly adapted coordinates, seems a pointless enterprise.

On the other hand, perhaps we are expecting some purely logical inconsistency within the theory itself. If that is the case, it certainly will not be identified by examining the question as it seems to be posed here: is there some way of ascertaining locally whether a charge is influenced only by gravity (and possibly its own fields) or whether some of its antics are due to our own motion? The answer seems irrelevant once we have built our principle so deeply into the theory that we can only attempt to answer that question by assuming an even more sophisticated version of the principle to formulate the problem at all.

But DeWitt and Brehme do appear to view this as the task at hand. At the very beginning of their introduction, they say:

> An important part of the development of any physical theory is the testing of its consequences against its original physical foundations. In the general theory of relativity this part plays an anomalously large role for the simple reason that the theory has progressed experimentally so little beyond these foundations. Two experimentally established principles, the principle of equivalence and the principle of the covariance of physical laws, form the basis for the general theory. The concept embodied in the principle of equivalence is a 'local' one which, when combined with the covariance concept, leads to the introduction of curvilinear coordinate systems and nonvanishing intrinsic curvature into the description of spacetime.

As an aside, it is interesting to note that the principle of covariance is considered to be an experimentally established principle. One might say that, in GR, the principle of covariance is just the insistence that the only description of spacetime and what happens in it be a generally covariant one, i.e., that there be no privileged frames of reference for the theory in its most general formulation. For surely one would expect any physical theory worth its salt to have a generally covariant formulation, in the sense that one can choose any coordinates one likes to describe spacetime and what happens in it. Naturally, a theory like SR will not look so nice with curvilinear coordinates as it does with Lorentz coordinates. (Of course, one might try to argue that it would not be the same theory if one changed to general curvilinear coordinates.) What distinguishes SR is the fact that there is a class of privileged coordinate systems in which it looks nice in the well known way. The principle of covariance in GR could perhaps be taken as saying that there is generally no such class of privileged systems (although, of course, there will be for specific, highly symmetrical situations, such as an empty spacetime). One may well wonder whether this is really an experimentally established principle.

In the continuation of the last quote, however, we are provided with a statement of the equivalence principle we are supposed to be examining:

> In its simplest form the principle of equivalence states that a gravitational force cannot be distinguished from an inertial force by any experiment which is conducted on a purely local basis.

As stated here, this looks very like the kind of experimental observation that people using NG or SR might well have taken as the starting point for geometrising away one of the very things it refers to, namely the gravitational force. But having done so, in the new theory without gravitational forces, one presumably needs to reformulate the idea. This is all the more urgent because it becomes one of the logical founding principles of the theory, albeit experimentally inspired, since it was in the previous theories only an experimental observation, whereas it has in GR become a genuine principle.

The other problem here is that we are given no help with the word 'local'. We shall take it to mean 'precisely at a given spacetime event, and to within experimental accuracy within some spacetime neighbourhood'. This is indeed the meaning it has when the idea is formulated in NG or SR. We do not expect to be fooled by a given test particle motion into thinking it is due to a gravitational force rather than our own motion, or vice versa, if we are allowed to examine the motions of test particles over a big enough neighbourhood, where the phrase 'big enough' is determined by the accuracy of our measurements.

The rest of the paragraph which begins with this statement of the equivalence principle continues as follows:

> [...] While this principle is certainly valid to a high degree of precision, and may even be valid with absolute precision for neutral matter, there is some question as to whether it can be absolutely valid for matter which carries an electrical charge. To put the question in physical terms, imagine a charged particle located in empty space at a great distance from any gravitating masses. If a force is exerted on the particle it will begin to accelerate, and we know from the laws of classical electrodynamics within the framework of the special theory of relativity, which is valid under the above circumstances, that the particle will radiate, producing a reactive damping force in addition to its mechanical inertial force. Let us next bring the particle to rest in a static gravitational field which exerts a force equal to that to which the particle was previously subjected in empty space. Although the particle experiences the same force in both cases it would be absurd to suppose that it continues to radiate under the latter conditions.

The reference to precision at the beginning does suggest that we are talking here about experimental comparison of inertial mass and passive gravitational mass.

If one views the ensuing discussion from the standpoint of SR, the first particle is being accelerated by an external force and it will radiate. The second particle is not being accelerated because there are two balancing forces on it, so it will not radiate. That is what Maxwell's theory tells us in SR, whether it is correct or not.

If one views this discussion from the standpoint of GR, the first particle is being accelerated by an external force in a flat (region of) spacetime, where we expect SR and Maxwell's theory to hold, so it will radiate according to GR+SEP and the minimal extension of Maxwell's theory, which happen to coincide with SR and Maxwell's theory here. The second particle, however, is now considered to be accelerated in a curved spacetime. GR strongly suggests that it will radiate, because it is accelerating in a local inertial frame and we have implemented a principle of equivalence which insists that the minimal extension of Maxwell's theory to GR will look like ordinary special relativistic Maxwell theory in a local inertial frame.

In fact, we have a very clear demonstration of the latter claim in the case of a static homogeneous gravitational field (SHGF). In that case, the situation described is precisely the situation of an accelerating charge in an inertial frame. (The local inertial frame happens to be globally inertial in that case.) Even when viewed from a frame accelerating with the charge, and one might say that we are using such a frame implicitly in the above example, this charge is radiating. The reason there has been some doubt about this particular demonstration is, as we have seen, that the so-called supported frames have uniform acceleration. In this case, the radiation terms in the Lorentz–Dirac equation (which is supposed to hold in this flat spacetime) happen to be zero, and there are other special, even pathological, features in this scenario.

At the end of the last quote, DeWitt and Brehme say that this conclusion, that a stationary charge in a static spacetime will radiate, is absurd. We have come to the problem raised by Bondi and Gold in [2]. Unfortunately, they do not explain on what grounds this eventuality is absurd. The fact that it follows from GR+SEP is presumably an argument in its favour. In fact it follows directly from the strong version of the equivalence principle whereby one obtains a minimal extension of Maxwell's theory to curved spacetime, and this is the very theory that DeWitt and Brehme are working with here, because they assume this minimal extension when they formulate electromagnetism in such a way that ordinary coordinate derivatives are replaced by covariant derivatives. So presumably the absurdity they feel arises from a comparison with physical intuition. But the intuition they are referring to looks very like the naive intuition of SR, wherein a charge supported against the effects of a gravitational field in such a way that it cannot move is actually an inertial particle. In GR, this same particle is no longer inertial.

At the same time, it is striking that DeWitt and Brehme do not even mention the question of whether a charged particle can appear to radiate when it is inertial, i.e., following a geodesic, but is viewed from an accelerating frame. Put briefly, does a charge with inertial acceleration radiate? Now the fact is that it does not, according to any theory, whether we consider things from the standpoint of SR or GR:

- In SR, we may try to imagine what an accelerating frame should be, e.g., a semi-Euclidean frame whose origin has some acceleration relative to an inertial frame, and transform the electromagnetic fields from the inertial frame to the accelerating frame in the way dictated by the usual tensor transformation laws. We have to extend SR slightly, because the transformations are not Lorentz transformations. In fact, we have to use the generally covariant version of SR. We find that the fields look strange, but there is no reason to consider them as radiation fields (see the opposite view expressed by Rohrlich in Sect. 16.10). If there happens to be a gravitational field, and if we wish the charge to be inertial, it must be supported, i.e., we must balance the gravitational field with some other force so that the charge does not accelerate relative to the inertial frame. But this changes nothing. The charge will not radiate, no matter how we look at it, according to this theory.

- In GR+SEP, and we consider the minimal extension of Maxwell's equations (MEME) to be part of this theory, an inertial charge is one that remains at the origin of a locally inertial frame so, to a first approximation (and the level of approximation is on a par with any vagueness in the term 'local'), the electromagnetic fields will look like those of an inertial particle in an inertial frame in SR, so there will be no radiation. This is only an approximation. Even the idea of a freely falling charge becomes an approximation in GR, as one can see from the complicated DeWitt–Brehme equation of motion (12.1). One might well expect some small differences between the EM fields in GR and those of an inertial particle in an inertial frame in SR. Now in the primitive version of the equivalence principle, this situation corresponds to a charge subject to a gravitational force. On the other hand, if we view the same particle and its fields in an accelerating frame, i.e., one that is not freely falling, we no more expect this to make the fields look like radiating fields than we expect the Coulomb field to look like a radiating field in SR when viewed from an accelerating frame (but see the opposite view expressed by Rohrlich in Sect. 16.10).

So much for what the theories say. The experimental question remains open. Note as before, however, that SR and GR give opposite predictions. SR says that the particle supported in a gravitational field, which is inertial in SR, does not radiate, no matter what frame we view it from. GR says that the particle falling freely (as far as possible) in a gravitational field does not radiate (on a significant scale), no matter what frame we view it from. The standpoint adopted in the present account is that the choice of frame to view the inertial particles is irrelevant to the question of radiation. To a certain extent, this is supported by the idea whereby one can view EM radiation as the giving off of photons, something which one does not a priori expect to depend on the frame adopted by the observer. What can depend on the observer in this view is the spatial pattern and timing of arrival of the photons. But if photons are given off in one frame, they are given off in another. There is more about this issue in Chap. 14 which considers Mould's radiation detector.

So only GR supports the primitive version of the equivalence principle we have been examining. In SR, the presence of radiation distinguishes the inertial particle viewed from an accelerating frame (no radiation) from the gravitationally accelerated, non-inertial particle (radiation). In GR, the presence of radiation does not distinguish the inertial (freely falling) particle viewed from an accelerating frame (no radiation) from the gravitationally accelerated, hence (locally) inertial particle (no radiation). This is because the gravitationally accelerated particle is the same as the (locally) inertial particle, and changing to an accelerating frame in one case makes no difference to the detection of radiation. All this is based heavily on the idea that, whether one considers SR or GR, a radiating field has to be considered as a radiating field in any frame, and a non-radiating field has to be considered as a non-radiating field in any frame.

DeWitt and Brehme seem to be only vaguely aware of what is involved in the switch as one moves from SR to GR, as attested by the following paragraph:

> There is, however, a catch here. It is well known that an accelerating charged particle does not in fact suffer a reactive damping force as long as its absolute acceleration is uniform, i.e., constant in magnitude and direction. It is therefore better to turn the problem round and imagine the particle in an unaccelerated state. If the particle is far from gravitating matter, it can be said to be in a state of uniform motion, in which it does not, of course, radiate. If the particle approaches a strong gravitational field, however, the notion of unaccelerated state is changed into a different notion, namely, that of a state of free fall. Does a change of notion now cause the particle to begin to radiate? If a charged particle does not radiate when at rest in a gravitational field, does it also refrain from radiating when falling freely? Or can a charged particle be used, at least in principle, as a local entity which distinguishes between gravitational and inertial forces?

The change of notion they refer to is the very thing that makes GR different from SR. But if one is working in GR all along, there is really no change of notion: the particle in uniform motion far from gravitating matter is following a geodesic of the spacetime, and the particle in free fall in a strong gravitational field is also following a geodesic of the spacetime. It is only our prejudice against the theory that suggests that these things might be different.

Now they concluded earlier as to the absurdity of the idea that a charged particle supported in a static gravitational field might radiate, whereas this is precisely the prediction of GR. The absurdity arises only if one is still thinking special relativistically, where gravity has no special role. And as to the question of whether a charge will refrain from radiating when falling freely, the answer is that it will (more or less) according to GR, although it would actually radiate according to SR.

The last question in the paragraph is a logical non sequitur, but the lead into the paragraph is worse, because the catch they mention is not spelt out at all. So what if a reactive damping force is zero? As we have seen, the charge is probably still radiating according to the theory. Furthermore, a uniform acceleration is only a certain kind of (pathological) acceleration anyway, whereas we are supposed to be considering general accelerations.

In the following paragraph, we begin to get the conclusions that DeWitt and Brehme draw from their analysis:

> On the basis of physical intuition it would seem reasonable to suppose that a charged particle does in fact radiate when deflected by a gravitational field, i.e., that bremsstrahlung can be produced by gravitational as well as electromagnetic forces. This is the problem that will be tackled in the present paper.

This is where DeWitt and Brehme mention that this is cheating, because the EM fields are nonlocal, and this was discussed earlier. In any case, we have here another intuitive idea that is based on a special relativistic intuition. There is no doubt that, in SR, where gravitational effects are just forces like any other, we expect the gravitationally deflected charge to radiate like any deflected charge. This is the prediction of the theory. But in GR, the theory predicts quite the opposite: the gravitationally deflected charge is not deflected. In a locally inertial frame, it follows a straight trajectory to a first approximation, so that to within a certain level of accuracy it will not radiate. Any radiation there is can be put down to the fact that this is only an approximation, or indeed the fact that a gravitationally deflected (i.e., freely falling) charge will follow a worldline that is affected by its own EM fields, so that in this sense it is never freely falling in the same way as a neutral particle.

The paragraph concludes (as quoted previously):

> [...] We should therefore not be surprised if, when the radiation reaction is included, we find the Riemann tensor entering explicitly into the dynamical equations of a charged particle moving in a gravitational field.

This is DeWitt and Brehme's conclusion from the fact that the fields are nonlocal in some sense. But we now continue:

> The surprising thing is that the Riemann tensor does not so enter, at least to the extent to which the essentially classical calculations of this paper are valid. This does not, however, mean that electrogravitic bremsstrahlung does not occur. It does. But it has its origin in a more subtle phenomenon having to do with the failure of Huygens' principle, when taken in the narrowest sense, in a curved spacetime. As has been pointed out by Hadamard, a plane or spherical sharp pulse of light, when propagating in a curved 4D hyperbolic Riemannian

manifold, does not in general remain a sharp pulse, but gradually develops a tail. It is this
phenomenon which is responsible for the electrogravitic bremsstrahlung.

Of course, we now know that this is wrong, due to the omission of the terms (12.5)
on p. 109, also spotted several years later by Hobbs. The Riemann tensor then occurs
explicitly in the equation of motion. We have seen the way these terms cause radiation: the geodesic, describing free fall of a neutral particle, is no longer a solution
of the equation of motion, so $\ddot{z}^{\alpha}$ will not be zero, and the radiation damping terms
in the equation of motion are no longer generally zero. We therefore conclude that
there will be radiation.

But these are usually very minor effects that do not come into a first approximation. To a first approximation, GR says that the (almost) freely falling charge
will produce fields in a locally inertial frame moving with it that look like those of a
charge with uniform motion in SR. This is the content of the very principle of equivalence that has been assumed to extend Maxwell's theory to GR! This prediction,
however, is at odds with the physical intuition one might have in SR, as mentioned
above. Indeed, SR predicts that the freely falling (or almost freely falling) charge
will radiate, because there are gravitational forces and they are no different from
any others.

DeWitt and Brehme conclude in further detail with this:

> The picture then is the following. The charged particle tries its best to satisfy the equivalence
> principle, and on a local basis, in fact, does so. In the absence of an externally applied elec-
> tromagnetic field, the motion of the particle deviates from geodesic motion only because of
> the unavoidable tail in the propagation function for the electromagnetic field, which enters
> the picture nonlocally by appearing in an integral over the past history of the particle. Phys-
> ically, the tail may be pictured as arising from a sort of scatter process, with the 'bumps'
> in spacetime playing the role of scatterers, which allows the radiation field originating in
> the particle, which normally 'outruns' the particle, to act directly back on the particle in an
> anomalous fashion.

We may go a little further here. Not only does the charged particle try its best to
satisfy the equivalence principle, but it wholly succeeds. Of course, this statement
depends on what equivalence principle we are referring to. The equivalence principle whereby GR physics looks just like SR physics (to a first approximation) in
a locally inertial frame is entirely satisfied. The equivalence principle whereby one
cannot distinguish gravitational from inertial forces is in danger for its life in GR,
where the gravitational force no longer exists. However, if we take a gravitational
force to mean one revealing itself through the free fall of test masses, then there is
absolutely no reason to expect a change of frame to make anyone think that they are
radiating. They are not radiating (to a first approximation) in the frame falling freely
with them, and there is no reason to consider them as radiating in any other frame.

The words 'to a first approximation' remind us that the equivalence principle
we use to extend Maxwell's theory to GR contains the word 'local'. In the second
approximation, DeWitt and Brehme have shown that the notion of free fall now has
to include the EM force that charged particles exert upon themselves, and that the
non-radiating field of the first approximation will contain other perturbations due to
curvature effects quite apart from those due to this deviation from geodesic motion.

One might in fact conjecture that, at the beginning of the last quote, DeWitt and Brehme actually mean to say that the charged particle does its best to satisfy the *geodesic* principle. It could be that they are assimilating this with the equivalence principle, since it follows, at least as an approximation, from the equality of inertial and passive gravitational mass, the experimental result that authorises the construction of a theory like GR on the basis of a principle like WEP (see Sect. 2.5). Of course, one only expects the geodesic principle to work if the particle cannot exert forces on itself. And even then, if it has some spatial extent and is spinning, for example, we have seen that it would not need to carry a charge distribution in order to be perturbed from a geodesic.

12.3 Conclusions

The aim here was to try to understand DeWitt and Brehme's problem. In short, their conclusions are as follows:

- Free fall is different for a charged particle and for a neutral particle, basically because the former produces EM fields that affect its own motion. In this process, the curvature of spacetime along the charge worldline directly enters the proposed equation of motion. So the geodesic 'principle' discussed at length in Sect. 2.5 does not apply to charged particles, although it will apply approximately in most cases.
- They agree with Bondi and Gold that a charge that is stationary relative to coordinates in which the metric is static should not radiate. Interestingly, they do not directly check this with the formulas they generate for the EM fields produced by a charged particle. They also think that there should be electrogravitic bremsstrahlung, i.e., a charged particle falling freely in a curved spacetime, subject only to its own EM self-interaction, should radiate.

The conclusions in the present chapter are as follows:

- General relativity and special relativity make radically different predictions about radiation from freely falling charged particles.
- Once in the context of GR, one should not try to formulate pre-GR equivalence principles referring to whether or not one can distinguish things.
- To a first approximation, remaining close enough to the charge for curvature effects to be negligible, in the sense that the metric components remain roughly constant, GR+SEP tells us that there should not be electrogravitic bremsstrahlung for a charge following a geodesic, although there will when the charge follows curves satisfying DeWitt and Brehme's equation of motion (12.1), due to its deviation from the geodesic. And it also tells us that a charge that is stationary relative to coordinates in which the metric is static will sometimes radiate.

So the key question in this whole debate is this: in GR, should charged particles that are stationary relative to coordinates in which the metric is static really be expected to radiate? At least in some cases, this is the prediction of GR+SEP and if it does not correspond to reality, then one must question the theory.

Chapter 13
Static Charge in a Static Spacetime

The time has come to have a closer look at what theory can tell us about charged particles that are stationary relative to coordinates in which the metric is static. This chapter is based on the paper [7, Appendix 3] by Parrott and uses his notation. To understand Parrott's own position on this issue, his appendix begins with:

> It is often stated in the literature [...] that a charged particle which is stationary with respect to the coordinate frame in a static spacetime generates a pure electric field in that frame; since the Poynting vector vanishes, [it is then concluded that] there is no radiation. However, we know of no proof in the literature, and the matter seems to us not as simple as it apparently does to the authors who make this assertion.

Note that Parrott blurs the distinction here between a static spacetime and a static spacetime in which one has given the metric a static form, viz.,

$$\mathrm{d}s^2 = g_{00}(x^1,x^2,x^3)(\mathrm{d}x^0)^2 + \sum_{I,J=1}^{3} g_{IJ}(x^1,x^2,x^3)\mathrm{d}x^I\mathrm{d}x^J \,, \tag{13.1}$$

by suitable choice of coordinates. He is referring to the latter case. The words 'it is then concluded that' have been inserted to make it clear that this is not what Parrott concludes, as we shall see later. A stationary particle is then one whose worldline has the form

$$x^0 \longrightarrow (x^0,c^1,c^2,c^3) \,, \quad c^1,c^2,c^3 \text{ constant} \,, \tag{13.2}$$

relative to these coordinates.

Parrott continues as follows:

> Implicit in such statements is that the field generated by the particle is the 'retarded field' for its worldline. The problem is that there is no generally accepted, mathematically rigorous definition of 'retarded field' in general spacetimes. In Minkowski space one can define the retarded field via the usual explicit formula, but no similar closed form expressions are known for general spacetimes.

The Green function analysis in [4] is certainly a step in that direction. In any case, the first task is to say what is meant by a retarded field construction, i.e., what would count as a retarded field for a given worldline. It seems reasonable to require that it should be a rule which assigns to the particle worldline $\tau \mapsto z(\tau)$, defined as a curve in spacetime with unit norm tangent $u(\tau) = dz/d\tau$, a 2-form $F = F(x)$ satisfying (the minimal extension to the curved spacetime of) Maxwell's equations with source the distribution current associated with the worldline:

$$dF = 0 , \qquad (*d*F)(x) = \int \delta\big[x - z(\tau)\big] q\bar{u}(\tau)\, d\tau , \qquad (13.3)$$

where δ is the four-dimensional Dirac delta, $\bar{u}$ is the 1-form associated with u, the operator d is the exterior derivative, and stars denote the Hodge dual (see [13] or [20] for an account of this formulation of electromagnetism). But one cannot accept just any solution to these equations. For the field to be a retarded field, the value of $F(x)$ at a spacetime point x off the worldline should depend only on the part of the worldline on or within the backward light cone with vertex x. Two worldlines that are identical within the past light cone at x should yield the same $F(x)$.

On top of this, there is another key assumption which turns out to be the lynchpin of Parrott's proof: we assume that, in a static spacetime, the retarded field for a stationary particle is time-independent. Later we shall see an example of this that is about as simple as one could hope for (see p. 231). Finally, we assume that the retarded field construction is unique.

The idea of the proof is to assume that one has this unique retarded field and project out the electric field part of it as viewed in the static coordinate system. One then shows that this projected electric field part is a retarded field solution for the same charge distribution, and in particular that it satisfies Maxwell's equations with the particle worldline as source, as specified by (13.3). The uniqueness of such a retarded field then does the rest. Even without the uniqueness assumption, one has shown that there exists a retarded field construction with zero magnetic field. One then requires further physical constraints to select the solution that actually occurs.

The following is taken more or less verbatim from the remarkable paper [7]. We consider a particle sitting at the origin of the spacetime in which the metric has been made to look like (13.1). The four-velocity of the charge is u^μ and the associated one-form is $\bar{u}$ with components $u_\mu := g_{\mu\nu}u^\nu$. More explicitly,

$$u = g_{00}^{-1/2}\partial_{x^0} , \qquad \bar{u} = g_{00}^{1/2} dx^0 . \qquad (13.4)$$

We suppose we have a time-independent distribution two-form F with components $F_{\mu\nu}$ satisfying the (minimally extended) Maxwell equations

$$dF = 0 , \qquad *d*F = +\delta_3\bar{u} , \qquad (13.5)$$

where the star denotes the Hodge dual, d denotes the exterior derivative, and δ_3 is the 3D Dirac delta distribution on the spatial coordinates. There is a lot more about this formulation in [13]. Time-independence means that the components $F_{\mu\nu} = F_{\mu\nu}(x^1,x^2,x^3)$ do not depend on the time coordinate x^0.

Now F can be uniquely expressed in the form

$$F = \overline{\mathbf{E}} \wedge \bar{u} + \beta \,, \tag{13.6}$$

where

$$\mathbf{E} = \sum_{I=1}^{3} E^{I} \partial_{x^{I}}$$

is a purely spatial vector field, $\overline{\mathbf{E}}_{\mu} := g_{\mu\nu} E^{\nu}$ is the associated one-form field, β is a purely spatial two-form field, and the wedge symbol denotes exterior product. The bold type notation indicates vectors that have zero time component in the given coordinate system. Capital Roman indices indicate spatial components. The statement that β is purely spatial means that it has the form

$$\beta = \sum_{I,J=1}^{3} \beta_{IJ} \mathrm{d}x^{I} \wedge \mathrm{d}x^{J} \,. \tag{13.7}$$

Physically, β is the 3-space Hodge dual of the one-form corresponding to the magnetic field three-vector $\mathbf{B}$ as it is usually defined from the electromagnetic two-form field. The expression (13.6) is obtained by writing F as a linear combination of $\mathrm{d}x^{\mu} \wedge \mathrm{d}x^{\nu}$, then using the fact that $\mathrm{d}x^{0}$ is proportional to $\bar{u}$, and collecting terms involving $\bar{u}$.

We can now see what is meant by projecting out the electric field part of F: we take this part to be $\overline{\mathbf{E}} \wedge \bar{u}$. The problem now is just to show that this satisfies (13.5):

- Consider the Maxwell equation $\mathrm{d}F = 0$. This tells us that

$$\mathrm{d}\!\left(\overline{\mathbf{E}} \wedge \bar{u}\right) + \mathrm{d}\beta = 0 \,. \tag{13.8}$$

It turns out that

$$\mathrm{d}\bar{u} = \bar{u} \wedge \bar{a} \,, \qquad a := \frac{\mathrm{D}u}{\mathrm{D}\tau} \,, \tag{13.9}$$

where the latter is the acceleration of the source charge, given by the covariant derivative of u along the worldline. This follows because

$$\mathrm{d}\bar{u} = \mathrm{d}(g_{00}^{1/2}\mathrm{d}x^{0})$$

$$= \mathrm{d}g_{00}^{1/2} \wedge \mathrm{d}x^{0}$$

$$= -\bar{u} \wedge \left(g_{00}^{-1/2}\mathrm{d}g_{00}^{1/2}\right) \,.$$

But

$$a^{\mu} = \frac{\mathrm{D}u^{\mu}}{\mathrm{D}\tau} = \frac{\mathrm{d}u^{\mu}}{\mathrm{d}\tau} + u^{\nu}\Gamma^{\mu}_{\nu\sigma}u^{\sigma}$$

$$= g_{00}^{-1}\Gamma^{\mu}_{00} \,,$$

for this motion, and the Levi-Civita connection is

$$\Gamma^{\mu}_{00} = \frac{1}{2}g^{\mu\nu}(g_{0\nu,0} + g_{0\nu,0} - g_{00,\nu}) = -\frac{1}{2}g^{\mu\nu}g_{00,\nu}\ ,$$

whence

$$\Gamma^{0}_{00} = 0\ ,\qquad \Gamma^{I}_{00} = -\frac{1}{2}g^{IJ}g_{00,J}\ ,$$

so one has

$$a^0 = 0\ ,\qquad a^I = -\frac{1}{2}g^{IJ}g_{00}^{-1}g_{00,J}\ .$$

Then $\bar{a}_0 = g_{00}a^0 = 0$ and

$$\bar{a}_I = g_{IJ}a^J = -\frac{1}{2}g_{00}^{-1}g_{00,I}\ ,$$

whence

$$\bar{a} = -g_{00}^{-1/2}\mathrm{d}g_{00}^{1/2}\ , \tag{13.10}$$

as required. This makes heavy use of the form of the metric given in (13.1) and the form of the charge motion given in (13.2) with c^1, c^2, c^3 all zero. We now have

$$\mathrm{d}\left(\overline{\mathbf{E}} \wedge \bar{u}\right) = -\mathrm{d}\bar{u} \wedge \overline{\mathbf{E}} + \bar{u} \wedge \mathrm{d}\overline{\mathbf{E}}$$

$$= \bar{u} \wedge \left(\overline{\mathbf{E}} \wedge \bar{a} + \mathrm{d}\overline{\mathbf{E}}\right)\ . \tag{13.11}$$

Since this has the form $\bar{u} \wedge X$ for some (two-form) X, it is orthogonal to any purely spatial three-form with respect to the usual inner product on three-forms induced by the spacetime metric [13, p. 39]. However, the first Maxwell equation (13.8) is telling us that

$$\mathrm{d}\left(\overline{\mathbf{E}} \wedge \bar{u}\right) = -\mathrm{d}\beta\ ,$$

and $\mathrm{d}\beta$ is a purely spatial three-form because we have made the key assumption that F and hence also β are time-independent. This means that $\mathrm{d}\left(\overline{\mathbf{E}} \wedge \bar{u}\right)$ and $\mathrm{d}\beta$ are both zero. But the relation

$$\mathrm{d}\left(\overline{\mathbf{E}} \wedge \bar{u}\right) = 0 \tag{13.12}$$

is precisely the first Maxwell equation for $\overline{\mathbf{E}} \wedge \bar{u}$.

- Now consider the Maxwell equation $*\mathrm{d} * F = +\delta_3\bar{u}$. Since the two-form β is purely spatial, its Hodge dual can be written

$$*\beta = \bar{u} \wedge \overline{\mathbf{S}}$$

for some purely spatial vector **S**. By the argument used to get the first Maxwell equation, one now has

$$\mathrm{d} * \beta = \bar{u} \wedge Y \,,$$

for some two-form Y. Taking the Hodge dual again, one finds that $*\mathrm{d} * \beta$ is a purely spatial one-form, i.e., $*\mathrm{d} * \beta$ is the one-form associated by index-raising with a vector orthogonal to u. We now consider the effect of $*\mathrm{d}*$ on $\overline{\mathbf{E}} \wedge \bar{u}$. We observe that $*\left(\overline{\mathbf{E}} \wedge \bar{u}\right)$ is a purely spatial two-form with time-independent coefficients, whence $\mathrm{d} * \left(\overline{\mathbf{E}} \wedge \bar{u}\right)$ is a purely spatial three-form, and finally $*\mathrm{d} * \left(\overline{\mathbf{E}} \wedge \bar{u}\right)$ is a multiple of $\bar{u}$. What we have shown is that, when $F = \overline{\mathbf{E}} \wedge \bar{u} + \beta$, the second Maxwell equation for F becomes

$$\delta_3 \bar{u} = *\mathrm{d} * \left(\overline{\mathbf{E}} \wedge \bar{u}\right) + *\mathrm{d} * \beta \,,$$

in which the first term on the right-hand side is a multiple of $\bar{u}$ and the second term is a one-form orthogonal to $\bar{u}$, which can only happen if

$$\delta_3 \bar{u} = *\mathrm{d} * \left(\overline{\mathbf{E}} \wedge \bar{u}\right) \,, \qquad *\mathrm{d} * \beta = 0 \,. \tag{13.13}$$

The first of these shows that $\overline{\mathbf{E}} \wedge \bar{u}$ satisfies the second Maxwell equation of (13.5).

This shows that the electric field part $\overline{\mathbf{E}} \wedge \bar{u}$ of the original solution F is also a solution of the Maxwell equations.

Together with the assumption that retarded field solutions are unique, one would also have $\beta = 0$. Even if there should exist several such solutions and $\beta \neq 0$, one still has a purely electric retarded field solution here. That is, if one writes the EM field two-form $F := \overline{\mathbf{E}} \wedge \bar{u}$ in component form for this coordinate system, one has

$$(F_{\mu\nu}) = \begin{pmatrix} 0 & E_1 & E_2 & E_3 \\ -E_1 & 0 & 0 & 0 \\ -E_2 & 0 & 0 & 0 \\ -E_3 & 0 & 0 & 0 \end{pmatrix} \,,$$

for some functions E_i, $i = 1, 2, 3$, of spacetime specifying the electric field vector in this coordinate system. The magnetic field entries in the matrix are zero.

There is a catch here though which is not discussed enough in the literature. One ought to ask in what sense the E_i are really components of an electric field as we know it, relative to these coordinates. Because the EM field tensor is a two-form (antisymmetric, covariant, and rank two), its component matrix can always be written in the form

$$(F_{\mu\nu}) = \begin{pmatrix} 0 & E_1 & E_2 & E_3 \\ -E_1 & 0 & B_3 & -B_2 \\ -E_2 & -B_3 & 0 & B_1 \\ -E_3 & B_2 & -B_1 & 0 \end{pmatrix} \,,$$

for some functions E_i and B_j on the spacetime, whatever coordinate system one chooses. Does this mean that $\mathbf{E}$ and $\mathbf{B}$ are electric and magnetic fields as we know them?

One could of course make this the definition, but it should be borne in mind that they only correspond to our usual understanding in a locally inertial frame, and then only locally, i.e., with decreasing reliability as one moves away from the spacetime event at which one has made the connection components equal to zero. This is the utility of the strong principle of equivalence: physics looks roughly as we know it in locally inertial frames. Without this principle, one could not even begin to interpret the constructions one has made on these manifolds. The weak and strong principles of equivalence taken together allow one to interpret the motion of test particles relative to arbitrary coordinates, and also to interpret other fields defined relative to arbitrary coordinates.

It is worth remembering that one could say nothing about the link between the spacetime manifold and the spacetime observed without having principles like these. Without SEP, one would require some other principle that made the link. To understand $F = \overline{\mathbf{E}} \wedge \bar{u}$, one needs to transform to normal coordinates at the event one is interested in and then look at the components of F.

This problem is illustrated by the debate over whether the $\mathbf{E}$ we have found here is a Coulomb field, as discussed so decisively in [7]. Of course, there is absolutely no reason to expect it to be a Coulomb field in any possible sense. Later we shall see what the EM fields of an accelerating charge in flat spacetime look like in semi-Euclidean coordinates for an observer accelerating with it, and we shall find that they do not look like a Coulomb field in those coordinates (see p. 231 ff). What is more, they are obviously not going to look like a Coulomb field in the happily available global inertial frame.

Parrott also gives an explicit example here, for the spacetime with interval

$$ds^2 = c(x)^2 dt^2 - dx^2 - dy^2 - dz^2 \,, \tag{13.14}$$

where c is some function of x alone. An object sitting at the spatial origin of this spacetime for the given coordinates then has four-acceleration

$$a = \frac{c'}{c} \partial_x \,,$$

where the prime denotes ordinary differentiation. Indeed it is a simple matter to show that the only nonzero connection coefficients are

$$\Gamma^0_{01} = \Gamma^0_{10} = \frac{c'}{c} \,, \qquad \Gamma^1_{00} = c'c \,. \tag{13.15}$$

This gives us the acceleration of a stationary particle, since it has four-velocity $u = (1/c, 0, 0, 0)$ and

$$a^i = (\mathrm{D}_u u)^i = u^j \frac{\partial u^i}{\partial x^j} + u^j \Gamma^i_{jk} u^k = \frac{1}{c^2} \Gamma^i_{00} \,, \tag{13.16}$$

whence

$$a = \mathrm{D}_u u = \frac{c'}{c}\partial_x \ . \tag{13.17}$$

As an aside, certain functions $c(x)$ lead to flat spacetimes. It turns out that the Riemann curvature is zero if and only if $c(x) = Ax + B$ for some $A, B \in \mathbb{R}$.

Parrott proposes

$$\mathbf{E} = \frac{x\partial_x + y\partial_y + z\partial_z}{(x^2 + y^2 + z^2)^{3/2}} \tag{13.18}$$

as a Coulomb field in this spacetime. This is of course just one idea of what a Coulomb field should be in such a spacetime. Relative to normal coordinates at some event, it will not look like this, so one ought to be wary of such a claim. In any case, this field cannot possibly satisfy

$$\mathrm{d}\left(\overline{\mathbf{E}} \wedge \bar{u}\right) = 0 \ , \tag{13.19}$$

the equation (13.12) one hoped to satisfy in the case of the retarded field solution. The reason is just that we have

$$\mathrm{d}\left(\overline{\mathbf{E}} \wedge \bar{u}\right) = \bar{u} \wedge \left(\overline{\mathbf{E}} \wedge \bar{a} + \mathrm{d}\overline{\mathbf{E}}\right) \ ,$$

by (13.11), whereas $\mathrm{d}\overline{\mathbf{E}} = 0$ (except of course at the spatial origin where the field is not specified), and

$$\bar{u} \wedge \overline{\mathbf{E}} \wedge \bar{a} = \frac{c'}{c^2}\mathrm{d}t \wedge \mathrm{d}x \wedge \overline{\mathbf{E}} \neq 0 \ .$$

There is no real surprise here. We had no reason to expect such a neat result relative to coordinates that we have not even tried to understand yet.

Returning to the result proven about the electric field part of any retarded field solution F being itself a retarded field solution, it does seem a shame that one has to assume the time-independence of F. This in itself seems tantamount to assuming that a charge that is stationary relative to coordinates in which the metric is static will not radiate. However, we noted earlier [see (2.16) on p. 9] that the metric for semi-Euclidean coordinates adapted to an accelerating observer in flat spacetime would only have the static form if the acceleration was eternally uniform, and in this case it turns out that the retarded electric field components of a charge moving with the observer are in fact constant with respect to semi-Euclidean time (and the retarded magnetic field components are identically zero). When we discuss this later (see p. 231), we shall see that it is not something immediately obvious, even in this very simple case where the spacetime is flat. On the other hand, it seems plausible enough to suppose that the retarded field solutions for a charge that is stationary relative to coordinates in which the metric is static should be time-independent in those coordinates, and that there may be some general way to demonstrate this.

In any case, this heavy assumption in the above proof is not the only problem with the idea that a charge that is stationary relative to coordinates in which the metric is static will not radiate. Another difficulty is that the time coordinate in one coordinate system may not much resemble the time coordinate in another. What is static for one system may not be for another. For example, one consequence of the result that $F = \overline{\mathbf{E}} \wedge \bar{u}$ is a retarded field solution is that the corresponding energy–momentum tensor, defined as usual by

$$T^{\mu\nu} = F^{\mu\sigma}F_{\sigma}{}^{\mu} - \frac{1}{4}F^{\sigma\tau}F_{\sigma\tau}g^{\mu\nu} \,,$$

has $T^{0J} = 0$ for $J = 1,2,3$, i.e., provided that one defines it in this way for arbitrary coordinates, the Poynting vector is zero. But does this mean that there is no energy radiation through a stationary closed surface surrounding the particle? Although Parrott would appear to be suggesting this [7, p. 27], the whole thesis of [7] turns around the idea that, when one integrates the energy–momentum tensor over a closed surface in this coordinate system, one is not actually finding an energy flow, as we shall discuss later in the context of Parrott's reply to Boulware's paper (see Sect. 16.8).

The last claim should come as no surprise either. What do the components $T^{\mu\nu}$ do for us in these non-normal coordinates? As usual, one could define radiation as the flux that is worked out in this way, but at some point one has to link this flux with something that could be measured in the real world, a point about which not enough is said in discussions on general relativity.

There is a last point that is logically problematic here and very easily overlooked. Equations (13.3) are not the Maxwell equations, but rather their minimal extension to this curved spacetime based on the assumption that the strong equivalence principle (SEP) holds. This is problematic because GR+SEP applied to the SHGF implied that a static charge in a static spacetime would sometimes radiate. But GR+SEP has shown in the present chapter that the Poynting vector relative to coordinates in which the metric is static is zero for a charge that is stationary relative to those coordinates (under certain other assumptions discussed above). If this does mean that a static charge in a static spacetime cannot radiate, then one might conclude that GR+SEP is self-contradictory.

It is not hard to see ways around this unpromising conclusion, e.g., questioning the other assumptions required for the argument at the beginning of this chapter, or considering the possibility that zero Poynting vector in some coordinates might not mean what we normally mean by no radiation. In fact, for the SHGF, we shall see later that the Poynting vector is zero relative to the semi-Euclidean frame, as expected from the above arguments (see Sect. 15.12). Indeed, the electric field components are time-independent (for semi-Euclidean time) and the magnetic field components are zero (see p. 231), so the argument discussed above is fully applicable. The question remains as to whether one should interpret this to mean that there is no radiation, or whether one can have radiation as viewed via one coordinate system and no radiation as viewed via another.

One point that should be retained, however, is the fact that, without SEP, we do not have any simple option for formulating electromagnetism in curved spacetimes. SEP is implicitly required to deduce that the Poynting vector is zero for a supported charge in an SHGF when expressed relative to the semi-Euclidean system. We have to ask what theory Bondi and Gold were proposing to use to show that a static charge in a static spacetime cannot radiate.

Do DeWitt and Brehme's calculations throw any light on the matter? They estimate the retarded fields due to a charge with arbitrary timelike worldline. One could look at those fields and decide whether they contain a radiative component. One way to do this would be to carry out the integration of the corresponding EM energy–momentum tensor over the walls of a world tube containing a portion of the worldline. This would lead to some of the terms in their proposed equation of motion (12.1) on p. 107. In fact, it leads to the terms they call the radiation reaction terms, and also to inertial mass renormalisation terms that have been absorbed into the parameter m. So one would like to assume that the worldline is one like (13.2) at the beginning of this chapter, in which the charge is stationary relative to some coordinates in which the metric has the static form (13.1), and see whether the proposed radiation terms in (12.1) are zero. Or again, one would like to consider the motion resulting when F_{in}^{α} is zero, i.e., the new kind of free fall including self-interaction effects, but also the old kind of free fall in which the charge follows a geodesic, and see whether the radiation terms are zero.

There are some problems with this. To begin with, in one very simple situation, the radiation reaction terms being zero does not mean there is no radiation. But treating this case as pathological, i.e., assuming that, if there is radiation, then the radiation reaction terms will generally signal it, what we observe is that there are always radiation damping terms in the general context. One hardly has to formulate the case of the stationary charge relative to coordinates in which the metric has static form in order to see that the situation is complex enough to generate some radiation damping terms. This may seem comforting for the claim made in the present book that such a charge will generally radiate. On the other hand, one can also see that the motion with $F_{\text{in}}^{\alpha} = 0$ and also the motion along a spacetime geodesic will generally generate radiation damping terms, and hence, one supposes, radiation.

We have to ask how these results fit in with the claim in the present book that a charge following a geodesic will not radiate (to a first approximation) and a charge that is stationary relative to coordinates in which the metric components are static can nevertheless radiate. These claims are based on a very simple observation: according to GR+SEP, which is the theory used by DeWitt and Brehme to establish all their results, there are (normal) coordinates near any chosen event on the charge worldline in which electromagnetism looks approximately as it would in flat spacetime.

The approximation here is just the approximation that one must remain close enough to the worldline for the connection coefficients to remain small enough. These grow roughly linearly with the coordinate separation from the relevant event, the constant of proportionality being roughly the value of the curvature at this spacetime locality, but in the normal coordinates for the chosen event on the worldline, the connection coefficients have been made equal to zero at the chosen event on the charge worldline. How far one can move from the chosen event on the worldline without the connection coefficients growing too big thus depends on the value of the curvature there. In the neighbourhood of this event, we are hoping to examine the EM fields for emergent radiative terms, and the strength of these depends principally on the charge of the particle. So what decides whether the connection coefficients are too big is basically the charge, source of EM effects.

The perfect situation for seeing the kind of thing that will happen in a first approximation, i.e., when the charge is strong and the curvature weak, is a situation in which the curvature is actually zero, and this is effectively a case like an SHGF. We know in this case that the stationary motion relative to SE coordinates does produce EM radiation. We also know that, for free fall, there is no radiation. This is why the case of the SHGF can be taken as an indicator of what can be considered a first approximation in cases of small curvature.

The contribution from DeWitt and Brehme is just to show how this approximation gets perturbed when the curvature starts to grow. Of course, with curvature, interacting as it does directly with the EM fields generated by the charge, and with those fields acting back on the charge, this particle will no longer follow a geodesic, and that in turn will affect the fields it generates. These are the curvature-generated effects, which mean that a charge that is stationary relative to coordinates in which the metric has static form will not just radiate in a first approximation that is curvature-independent, but will also radiate curvature-dependent terms; and likewise which mean that a charge compelled by an external field to follow a geodesic, even though it does not radiate in the first approximation, will nevertheless radiate curvature-dependent terms.

At the end of Chap. 10 (see p. 92), we asked the following question: is it possible that curvature effects could dominate the first-order expectation implied by GR+SEP? For example, could the radiation effect expected for a supported charge in a real (non-homogeneous) gravitational field be somehow cancelled by curvature effects? One can already guess the answer: there is no reason why it should, because EM effects depend on the strength of their sources, while gravitational effects like curvature depend on the strength of their sources, and these strengths are physically independent. Of course, the effects might perfectly cancel to give no radiation in this case, but what is evident is that they will not generally do so.

Chapter 14
A Radiation Detector

In 1964, Mould wrote a remarkable paper [5] which really does not seem to have done anything to help things along, although it purports to deal with precisely the issue raised in the last chapter, viz., how would one link the theoretical energy flux with something that could be measured in the real world? The aim here is once again to give a brief summary and concentrate on the question of what one should or should not expect on the basis of GR and its equivalence principles.

14.1 Equivalence Principle According to Mould

One should start by quoting the abstract for this paper:

The existence of electromagnetic radiation from a uniformly accelerated charge has appeared in some recent work to present a problem suggesting the inadequacy of the equivalence principle. For a proper treatment of the problem, it is desirable to show how the absorption properties of detectors are affected by being physically attached to noninertial frames of reference. An invariant criterion of internal absorption is formulated, and is identified with the observable behavior of an elementary detector. It is shown that the properties of a detector fixed in a uniformly accelerated frame are different from the properties of an inertial detector of similar construction, and that this difference is consistent with the usual form of the equivalence principle.

One hopes to find now a clear statement of the relevant equivalence principle and the problem with it. If there is a difficulty with the strong principle of equivalence, then we can no longer say anything about electromagnetic effects in the context of GR. If there is a difficulty with the weak equivalence principle, then GR is out and we no longer have a theory of motion for bodies subject to gravitational influences. Unless of course, in each case, one uses an older theory, or invents a new one. The other possibility is that one might be talking about some other version of the equivalence principle. This is really what never seems clear in any of these papers.

The introduction begins with a rather clear statement of Pauli's conclusion with regard to uniformly accelerated charges. He concluded that such charges would not

radiate, and we have already discussed this. Mould then mentions the papers by
Bondi and Gold [2] and Fulton and Rorhlich [3], which would appear to show (successfully) that such a charge would radiate. He then moves on to Bondi and Gold's
problem with the equivalence principle:

> Bondi and Gold draw the further conclusion that the gravitational principle of equivalence
> cannot be generally applied to radiation phenomena. Their argument proceeds in two steps.
> First, it is recognized that a static charge in a static gravitational field cannot radiate, for oth-
> erwise energy would have to be transferred through the surrounding space (including the flat
> space at large distances from real gravitating masses) by time-independent fields. Second,
> we are asked to restrict attention to the local uniform gravitational field of the charge, and
> consider a freely falling coordinate frame relative to which the charge is uniformly accel-
> erating. According to the principle of equivalence, this falling frame is an inertial system
> in which Maxwell's equations are valid; and so the accelerating charge must be radiat-
> ing energy to its surroundings. We therefore have a case in which radiation exists in one
> (inertial) frame but not in another (gravitational) frame. This is claimed to be paradoxi-
> cal because the observation of radiation should be invariant. To solve the paradox, these
> authors suppose that the local gravitational field (the first order limit of the equivalence
> principle) must be contained in the local electromagnetic field where radiation of this ori-
> gin is not detectable. In the wave zone of this radiation, it is supposed that higher order
> gravitational effects just compensate the electromagnetic field in such a way as to prevent
> detection.

The first thing to note is that there is no statement of the equivalence principle we
are talking about here. So let us recall what Bondi and Gold themselves said:

> The principle of equivalence states that it is impossible to distinguish between the action on
> a particle of matter of a constant acceleration or of static support in a gravitational field.

We have already discussed the problem with versions of the equivalence principle (or one should perhaps say, with equivalence principles) which refer to what
can or cannot be distinguished. Such a principle survives in the context discussed
in Chap. 2: an observer cannot say whether she is uniformly accelerating in a flat
spacetime (and using semi-Euclidean coordinates) or sitting still relative to coordinates that describe a spacetime containing a static, homogeneous gravitational field,
or even a mixture of both. In realistic spacetimes, what remains of the equivalence
principle has to contain the word 'local', but in a flat spacetime described using
semi-Euclidean coordinates for a uniformly accelerating observer, one can dispense
with it because the locally inertial frame is globally inertial.

Now we can still construe the last quoted statement as a version of the strong
equivalence principle by inserting the word 'local' somewhere. But then that does
mean that, if we look carefully enough around our neighbourhood of spacetime, we
will be able to distinguish anything we want. If one uses a wording of the equivalence principle that speaks of what can or cannot be distinguished, it looks as though
Bondi and Gold's problem only remains in the flat spacetime described using semi-
Euclidean coordinates for a uniformly accelerating observer. But if one is using the
strong equivalence principle which allows one to extend electromagnetism to GR,
and if one still believes for some reason that a static charge in a static gravitational
field should not be able to radiate, then one is in trouble because GR+SEP tells us
that such a charge will radiate. For example, it does so in the highly idealistic case of

a spacetime containing an SHGF, and DeWitt and Brehme's sophisticated radiation reaction terms suggest that it will do so in a general spacetime. But just SEP and a locally inertial frame centered on the charge at some event already tell us that, if the charge is accelerating relative to the coordinates of that frame, then it will radiate.

Returning to the last quote from Mould, we do have an attempt to explain why a static charge in a static gravitational field cannot radiate. In the light of what has just been said, one wonders what theory it could be based upon, and it is interesting to read it again with this in mind. If it is something to do with physical intuition, then one ought to be skeptical. Take, for example, the reference to time-independent fields. This means that they are independent of the particular time coordinate used, just as the electric field of the uniformly accelerating charge in flat spacetime turns out to be independent of the semi-Euclidean coordinate y^0 of (2.12) on p. 9, when one works out what the electric field part of the EM field tensor should be for these coordinates. But time is a coordinate-dependent notion (and so is the electric field). Are we applying physical intuition to these things to understand them? Ought we not to be using WEP and SEP to understand what is happening physically by referring everything to a locally inertial frame? This same case of a uniformly accelerating charge in flat spacetime shows that in the (globally) inertial frame, the electric field is not time-independent.

That is the import of Mould's second step in the above quote. Furthermore, he appears to be applying SEP in that second step. Once again, one has to ask how one could say anything about electromagnetism in GR without SEP, and what principle it is that allows one to conclude something about electromagnetism for a static charge in a static gravitational field if it is not SEP. But in any case, if one has reason to think that a stationary charge in a static gravitational field cannot radiate, then there is a problem. It looks as though maybe radiation can exist in one view of the situation and not in the other, as Mould puts it. So the paradox is that the observation of radiation should be invariant, meaning presumably that, if one observes radiation in one frame, one ought to observe it in any other. And this is the question tackled by Mould.

The last two sentences of the above quote deserve a little attention. They purport to explain Bondi and Gold's solution to their problem. Despite the oddly jargonistic way of putting things, let us try to understand it. Could it be that, when Mould speaks of the local gravitational field or the first order limit of the equivalence principle, he is referring to a region in which the equivalence principle is supposed to hold sway, i.e., a region in which the equivalence principle in his argument just prior to this statement would have implied the presence of radiation? If we then decree that this region must be contained in the region where electromagnetic radiation is not detectable, we are saying that, as far out as the equivalence principle might have imposed a radiation field on us, there will be no observable radiation. The wave zone of this radiation presumably refers to a more distant region in which radiation terms dominate the EM field, but it is suggested that, in such a large region, we will be able to measure higher order gravitational effects, i.e., curvature. The solution to the problem is that curvature effects will just compensate the electromagnetic field in such a way as to prevent detection of radiation.

The idea seems to be that there is radiation, but that it is condemned to go unnoticed. This fits with the kind of equivalence principle which talks about whether or not two situations can be distinguished. To satisfy such a principle, it is sufficient that the radiation, which is actually there, should not be detectable. If one feels that this kind of equivalence principle is important in relativity, then there is perhaps some hope. But the real stumbling block for this argument is the claim that curvature effects can compensate the electromagnetic field. If this means that they can make it disappear, then there really is no future for this argument, because EM fields are very different from gravitational effects as expressed via the curvature, so there is no way they could cancel one another out mathematically. Worse, the two effects have wildly different sources, viz., active gravitational mass (or energy) and electric charge, so the extent of the various regions over which their influence may be significant will not generally be comparable.

Anyway, as just concluded, the paradox for Mould is that the observation of radiation would appear to depend on one's frame of reference, whereas he feels that it should be invariant, i.e., if radiation is observed in one frame, it should be observed in another. Now Mould's statement of intent is this:

> In the present paper, we assume that a uniformly accelerated charge in an inertial frame does radiate relative to that frame, and that the equivalence principle can be extended into the wave zone as far as the usual gravitational approximation permits. These assumptions do not imply conflicting detector observations, for we are able to show that detectors physically attached to different reference frames have different absorption properties. In particular we show that a detector at rest in a static uniform gravitational field will not absorb internal energy from an electric charge held static in that frame, whereas a free falling detector may (generally) absorb internal electromagnetic energy from such a field.

The talk here about extending the equivalence principle into the wave zone is presumably a way of dealing with the fact that locally inertial frames are only locally inertial, so that one does not expect electromagnetism in such a frame to look exactly as it would in a flat spacetime. One has to ask, however, what theory Mould is going to use to show anything about a radiation detector at rest in an SHGF. And one should remember that this is a purely theoretical paper, because of course no such experiment was ever carried out.

In fact, the problem is much more serious than Mould seems to realise. GR+SEP contradicts whatever theory it is that says that a static charge in a static gravitational field cannot radiate. Since he does in fact use GR+SEP (what else could one use?) to model his detector, and since it does not pick up the radiation predicted by GR+SEP for the static charge in a static gravitational field, then either it is not actually functioning as a radiation detector when it is fixed relative to the charge, or GR+SEP is self-contradictory. The logical implications of the second possibility are disastrous.

One way to save the situation might be to say that detection by this detector should define radiation, although a suitable redefinition of radiation could have done that without making any theoretical model of the detector, and in that case it is not clear that any paradox has been solved. The idea is tantamount to admitting that radiation in one frame might not be radiation in another frame. This reminds us of the qualitative argument from quantum theory mentioned by Fulton and Rohrlich

[3] that, when a photon is emitted, it can be registered by a counter located at a great distance from the source, and that such an effect can be seen by every observer. Indeed this would appear to show that the existence of radiation should be independent of inertial observer (although of course it does not show that the rate of radiation should be independent of inertial observer since the energy of the photons and the time scale used to count their rate of arrival both depend on the frame). In this view of things, Mould is saying that accelerating detectors will not always pick up radiated photons.

It is a surprising claim, confirmed in the summary at the end of Mould's paper: a static charge in the semi-Euclidean frame adapted to a uniformly accelerating observer will not induce internal transitions in a detector that is stationary relative to those coordinates, i.e., a detector that is uniformly accelerating with the charge or sitting at a fixed semi-Euclidean distance from it. (Note that the second case actually means that the detector has a specific fixed uniform acceleration that depends on its semi-Euclidean distance from the charge, but is generally different from that of the charge.) Now we know that the uniformly accelerated charge just mentioned will radiate, so photons are emitted by the charge but they are not picked up by detectors with this motion.

Worse still, Mould finds that, if a detector is accelerated through a static EM field, e.g., the Coulomb field of an inertially moving charge in flat spacetime, it can in general absorb internal energy. We know that this charge is not radiating, but the detector is excited. So in this view the charge is not now emitting photons, but the detector would appear to pick up some photons. The quantum picture certainly makes this look odd. Presumably, however, a detector moving inertially with the charge will not be excited.

In his book [17], Zeh would appear to aquiesce, claiming that the problem of how to apply Larmor's formula to a charged particle in a gravitational field was not understood until Mould demonstrated that the response of a detector to radiation depends on its acceleration:

> In general relativity, the principle of equivalence is only locally valid. However, a homogeneous gravitational field (as would result from a massive plane) is described by a flat spacetime, and thus globally equivalent to a rigid field of uniform accelerations a^μ on Minkowski spacetime. [...] The equivalence principle can therefore be globally applied to a homogeneous gravitational field. It requires that an inertial (freely falling) detector is not excited by an inertial charge, while a detector 'at rest' is. The latter would remain idle in the presence of a charge being 'equivalently at rest' (that is, at a fixed distance). A detector-independent definition of total radiation also turns out to depend on acceleration (as it should for consistency) because of the occurrence of spacetime horizons for truly uniform acceleration (see Boulware 1980).

The reference to Boulware is [6], which will be discussed later. We shall find that the detector-independent definition mentioned in this quote is not actually a definition of energy radiation (see Sect. 16.8) [7]. Further support for the view expressed in the last quote comes from Rohrlich [22] (discussed in Sect. 16.10).

For the moment, let us note that the quote confirms the idea that one is here redefining radiation, since Zeh refers implicitly to a detector-dependent definition. Since there is no problem with WEP or SEP (unless Bondi and Gold were right to say that a charge that is stationary relative to coordinates in which the metric is static should not radiate), and indeed since these principles are abundantly used to prove Mould's claims, or indeed to make any kind of deduction in this domain, one wonders if he is not trying to save an equivalence principle of the kind that says that one can or cannot distinguish something.

With this in mind let us examine the idea that these detectors confirm the following claim: one cannot tell whether the charge is moving freely (freely falling) under the effects of an SHGF, or whether one is oneself uniformly accelerating in a flat spacetime, with no gravitational effects, not even non-tidal ones, in which the charge is sitting still relative to inertial coordinates. We know that this is imposed by SEP. We put it to the test by considering radiation by the charge, as predicted by GR+SEP. In fact there is no radiation according to GR+SEP in either case. Mould's detector will pick up radiation if supported against the effects of the SHGF in the first case and if uniformly accelerating with the observer in the second case. But it will not pick up radiation if falling freely with the charge in the first case and if sitting still relative to inertial coordinates in the second case. It does not distinguish the two cases, even though it says there is radiation when there is not in two of the configurations considered.

So what are these detectors doing for us? The point is of course that the detector that is stationary relative to coordinates in a curved spacetime in which the metric is static will not detect radiation from a charged particle that is stationary relative to those same coordinates, and this was precisely the particle that Bondi and Gold did not want to be radiating, even though it is in fact radiating according to GR+SEP. So Mould's solution to this paradox is that one can (in principle) make a detector that will not detect the radiation. If one then defines radiation as what such a detector will detect, it looks at first glance as though one has quite simply defined away the problem.

What about the extraordinary idea that these detectors would pick up photons from a Coulomb field by accelerating through it? Was anyone ever worried about the idea that inertially moving charges in flat spacetime might be expected to radiate to some observers? Later we shall see what the EM fields of such charges look like in semi-Euclidean coordinates, and we shall of course find that they take on a rather complex appearance [see (14.15) and (14.16) on p. 154]. But does that mean that they are radiating fields? After all, this is only one possible choice of coordinates that the accelerating observer could adopt. Should the question of whether a charge is radiating or not depend on a choice of coordinates? In Sect. 16.10, we shall examine this point of view, as expressed by Rohrlich.

Perhaps there is a link with the idea expressed by DeWitt and Brehme in the quote on p. 123, where they claim that, on the basis of physical intuition it would seem reasonable to suppose that a charged particle does in fact radiate when deflected by a gravitational field, i.e., that bremsstrahlung can be produced by gravitational as well as electromagnetic forces. This is precisely what GR+SEP says will not happen, except possibly for rather small curvature-dependent effects, as DeWitt and Brehme in fact confirmed. But there are no curvature-dependent terms in this application *par excellence* of SEP, in which the locally inertial frame turns out to be globally inertial.

When the observer sitting at the origin of semi-Euclidean coordinates interprets what she observes with reference to an SHGF, GR+SEP must make exactly the same prediction as the one obtained for an inertial charge in a flat spacetime with no gravitational effects, not even non-tidal ones, viz., it will not radiate.

There does of course remain the question as to whether any radiation would be detected by a uniformly accelerating observer in the latter case. The present view is that, at least for the purposes of assessing the success of WEP and SEP, one ought to refer to a locally inertial frame in order to decide whether there is radiation or not, but one needs to bear in mind the following possibilities:

- It may be a good idea for some purposes to define radiation for accelerating observers, but it does seem likely that this definition must depend on the coordinates chosen by that observer. The problem here is that there are no canonical coordinates for such an observer, although one can find quasi-canonical coordinates (defined by DeWitt and discussed in Sect. 2.4). It is not obvious that such a concept would be useful. But one thing is certain: it is not necessary for the assessment of WEP and SEP, and if it is purely motivated by the desire to save an obsolete version of the equivalence principle which speaks about what can or cannot be distinguished then it serves no purpose.
- A real physical detector that always gives the right measurement results when moving inertially might behave as Mould claims when accelerating. This would give a physical reason for defining radiation as he suggests. It may be that the Mould detector does this. But one does need to consider the possibility that other real physical radiation detectors of a different construction, which always give the right measurement results when moving inertially, might give different results to Mould's detectors when accelerating. Presumably, Mould's thesis is that his theoretical basis for the detector described in [5] is general enough to be all-encompassing, so that this could not happen.

So the claim being made in the present book is certainly not a complete rejection of Mould's work, only a criticism of its relation to the question of equivalence principles.

In this respect, however, it is worth noting the following. The prediction that a freely falling charge should radiate as observed in a frame that is not allowed to fall relative to some gravitating body would be made by a special relativistic model of gravity, where gravity is a force like any other. But in that model, the freely falling charge is not inertial. The odd thing about Zeh's claim that an inertial (freely falling) detector is not excited by an inertial charge, while a detector at rest is, is that he clearly identifies the freely falling detector and charge as inertial, which implies that he is using GR. Furthermore, if the inertially moving charge is not radiating, why should a supported detector be excited by it? We shall see Rohrlich's argument in favour of this idea in Sect. 16.10 [22].

Before moving on to Boulware's paper [6], which is often referred to as definitive with regard to these issues (see the next chapter), it is worth making some general points about calculations in curved spacetimes, or using non-inertial coordinates in flat spacetimes. This will also be an opportunity to understand the strange behaviour of Mould's detectors when accelerated.

14.2 Construction of the Detector and Calculations in General Coordinates

Here is the definition of the detector, made relative to inertial coordinates in a flat spacetime:

The elementary detector consists of a pair of positive and negative charges which are constrained to move over some specified closed path in the region of interest. The charges are assumed to begin together at some spacetime point A, and to move over independent paths from there to a second point of coincidence B, thereby closing a 'detector loop'. If one observes such a detector loop from a covering inertial system, then the total electromagnetic four-momentum absorbed by the charges $+q$ and $-q$ in a single cycle is given by

$$\Delta p_\mu = q \int_A^B F_{\mu\nu} \mathrm{d}\lambda^\nu (\text{path } {}^+q) - q \int_A^B F_{\mu\nu} \mathrm{d}\lambda^\nu (\text{path } {}^-q)$$

$$= q \oint F_{\mu\nu} \mathrm{d}\lambda^\nu , \tag{14.1}$$

where $F_{\mu\nu}$ is the electromagnetic field tensor. The positive direction around the loop is given by the motion of the positive charge.

We shall suppose that we understand this construction and see how it fares when generalised to the kind of non-inertial coordinates one might adopt when accelerating through a flat spacetime, or when moving through a curved spacetime.

But first, briefly, let us say how Δp_μ is used to decide whether the detector has absorbed energy or not, in a way that does not depend on the inertial frame one happens to be using. This is what Mould calls an invariant criterion for deciding whether detection has occurred or not. We say that internal energy has been absorbed if Δp_μ is timelike, and that internal energy has not been absorbed if Δp_μ is spacelike. When Δp_μ is null, we declare absorption if and only if the time component is nonzero. We shall not question the integrity of this proposal here but consider only the way it leads to the strange detection results found theoretically in Mould's paper.

One can apply Stokes' theorem to (14.1) in the usual way to express Δp_μ as an integral over any surface Σ spanning the detector loop:

$$\Delta p_\mu = \frac{1}{2} q \int_\Sigma (F_{\mu\beta,\alpha} - F_{\mu\alpha,\beta}) \mathrm{d}\sigma^{\alpha\beta}$$

$$= \frac{1}{2} q \int_\Sigma F_{\alpha\beta,\mu} \mathrm{d}\sigma^{\alpha\beta} , \tag{14.2}$$

where $\sigma^{\alpha\beta}$ is some kind of measure on the surface, antisymmetric in its superscripts, and the second step uses the source-free Maxwell equations. We are still using inertial coordinates in a flat spacetime. (In fact, Mould only considers flat space-times.) Commas denote ordinary coordinate derivatives.

We can already see from (14.2) why it is that, if the field $F_{\alpha\beta}$ is time-independent, then the time component of Δp_μ, viz., the component Δp_4 in Mould's notation, must be zero, simply because $F_{\alpha\beta,4}$ is then zero. So these detectors will never absorb internal energy from electromagnetic fields that are time-independent relative to some inertial frame. Of course, this does not mean that $\Delta p_4 = 0$ in every inertial frame. In fact, it will have a nonzero 4-component in most inertial frames, as can be seen by ordinary Lorentz transformation. But it must be spacelike if there is some inertial frame in which the EM fields are time-independent for the time in that frame; and then in that frame, its 4-component is zero, so according to Mould's invariant criterion, it has not absorbed internal energy.

But is this conclusion independent of what the detector is doing? Indeed, how do we know what the detector is doing? This must in fact be determined by A and B, and the paths followed by the positive and negative charges when they move between them, and hence also by the vector-valued measure $d\lambda^\nu$. These are the things Mould refers to as the geometry of the loop, where the word 'geometry' includes its motion in the spacetime context. Since we have not specified anything particular about these things, it would appear that the conclusion of the last paragraph is in fact independent of what the detector is doing.

Now here is an odd thing, for one of the conclusions in Mould's paper is that a detector accelerating through an inertial frame in which the EM field is Coulomb will make a detection. This should make us look very carefully indeed at the next step in the theory, where one extends the absorption criterion to non-inertial frames of reference (still in flat spacetime). This is done by introducing a displacement bitensor, an object that generalises to curved spacetimes and is used abundantly by DeWitt and Brehme in [4]. Let us see how this works, but introducing it in a rather more explanatory way than we find in [5].

We shall use primes to distinguish arbitrary coordinates $\{x'^\mu\}$ from inertial coordinates $\{x^\mu\}$. The problem with (14.1) is that we are summing vectors (in fact, covectors) at different spacetime points when we calculate the integral. This nevertheless gives a Lorentz covariant quantity, i.e., one that always looks like the expression in (14.1) when expressed relative to an inertial frame. This in turn happens because the Lorentz transformations used to change from one inertial frame to another act on vectors, tensors, etc., everywhere in the same way, i.e., they are independent of the spacetime point with which the vector, tensor, etc., is associated, and so can be brought out of the integral. But this is not the case for a general coordinate transformation giving the arbitrary coordinates x'^μ as a function of the inertial coordinates x^ν.

There is in any case a question of how a covector like Δp_μ should transform in going to an arbitrary coordinate frame. For someone used to working with GR, this is obvious: one should define

$$\Delta p'_\mu(w) := \left.\frac{\partial x^\nu}{\partial x'^\mu}\right|_w \Delta p_\nu \,. \tag{14.3}$$

There are three things to note here:

- Firstly, this is a definition. Special relativity expressed entirely in terms of inertial frames says nothing about what such a thing should look like in an arbitrary coordinate frame. However, applying the general principle of covariance to SR, in the form which says that any viable physical theory must be expressible in a generally covariant way, this would certainly appear to provide a suitable way of proceeding, if only because it fits SR into the pattern used in GR.
- Secondly, the momentum impulse is now a function of the spacetime argument w, which is perfectly arbitrary.
- Thirdly, $\Delta p'_\mu(w)$ is spacelike or timelike if and only if Δp_ν is spacelike or timelike, respectively.

There is cause to wonder here. What does the vector field $\Delta p'_\mu(w)$ tell us about the physical world? What value of w should we be concerned with? The object Δp_μ was not a vector field, but just a vector associated with the world as a whole, but we have had to treat it as a constant vector field in spacetime in order to obtain the new object. In applications one has to make a choice for w.

Let us see how to obtain Mould's formula for $\Delta p'_\mu(w)$ from (14.1). One has

$$\Delta p'_\mu(w) := \left.\frac{\partial x^\nu}{\partial x'^\mu}\right|_w \Delta p_\nu$$

$$= q \oint \left.\frac{\partial x^\nu}{\partial x'^\mu}\right|_w F_{\nu\sigma} d\lambda^\sigma$$

$$= q \oint \left.\frac{\partial x^\nu}{\partial x'^\mu}\right|_w \left.\frac{\partial x'^\alpha}{\partial x^\nu}\right|_x \left.\frac{\partial x'^\beta}{\partial x^\sigma}\right|_x F'_{\alpha\beta} \left.\frac{\partial x^\sigma}{\partial x'^\kappa}\right|_x d\lambda'^\kappa$$

$$= q \oint \left.\frac{\partial x^\nu}{\partial x'^\mu}\right|_w \left.\frac{\partial x'^\alpha}{\partial x^\nu}\right|_x F'_{\alpha\beta} d\lambda'^\beta$$

$$= q \oint \Lambda'_\mu{}^\alpha(x,w) F'_{\alpha\beta} d\lambda'^\beta \,, \tag{14.4}$$

where x is the field point or integration variable. The object

$$\Lambda'_\mu{}^\alpha(x,w) := \left.\frac{\partial x^\nu}{\partial x'^\mu}\right|_w \left.\frac{\partial x'^\alpha}{\partial x^\nu}\right|_x \tag{14.5}$$

is the parallel displacement bitensor for Minkowski spacetime, with index α at x and index μ at w, transforming as a contravector at x and a covector at w under general coordinate transformations.

In inertial coordinates, the components of Λ are constant over spacetime and equal to δ^α_μ. This could already have been inserted into the expression (14.1) for Δp_μ in inertial coordinates. So in fact, in one sense, we have not changed anything here, merely written it in a form that can be associated with an arbitrary coordinate frame. In another sense, however, the model has changed, because we have extended to a generally covariant view and the physically relevant quantity has become a covector field.

The expression (14.2), which we just interpreted as telling us that these detectors will never absorb internal energy from electromagnetic fields that are time-independent relative to some inertial frame, has a much more complicated form now. For (14.4), Stokes' theorem gives (dropping primes on the arbitrary coordinates now)

$$\Delta p_\mu = \frac{1}{2} q \int_\Sigma \left[(\Lambda_\mu{}^\kappa F_{\kappa\beta})_{\cdot\alpha} - (\Lambda_\mu{}^\kappa F_{\kappa\alpha})_{\cdot\beta} \right] d\sigma^{\alpha\beta} \; ,$$

where dots denote covariant derivatives using the Levi-Civita connection for the Minkowski metric relative to these coordinates. The covariant derivatives are taken with respect to the field point x over which integration is carried out. Maxwell's source-free equation can be written

$$F_{\kappa\beta\cdot\alpha} + F_{\alpha\kappa\cdot\beta} = F_{\alpha\beta\cdot\kappa} \; , \tag{14.6}$$

and one soon obtains Mould's formula

$$\Delta p_\mu(w) = \frac{1}{2} q \int_\Sigma M_{\mu\alpha\beta}(x,w) d\sigma^{\alpha\beta} \; , \tag{14.7}$$

where

$$M_{\mu\alpha\beta}(x,w) := \Lambda_\mu{}^\kappa(x,w) F_{\alpha\beta\cdot\kappa} + \Lambda_\mu{}^\kappa(x,w)_{\cdot\alpha} F_{\kappa\beta} - \Lambda_\mu{}^\kappa(x,w)_{\cdot\beta} F_{\kappa\alpha} \; . \tag{14.8}$$

The object M is a bitensor with one index μ associated with the point w and two indices α and β associated with the point x.

At least, this is the manifestly covariant form of Mould's equation, which he writes using ordinary coordinate derivatives (denoted by commas). One can replace the dots by commas in Stokes' theorem, and everywhere thereafter, because the Levi-Civita connection is symmetric in its lower indices. This is true even for Maxwell's source-free equation (14.6). Mould's expression for $\Delta p_\mu(w)$ is covariant, but not manifestly so.

Mould now deduces from this that time-independent EM fields in conjunction with a static gravitational field do not generally guarantee a spacelike result for Δp_μ. But it is quite impossible to make any deduction by noting naively that the four-component of Δp_μ as given by (14.7) and (14.8) is obtained in part from the term $\Lambda_\mu{}^\kappa(x,w) F_{\alpha\beta\cdot\kappa}$, and that there is no reason to expect the displacement bitensor to be diagonal. The situation is much worse if one keeps the covariant derivative $F_{\alpha\beta\cdot\kappa}$ of the EM field, which generally introduces nonzero connection components, and in any case there are other terms in (14.8).

One should note also that the notion of a time-independent EM field has been transformed along with the arbitrary coordinate transformation. Worse still, even if some coordinate can still be associated with the notion of time, e.g., because the corresponding coordinate axes have timelike tangent vectors, there is no reason to expect it to be simply related to any inertial time. The notion of time-independence is coordinate-dependent. On the other hand, we did observe above that the timelike or spacelike attribute of Δp_μ is coordinate-independent. Indeed, this is what made Mould's detection criterion invariant under change of inertial frame, and it makes it invariant under change of arbitrary coordinate frame too.

These comments are intended to paint a rather black picture of what can be done with this kind of covariant expression, an issue that is often avoided in the textbooks. The latter concentrate on covariance because it seems to be a prime feature of GR. Indeed, covariant expressions must always be possible, even in SR, but the fact is that they really do not help one to solve specific problems. In fact they raise a new problem that is rarely discussed: what do the objects represent physically when expressed relative to arbitrary coordinates?

But there is an even more important issue here: Mould is now talking about gravitational fields. Indeed, when introducing the idea of allowing arbitrary coordinate transformations, he refers to transformations from the inertial frame to the gravitational reference frame at the point w, an interpretation that is not easy to understand in the context. One can only presume that he is referring here to an interpretational transition of the kind outlined in Chap. 2.

Mould applies his formula (14.7) to the case of what he refers to as a uniformly accelerating frame. We have the standard transformation from inertial coordinates to new coordinates adapted to an observer with eternal uniform acceleration (see Chap. 2), at least over the region of spacetime where this is possible, and the Born expression for the EM fields generated by a charge sitting permanently at the origin of such coordinates, once again in the region where this expression is considered to be valid. These fields are first given in the inertial frame, then transformed by applying the principle of general covariance to the semi-Euclidean coordinates for this flat spacetime. It is noted that the electromagnetic field is time-independent and that its axial components are zero.

Once again, one ought to be cautious here. We no longer understand the field components when they are expressed relative to arbitrary coordinates. The axial components can always be interpreted as representing a magnetic field, but this does not mean that the things we learnt at school about magnetic fields apply to these objects. The components, whatever they represent physically, can be decreed static, but they are only time-independent with respect to the semi-Euclidean time coordinate. The components that we do understand from school, in the inertial frame, are not static at all with respect to the time that we all have some feel for (despite certain oddities in SR) in the inertial frame. It should also be remembered that the semi-Euclidean coordinates have no particular significance for a uniformly accelerating observer, who could presumably find other ways of adapting coordinates to her motion.

We now reach one of the most problematic parts of Mould's paper: the construction of the detector loop, presented without diagram. Rewards go out to anyone who can draw a diagram that corresponds to his description. Still, one understands that the aim of the construction is to simplify the calculation of the integral in (14.7) by arranging for a lot of terms to cancel. This is done by contriving the geometry of the detector loop so that it has two subloops labelled 1 and 2 with symmetries with respect to the differential area measure, viz.,

$$d\sigma^{ij}(2) = d\sigma^{ij}(1)\,, \qquad d\sigma^{i4}(2) = -d\sigma^{i4}(1)\,, \qquad i,j = 1,2,3\,,$$

which certainly looks feasible. A convenient choice of w is then made and we are told that the values of the bitensor $M_{\mu\alpha\beta}(w,x)$ satisfy the relations

$$M_{ijk}(w,2) = -M_{ijk}(w,1)\,, \qquad M_{ij4}(w,2) = M_{ij4}(w,1)\,,$$

$$M_{4ij}(w,2) = -M_{4ij}(w,1)\,, \qquad M_{4i4}(w,2) = -M_{4i4}(w,1)\,,$$

on subloops 1 and 2, where $i,j,k = 1,2,3$. Given the complexity of the object $M_{\mu\alpha\beta}(w,x)$ as specified by (14.8), with the non-trivial displacement bitensor components in the semi-Euclidean coordinate frame, this is quite an achievement.

Supposing that we can understand how this works, we then obtain an impulse $\Delta p_\mu(w)$ with $\Delta p_i = 0$ for $i = 1,2,3$, and $\Delta p_4 \neq 0$, for the chosen value of w. The latter apparently has no particular significance except that it is convenient. However, another choice of w would not change the timelike character of $\Delta p_\mu(w)$, as can be seen from the definition (14.3): if the vector calculated by (14.2) in an inertial frame is timelike, then the whole vector field $\Delta p_\mu(w)$ is timelike.

Mould's loop apparently brings the charges back to their starting points relative to the spatial coordinates of the accelerating frame so that the process can be repeated indefinitely. So this detector is considered to be moving with the accelerating frame. It is an accelerating detector moving with an accelerating charge and it continuously absorbs a timelike four-momentum from the EM field. Here we have Bondi and Gold's paradox, because we are not supposed to expect a charge that is stationary relative to coordinates in which the metric is static to radiate EM energy, whereas this detector is picking up energy from the EM field. Of course, we are assuming that the detector is a radiation detector when accelerating, something that has not been analysed yet; or should one say, something that has not been justified yet. Finally, it is noted that the effect here is local, meaning that the detector may be right beside the field-generating charge and it will still make a detection, whereas radiation had been considered to be a long-range phenomenon in the earlier part of the debate.

We now observe that there is a net displacement in the detector's center of mass and that the accompanying change in gravitational energy is not included in the calculation of Δp_4. So we have surreptitiously made the switch from an empty spacetime with no gravitational effects, not even non-tidal ones, in which the strange appearance of geodesic motions (relative to semi-Euclidean coordinates) is attributed to the fact that the observer is accelerating, to a spacetime containing an SHGF, in which the strange appearance of geodesic motions is attributed to a non-tidal gravitational field. As pointed out in Chap. 2, one must apply the strong equivalence principle to be able to interpret the EM field tensor now. Likewise to interpret the four-momentum impulse given by (14.7).

Mould calculates the gravitational potential energy gained by the detector during the detection and finds that it exactly cancels the energy that was apparently absorbed by the detector from the EM fields. This remarkable cancellation resolves Bondi and Gold's paradox. Let us quote:

> It is suggested by this timelike result [namely $\Delta p_i = 0$ for $i = 1, 2, 3$, and $\Delta p_4 \neq 0$ for the case considered] that a stationary detector in static gravitational and electromagnetic fields will show time-dependent behaviour of a kind associated with internal transitions to higher energy states. We can furthermore identify this behaviour with a continuous lowering of the detector's center of mass. But then it is also clear that we have not fully satisfied the claimed conditions, for a detector cannot be said to be stationary if its center of mass is continually falling.

It is a great pity that it is so difficult to understand Mould's detector configuration, because it is obviously crucial in order to see how he gets out of the supposed paradox. What is the detector doing relative to the semi-Euclidean coordinates? Is it moving with the semi-Euclidean frame or not? Presumably it was shifting slightly with respect to these coordinates when one obtained the positive detection, and what he is claiming is that any energy it absorbed would then be required to shift it back. Apart from the striking coincidence that these apparently very different energies cancel each other out perfectly, suggesting some deep connivance between our theories of electromagnetism and gravity, the same questions remain: Is it detecting EM radiation? And if detection by his detector is supposed to redefine EM radiation, is that really answering the Bondi/Gold paradox? And is there a paradox anyway?

Let us see how his argument proceeds. He now identifies the electric and magnetic fields as they would be defined from the components of the EM field tensor relative to the semi-Euclidean frame. This is an absolutely standard procedure, even though there are rarely any comments about how one is to interpret these three-vectors relative to such a frame. He identifies what he calls the quantity associated with energy conservation in the accelerated frame, by writing

$$\frac{c}{4\pi}\operatorname{div}(\mathbf{E} \times \mathbf{H}) + \frac{1}{8\pi}\frac{\partial}{\partial t}\left(\frac{E^2 + H^2}{az}\right) + \mathbf{J} \cdot \mathbf{E} = 0 \,, \tag{14.9}$$

where $\mathbf{J}$ represents the current density in some way. All the quantities here are the ones associated with the accelerating frame, so we do not really know what they correspond to physically, but the above equation, which Mould calls the Poynting relation, is indeed just a transformed version of a similar relation that is known and understood from the ordinary physics of inertial frames.

The first two terms are going to be zero because, in the case of the fields generated by a uniformly accelerating charge, $\mathbf{H} = 0$ and $\mathbf{E}$ is independent of semi-Euclidean time, so this relation reduces to

$$\mathbf{J} \cdot \mathbf{E} = 0 \,. \tag{14.10}$$

From this Mould deduces that the net impulse on the detector loop, denoted by $\Delta p_4^{(\text{accel})}$, presumably including change of gravitational potential, is zero. He claims that including the gravitational potential means that there is no violation of energy conservation. Rewards go out again for a clear version of these arguments. What would really make them clear is to show the parallel arguments when everything is kept in the inertial frame. Since everything in the semi-Euclidean frame is just transformed from the inertial frame where one can understand the physical significance of the quantities, Mould's arguments must be a transformed version of conservation laws in the inertial frame.

The advantage of working in the semi-Euclidean frame is presumably just mathematical, in the sense that the field components are simpler. One needs to be careful about physical interpretation, however, because the zero component of the transform of an energy–momentum four-vector is not really energy as we know it. Worse, we find that $\Delta p_\mu^{(\text{accel})}$ as Mould calculates it is not a four-vector anyway, presumably because we are told to get it by integrating over the loop without the parallel displacement ploy, as in (14.1). This makes it hard to imagine that it could really represent any physical quantity.

In any case, we come to the point where the criterion based on the complicated expression (14.7) has to be adapted to take into account what Mould calls additional supports or constraints. This appears to mean that one is going to add a change in energy and momentum due to a shift in position relative to the gravitational field:

> To satisfy definition, additional supports or constraints must be added to the detector system in the presence of a Coulomb source, where these perform the function of raising the center of mass in each cycle. Since these forces are not applied to the charge q directly, but only to the mass absorbed in each cycle, they are not formally included in (14.7). However, they are crucial to recognize, for they oppose the internal transitions induced by the electric field, thereby establishing the required equilibrium. To the extent that they fulfill this requirement, their effect plus the effect of the electromagnetic forces must on average transfer either spacelike impulses to the detector system, or null impulses with zero components.

The opening clause refers to the fact that one has not satisfied the condition that the detector is stationary if one finds that its center of mass is continually falling (Mould's own words). This explanation strongly suggests that we are now just adding in a change in gravitational potential, treating the observed effects in this view of things as a gravitational effect in the way described in Chap. 2. One supposes that one could equivalently consider this as happening in a flat spacetime without (non-tidal) gravitational effects and that energy and momentum have to be transferred to the detector to keep it stationary relative to the semi-Euclidean coordinate frame. What is not obvious is whether this is at all relevant to the question of whether the accelerating charge is radiating. Since one finds (by most accounts) that it is radiating when viewed from the inertial frames, the relevance of Mould's detector is at stake here.

The reference to the required equilibrium in the last quote also suggests that this detector is not expected to detect anything in this configuration, which would perhaps mean that it solves the Bondi/Gold paradox in some sense. But this fact alone also means that it is not a radiation detector in the sense in which radiation was meant in the Fulton/Rohrlich discussion, for example.

The section of [5] explaining how the additional constraints are to be included is a challenge, partly because it appears to contain many typographical errors (as does the rest of the paper), and partly because there is no real explanation of what is happening physically. Superficially, it certainly looks plausible, and it leads to the remarkable result that (14.1) and (14.2) were really right all along. In other words, the need to parallel displace using the bitensor Λ, leading to the complex expressions (14.7) and (14.8), is somehow cancelled out by including the change in gravitational potential, or impulses used to accelerate the detector, depending on whether one is discussing an SHGF or the observations of an accelerating observer in a spacetime with no gravitational effects.

Except for the annoying fact that this final result is not covariant (at least not in the obvious sense of the term), this almost seems too good to be true. It suggests either some deep physical conspiracy between gravity and electromagnetism, or some trivial relationship that has taken on a sophisticated disguise. One notable thing about the new result

$$\Delta p_\mu = q \oint F_{\mu\nu} \mathrm{d}\lambda^\nu = \frac{1}{2} q \int_\Sigma F_{\alpha\beta,\mu} \mathrm{d}\sigma^{\alpha\beta} \tag{14.11}$$

is that Δp_μ on the left is evaluated at a specific point w, described by Mould as the point where the loop momentum has been physically transported by the added constraints, whatever that means. The fact that the right-hand side is independent of w is explained as follows:

> [...] different lifting constraints are required for different choices of w, and in such a way that the numerical value of the resulting four-vector is the same for all w. That is, changing w here implies a physical change and not just a formal one.

On the face of it, this sounds plausible, but what exactly is meant by 'lifting constraints'? And is this object really a four-vector?

The conclusion is:

> In the special case of the static charge in a [semi-Euclidean] frame, we have seen that the fields $F_{\alpha\beta}$ are time independent. Therefore, the fourth component of (14.11) will be $\Delta p_4 = 0$, assuring that a static elementary detector in such a field will not absorb internal energy in the sense of our criterion. The observational paradox is then removed by the addition of constraints which insure that the detector is stationary in the given reference frame.

Once again, put like this, it looks plausible, but a lot of questions remain. The principal question is:

- Are we detecting the right thing here?

This leads to the other questions we wish to raise:

- Are we redefining radiation as positive detection by these detectors, and if so, does that really resolve the original paradox?
- Is there really an observational paradox anyway? In other words, should we really expect to detect nothing in this configuration? Is there some sense in which a charge that is stationary relative to coordinates in which the metric is static would not be able to radiate?

With regard to the last question, we seem to have an example *par excellence* in which we do expect such a charge to radiate. The Bondi/Gold question is whether one must not therefore begin to doubt the strong equivalence principle whose application to the uniform acceleration case leads to the conclusion that a charge can radiate in such a situation. And the answer proposed in the present book is that it is better to sacrifice one's intuition about static charges in static spacetimes than to renounce the strong equivalence principle, because one has no other (simple) way to use GR.

14.3 Detecting Radiation Where There Is None

One has to consider the penultimate section of [5], because it deals with a freely falling charge in the non-tidal gravitational interpretation of the flat spacetime and it reaches the rather surprising conclusion already mentioned. In this case, one finds the locally (and happily global) inertial frame falling freely with the charge and observes that, by Maxwell's equations which are supposed to hold there according to GR+SEP, the charge does not radiate. Its EM field in this frame is just the Coulomb field.

Can this field look like a radiating field to an accelerating observer? To answer that, one ought to choose some frame for the latter, but no such frame is laid down by the theory. The situation here is different from SR, because when the observer is inertial, a preferred frame is in fact laid down by the theory. But in GR one is stuck with all frames, a paradoxically unlucky situation. In reality, the best thing the accelerating observer could do would be to get hold of the description relative to some (locally) inertial frame. The fact that one can find semi-Euclidean coordinates adapted to the motion of the accelerating observer means nothing. They have no particular relevance according to GR, or even according to GR+SEP. The theory says precisely that no coordinates have any particular relevance. There are no favoured coordinate systems. The best we can do, with the help of SEP, is to refer to a locally inertial system. For this reason, one ought to be suspicious of glib calculations in accelerating frames.

Mould's paper ends with just such a calculation. One writes down the EM field in the inertial frame and transforms to the semi-Euclidean system. One then constructs another remarkably convenient detector loop comprising two subloops in such a way that almost everything in (14.11) cancels out. Once again there is no diagram to help us understand this crucial step. We obtain a purely timelike Δp_μ, in the sense that $\Delta p_i = 0$ for $i = 1, 2, 3$, while $\Delta p_4 \neq 0$. Although the charge here is considered to be moving inertially, because freely falling in a GR model, it is apparently radiating as one would expect a freely falling charge to radiate in the SR model where it would not be considered to be inertial. Are we just trying to save a special relativistic intuition that there ought to be electrogravitic bremsstrahlung, as DeWitt and Brehme put it? Even DeWitt and Brehme with their highly sophisticated calculation cannot back up this idea, because in this spacetime the curvature is zero.

In a certain sense, it is a useful exercise to take the Coulomb EM fields in Minkowski coordinates for flat spacetime, viz.,

$$\mathbf{E} = -\frac{e}{4\pi\varepsilon_0}\frac{\mathbf{r}}{r^3}\,, \qquad \mathbf{B} = 0\,, \tag{14.12}$$

with covariant electromagnetic field tensor

$$F_{\mu\nu} = -\frac{e}{4\pi\varepsilon_0 r^3}\begin{pmatrix} 0 & x & y & z \\ -x & 0 & 0 & 0 \\ -y & 0 & 0 & 0 \\ -z & 0 & 0 & 0 \end{pmatrix}\,, \tag{14.13}$$

and transform the tensor to its semi-Euclidean form. Not surprisingly, one obtains quite a complicated result for the 'electric' and 'magnetic' fields, $\mathbf{E}^{\mathrm{SE}}$ and $\mathbf{B}^{\mathrm{SE}}$, read off from the definition

$$F_{\mu\nu}^{\mathrm{SE}} = \begin{pmatrix} 0 & E_1^{\mathrm{SE}} & E_2^{\mathrm{SE}} & E_3^{\mathrm{SE}} \\ -E_1^{\mathrm{SE}} & 0 & -B_3^{\mathrm{SE}} & B_2^{\mathrm{SE}} \\ -E_2^{\mathrm{SE}} & B_3^{\mathrm{SE}} & 0 & -B_1^{\mathrm{SE}} \\ -E_3^{\mathrm{SE}} & -B_2^{\mathrm{SE}} & B_1^{\mathrm{SE}} & 0 \end{pmatrix}\,. \tag{14.14}$$

With the conventions used in Chap. 2, one finds

$$\mathbf{E}^{\mathrm{SE}} = -\frac{eCg^2}{4\pi\varepsilon_0 c^4}\frac{1}{\left(1+\frac{gy^1}{c^2}\right)^3\left\{\left[\left(1+\frac{gy^1}{c^2}\right)C-1\right]^2+\left(\frac{gy^2}{c^2}\right)^2+\left(\frac{gy^3}{c^2}\right)^2\right\}^{3/2}}$$
$$\times \begin{pmatrix} 1+\dfrac{gy^1}{c^2}-\dfrac{1}{C} \\ \dfrac{gy^2}{c^2} \\ \dfrac{gy^3}{c^2} \end{pmatrix}\,, \tag{14.15}$$

$$\mathbf{B}^{\mathrm{SE}} = \frac{eSg^3}{4\pi\varepsilon_0 c^6}\frac{1}{\left\{\left[\left(1+\frac{gy^1}{c^2}\right)C-1\right]^2+\left(\frac{gy^2}{c^2}\right)^2+\left(\frac{gy^3}{c^2}\right)^2\right\}^{3/2}}\begin{pmatrix} 0 \\ y^3 \\ -y^2 \end{pmatrix}\,, \tag{14.16}$$

with the shorthand

$$S := \sinh\frac{gy^0}{c^2}\,, \qquad C := \cosh\frac{gy^0}{c^2}\,.$$

The details of this complexity come from the coordinate transformation, obviously. With different coordinates adapted to the observer worldline, e.g., describing a frame with some rotation, one would get a different complexity. But which coordinates should the observer use?

Another question here is this: how would one use the above expressions to support the claim that the accelerating observer would observe radiation? Or not observe radiation, for that matter? What is our criterion for there being radiation? On the face of it, it does seem possible that the accelerating observer might set up these coordinates in some natural way and then find that the above 'electric' and 'magnetic' fields, treated naively as though expressed relative to Minkowski coordinates, look like a field with a radiation component. One might calculate the Poynting vector and observe that there is a flow of 'energy–momentum'. This is exactly what Rohrlich proposes to do in [22] (see the discussion in Sect. 16.10).

One could take this as a definition of radiation. But the view adopted in the present book is that this should not be taken as EM radiation; that EM radiation is in fact something that should be coordinate-independent. This point came out in Chap. 10, in the analysis of the different predictions made by special relativity and general relativity with regard to EM radiation from freely falling charges and charges prevented from free fall in the presence of gravitational fields. In each case, it was taken that the motion of the observer had no influence on whether the charge should be considered as radiating.

Mould is clearly proposing something very different. It could be that his detector really is theoretically sound and could even be built. In that case, there would appear to be hope for an observer-dependent definition of EM radiation.

Note finally that the question of the Poynting vector is discussed further in Sect. 16.8 in the complementary situation, where the charge is uniformly accelerating and the observer co-accelerating, and it is argued that the SE Poynting vector is not giving us the flow of energy–momentum as it is usually defined, but the flow of something else.

14.4 Conclusion

For the present purposes, Mould's discussion of the freely falling charge in an SHGF is the key section in his paper. Perhaps there are two contrasting ways of looking at this:

- We have the proof that Mould's detector is not a radiation detector. According to GR+SEP, the freely falling charge does not radiate. It produces a Coulomb field in this globally inertial freely falling frame. If the detector absorbs energy by accelerating through the Coulomb field, then it is not detecting radiation, because there is no radiation, at least according to the usual definition in terms of which Bondi and Gold's question was originally raised. This in turn shows the irrelevance of the fact that it does not absorb energy when 'stationary' in 'the' accelerating frame moving with a uniformly accelerating charge.

- Mould's detector may detect radiation correctly whenever it is moving inertially, but obtain the claimed measurement results when it is accelerating. In this case, it could be physically useful to define radiation for accelerating observers by the results of a Mould detection. However, WEP and SEP do not require these results for their well-being. Furthermore, these results would suggest a kind of conspiracy between electromagnetism and gravity that is really not required by any standard aspects of these theories.

The fact that calculations are carried out in the difficult semi-Euclidean coordinates of the accelerating frame should make us suspicious. It would have been easier to do everything in the inertial frame that also happens to be available here, because we then understand the meaning of the quantities. After all, why choose this particular accelerating frame? One could find other coordinates adapted to the observer motion.

Chapter 15
The Definitive Mathematical Analysis

The main achievement of the paper [6] by Boulware is a very careful analysis of the EM fields of a uniformly accelerating charge and the energy–momentum of those fields, with some interesting deductions about the flow of energy in spacetime for this scenario. Although there will be criticisms in what follows, this paper is highly recommended for the mathematical aspects of the fields, and in fact all the details are spelt out in the following. Criticism is aimed particularly at deductions made about the impossibility of measuring radiation if one is comoving with a uniformly accelerating charge, but also about the idea that one can, or should, try to save the claim that a charge that is stationary relative to coordinates in which the metric is static cannot radiate. Since the discussion here is very detailed, this chapter is divided into sections and subsections.

As an introduction, let us quote the abstract for [6], which first confirms what has just been said, then states what Boulware calls the equivalence principle paradox:

The electromagnetic field associated with a uniformly accelerated charge is studied in some detail. The equivalence principle paradox that the co-accelerating observer measures no radiation while a freely falling observer measures the standard radiation of an accelerated charge is resolved by noting that all the radiation goes into the region of spacetime inaccessible to the co-accelerating observer.

We shall return to a detailed discussion of Boulware's introduction only when we have considered how he derives the various results. Suffice it to say here that the paradox has changed slightly since Bondi and Gold first described it. Their problem was that, for them, a charge that is stationary relative to coordinates in which the metric is static cannot radiate, whereas in the case of an SHGF, the strong equivalence principle implies that a charge that is stationary relative to coordinates in which the metric is static does actually radiate. Their solution to this was that an SHGF is unphysical. The last quote, however, accepts that there is radiation for the freely falling observer, but not for the co-accelerating observer in that very case of the SHGF, or at least that one will measure it while the other will not.

This is a subtle but important transformation of the issue. Boulware does not claim that a charge that is stationary relative to coordinates in which the metric is static does not radiate, only that an observer who is also stationary relative to these coordinates will not be able to measure any radiation there is. We shall find that the radiation escaping over this observer's event horizon is not the only thing preventing her from measuring it, according to Boulware. The fact of being restricted to data from the region of spacetime within that event horizon also plays an essential role for this author. So this stance seems to some extent to support what has been claimed in the present account: there is no particular reason to say that a charge that is stationary relative to coordinates in which the metric is static does not radiate. But it does try to contradict another claim made in the present account: there is no justification for claiming that one has to be far from a source charge in order to detect its radiation.

But how will the fact that, although the radiation is there, one cannot detect it, allow us to resolve any paradox? In fact, what exactly is the paradox now? In the above quote, the paradox is supposed to be that a co-accelerating observer measures no radiation while a freely falling observer measures the standard radiation of an accelerated charge. But why would this be paradoxical? After all, Boulware's paper purports to explain and justify it, so there cannot be any internal contradiction there. One must also ask why it should even be true that a co-accelerating observer measures no radiation while a freely falling observer measures the standard radiation of an accelerated charge. Could it be Boulware's version of the equivalence principle that implies this, since he refers to it as the equivalence principle paradox?

We find from Boulware's introduction that he is taking the equivalence principle to say that a uniformly accelerated frame must be indistinguishable from a gravitational field. This does indeed appear to be close to the idea discussed in Chap. 2 that, in the spacetime described by (2.45) on p. 15 or (2.48) on p. 17, an observer cannot say by observing the geodesic motion of test particles whether she is uniformly accelerating in a flat spacetime (and using semi-Euclidean coordinates) or sitting still relative to coordinates that describe a spacetime containing a static, homogeneous gravitational field, or even a mixture of both. We have also seen that the adjunction of the strong equivalence principle (SEP) allows one to go further, extending the scope of the indistinguishability claim to EM fields, or other fields. In fact, EM fields can only be formulated theoretically in GR with the help of SEP. It is the thesis of the present book that, whatever else one says about what can or cannot be measured, the two situations of a uniformly accelerated frame and a non-tidal gravitational field are scrupulously the same according to this theory, simply because the mathematics is identical.

So here is an idea: perhaps Boulware still believes Bondi and Gold's claim that, in the case of a gravitational field, a charge that is stationary relative to coordinates in which the metric is static cannot radiate, and by the above indistinguishability (referred to loosely as 'the' equivalence principle), concludes that an observer co-accelerating with the uniformly accelerating charge (in flat spacetime with no gravitational effects) should not detect any radiation. Strictly speaking, the above indistinguishability actually implies that, if in the case of a gravitational field a charge that is stationary relative to coordinates in which the metric is static cannot radiate, then there really is no radiation for an observer co-accelerating with the uniformly accelerating charge, not just that the observer would be unable to detect it.

The problem is of course that the uniformly accelerating charge is known to radiate according to the theory. It radiates in the inertial frame (Boulware confirms that he accepts this), and the view adopted in the present account is that it therefore radiates in any frame, since radiation is an objective, observer-independent thing. So we still have Bondi and Gold's paradox, assuming that one can uphold their opinion that, in the case of a gravitational field, a charge that is stationary relative to coordinates in which the metric is static cannot radiate. If Boulware achieves anything it is just that he manages to assuage the disquiet of those who believe that opinion.

Put another way, in Bondi and Gold's version of the paradox, one appeared to be saying that radiation would look like radiation however one viewed it and this was really necessary in order to make difficulties for SEP, provided one accepted their opinion that, in the case of a gravitational field, a charge that is stationary relative to coordinates in which the metric is static cannot radiate. But in Boulware's version, one is saying that the radiation observed in the global inertial frame is inaccessible to the supported observer moving with the charge, i.e., it is still there, but not in the right place for the supported observer to measure. So the strict application of WEP and SEP described two paragraphs ago would then still contradict Bondi and Gold's opinion, meaning that one or the other would have to be sacrificed. However, Boulware is working, not with the strict application of WEP and SEP as formulated in the present account, but with a vague version of the equivalence principle that interprets the word 'indistinguishable' in a purely subjective way: it is enough for a particular observer to be unable to make the difference for some quite disparate reason, e.g., the radiation is only clearly detectable beyond an event horizon.

One can sense that the problem has been transformed. Up to now, one was more concerned with the safety of SEP, but here one is trying to save the idea that, at least in the case of a gravitational field, there should be no radiation from a charge that is stationary relative to coordinates in which the metric is static (and of course, using SEP to do so, because one cannot say anything about EM fields in spacetimes like this without applying SEP). The idea of no radiation from the charge in this case has now become just bad luck for that particular observer. But this idea would certainly not have satisfied Bondi and Gold: the idea that one particular observer would be expected not to observe radiation even if it was there. When Bondi and Gold suggest that a stationary charge in a static spacetime should not radiate, they are saying that no observer should detect radiation. They are actually saying that there is no radiation in some objective sense.

So it looks as though the whole thesis of Boulware's paper, viz., that the radiation from a uniformly accelerating charge will be inaccessible to a comoving uniformly accelerating observer, will prove irrelevant to Bondi and Gold's original paradox. On the other hand, it is certainly worth considering in its own right.

15.1 Static Gravitational Field

Boulware's account of the metric suffers a little from mathematosis. Although it is interesting in some respects, particularly in showing the limitations that simple constraints can put on the metric, it is generally too far removed from physical interpretation to really clarify the problems at hand, as we shall see. Note at the outset that the word 'homogeneous' should be appended as an attribute of the gravitational field, and we may sometimes use the abbreviation SHGF.

The spacetime is assumed to be invariant under time translations (static) and under the Euclidean group E_2 of translations and rotations in a plane perpendicular to the gravitational field (homogeneous). The most general metric allowed by these restrictions is given by the interval

$$\mathrm{d}s^2 = -\phi^2(z)\mathrm{d}\tau^2 + A^2(z)\mathrm{d}z^2 + B^2(z)(\mathrm{d}x^2 + \mathrm{d}y^2)\,. \tag{15.1}$$

The homogeneity of the gravitational field only refers to the x and y coordinates. It seems to vary in the z direction (but see note below). With the change of coordinate

$$z \longmapsto \zeta := \int^z \mathrm{d}\bar{z}A(\bar{z})\,, \tag{15.2}$$

we obtain the same metric but with $A \equiv 1$. The nonzero curvature components are

$$R^{\tau}_{\zeta\tau\zeta} = -\frac{\phi''}{\phi}\,, \qquad R^{\tau}_{x\tau x} = -\frac{\phi'B'B}{\phi} = R^{\tau}_{y\tau y}\,, \tag{15.3}$$

$$R^{x}_{\zeta x\zeta} = -\frac{B''}{B} = R^{y}_{\zeta y\zeta}\,, \qquad R^{x}_{yxy} = -B'^2\,. \tag{15.4}$$

We now require the spacetime to be flat, another thing not mentioned in the title of this section. In fact, we shall see that it is precisely this that makes the gravitational field homogeneous. This means that $B' = 0 = \phi''$, whence, with suitable scaling of coordinates,

$$\phi = 1 + g\zeta\,, \qquad B = 1\,, \tag{15.5}$$

where g is an arbitrary constant. We now have

$$\mathrm{d}s^2 = -(1 + g\zeta)^2\mathrm{d}\tau^2 + \mathrm{d}\zeta^2 + \mathrm{d}x^2 + \mathrm{d}y^2\,. \tag{15.6}$$

It turns out then that g is the proper acceleration of a body sitting at $\zeta = 0$. Bodies sitting at other values of ζ have different proper accelerations, so in this sense, the gravitational field would not appear to be homogeneous in the ζ direction.

As discussed in Chap. 2, it is not obvious that this should be called a gravitational field at all. Another problem with the kind of derivation that Boulware gives is that we have no idea how to interpret these coordinates physically. Indeed, this question is not even considered! The problem also arises in the derivation of the Schwarzschild metric as a solution of Einstein's equations under certain symmetry conditions, as found in any standard textbook on GR. The notation used for the coordinates is highly suggestive in this case, but little is usually said about how to relate coordinates to measurement. In fact, one does need to make use of the weak and strong equivalence principles to say how these coordinate versions of spacetime manifolds are to be understood physically.

The form of the interval in (15.6) shows that we have what we called semi-Euclidean coordinates in Chap. 2. Boulware now makes the coordinate translation

$$Z = \frac{1}{g} + \zeta \,, \tag{15.7}$$

which brings the interval to the form

$$ds^2 = -g^2 Z^2 d\tau^2 + dZ^2 + dx^2 + dy^2 \,. \tag{15.8}$$

We note immediately that $Z = 0$ is a singularity of this metric and that we shall therefore restrict to $Z > 0$. This problem is not mentioned by Boulware.

As noted above, it is not at all obvious that the gravitational field in question here is actually homogeneous, and perhaps this is why Boulware simply describes it as static. Indeed, on his rather mathematical approach, there is little hope of understanding what the gravitational field is like physically. Approaching via a uniformly accelerating observer in Minkowski spacetime, and understanding how she sets up semi-Euclidean coordinates, perhaps one gets a better grasp of these points. For one thing, one can see why the above spacetime is required to be flat. Otherwise, why should the spacetime of a static gravitational field be flat? There are of course curved spacetimes for which there exist coordinates such that the metric has the static form (13.1) specified on p. 127. An obvious example is the Schwarzschild spacetime. But any spacetime has coordinates such that the metric does not have the static form given in (13.1).

Now the uniformly accelerating observer who sets up the semi-Euclidean coordinates has constant space coordinates in the semi-Euclidean frame (in fact, zero). This is one of the requirements for a frame adapted to a moving observer discussed in Chap. 2. If we consider another observer with constant semi-Euclidean spatial coordinates, it turns out that this observer also has uniform acceleration, but generally a different one. This is actually because such an observer has a very artificial motion from a physical standpoint.

What is the motion of this second observer? At semi-Euclidean coordinate time τ, the first observer uses the hyperplane of simultaneity (HOS) of an inertial observer who is instantaneously comoving with her. At a later coordinate time τ', the HOS has changed, but we require the second observer to be at the same spatial coordinate relative to this new HOS. It is perhaps rather a coincidence (in this approach, at least) that the second observer has a uniform acceleration, but it is certainly no surprise to find that this uniform acceleration is a different one. So if we now consider the motion of the second observer relative to the Minkowski frame with an SHGF in it,

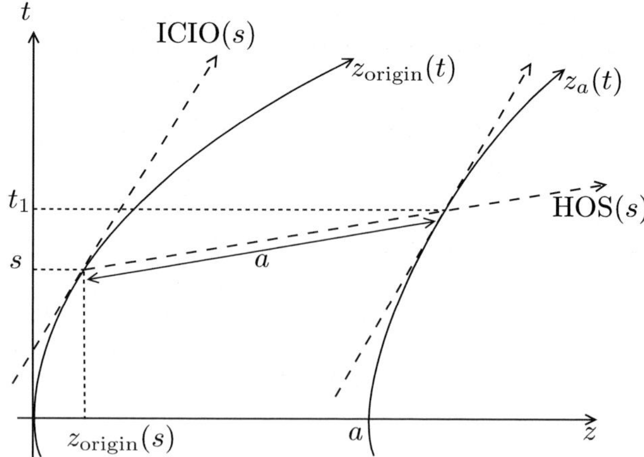

Fig. 15.1 Motion of an observer O_1 with constant semi-Euclidean coordinates relative to a uniformly accelerating observer AO, viewed from an inertial system with coordinates (z,t). The worldline of AO is $z_{\mathrm{origin}}(t)$ and the worldline of O_1 is $z_a(t)$. We consider AO at some inertial time s, when it is located at $z_{\mathrm{origin}}(s)$. Then ICIO(s) indicates the worldline of the instantaneously comoving inertial observer for AO at this event, and HOS(s) the hyperplane of simultaneity of this ICIO. The inertial time t_1 is such that the event $\left(z_a(t_1),t_1\right)$ on the worldline of O_1 is considered to be simultaneous with the event $\left(z_{\mathrm{origin}}(s),s\right)$ on the worldline of AO by either ICIO(s) or AO with her SE coordinates, which borrow simultaneity from instantaneously comoving inertial observers. The worldline of O_1 is defined by saying that ICIO(s) reckons the instantaneous separation of OA and O_1 to be a, and this for all s [not forgetting that ICIO(s) is a different observer for each s]. Note that O_1 happens to have the same instantaneous speed at the event $\left(z_a(t_1),t_1\right)$ as AO has at the event $\left(z_{\mathrm{origin}}(s),s\right)$. This is not obvious (see Sect. 15.5)

we find something rather paradoxical. For if the second observer is just supported under the influence of the SHGF, why does this require a different four-acceleration? Figure 15.1 shows a picture of the two observers as seen from an inertial system.

One might be tempted to say that the second observer moves under the influence of a different SHGF. But one cannot have the first observer moving under one SHGF and the second moving under another. One would like to say that there is only one SHGF permeating through our spacetime, physically speaking. In fact, there is no particular problem here, once we have understood that a point fixed at some spatial coordinate position does not have to correspond to a physically moving object. This is perhaps the understanding one gains from the construction of such coordinates for a uniformly accelerating observer in a Minkowski spacetime with no gravitational effects. One should be asking the following question: is the fact that a four-acceleration a^μ is required to hold an object at some fixed coordinate position a measure of the gravitational field?

In the same context, Parrott has this to say about this coordinate system [7, p. 8]:

It would be misleading to call [the Rindler frame] a uniformly accelerating elevator. Although every point on it is uniformly accelerating, the magnitude $1/X$ of the uniform acceleration is different for different points. Because of this, the everyday notion of a uniformly accelerating elevator gives a potentially misleading physical picture. A more accurate picture is obtained by thinking of each point of the elevator as separately driven on its orbit through Minkowski spacetime by a tiny rocket engine. Observers moving with the elevator experience a pseudo-gravitational force which increases without limit as the floor of the elevator at $X = 0$ is approached. Observers nearer the floor need more powerful rockets than those farther up.

He is referring here to a flat spacetime with no gravitational effects, not even non-tidal ones. We can see the risk with this. Talk of pseudo-gravitational forces increasing without limit makes it look as though, in the gravitational interpretation of these coordinates, there must be a black hole at the bottom of the lift! Boulware seems to have this view, as we shall see later, when he tries to explain the future event horizon at $Z = 0$ ($z = t$). But this cannot be the case, because the curvature is zero in this spacetime.

15.2 Relation with Minkowski Spacetime

We now discuss how the spacetime with coordinates (τ, Z, x, y), where $Z > 0$, and interval (15.8) is related to Minkowski spacetime. This discussion is also highly mathematised. The first observation is that (t, z, x, y) with t, z defined by

$$t := Z \sinh g\tau \,, \qquad z := Z \cosh g\tau \,, \qquad (15.9)$$

wherever these are actually defined, lead to a Minkowski expression for the interval:

$$\mathrm{d}s^2 = -\mathrm{d}t^2 + \mathrm{d}z^2 + \mathrm{d}x^2 + \mathrm{d}y^2 \,. \qquad (15.10)$$

Since we required Z to be strictly positive, the region of Minkowski spacetime actually covered has $z > 0$ and $|t| < z$, the latter arising because $|\sinh| < \cosh$.

This is not entirely surprising if we think about one of the uniformly accelerating observers as viewed in the newly found Minkowski spacetime. Her worldline lies entirely within the region $z > 0$, $|t| < z$ and approaches the null lines $t = \pm z$ in this region asymptotically as $\tau \to \pm\infty$, respectively. Boulware defines the regions I, II, III and IV of Minkowski spacetime by moving round the t, x plane anticlockwise and dividing it into four regions with the null lines (see Fig. 15.2). Then the region containing the uniformly accelerated observer is region I, and we see that she can receive no signal from regions II and III. This means that the line $t = z$ is an event horizon, as stated by Boulware. Further, she can never send a signal to regions III and IV. In this sense then, our observer is only fully concerned with region I, and it is this region that Boulware's coordinates τ, Z, x, y actually cover.

We note, however, that we can map the original spacetime with coordinates τ, Z, x, y to each of the other three regions of Minkowski spacetime in the following way (including the above mapping for completeness):

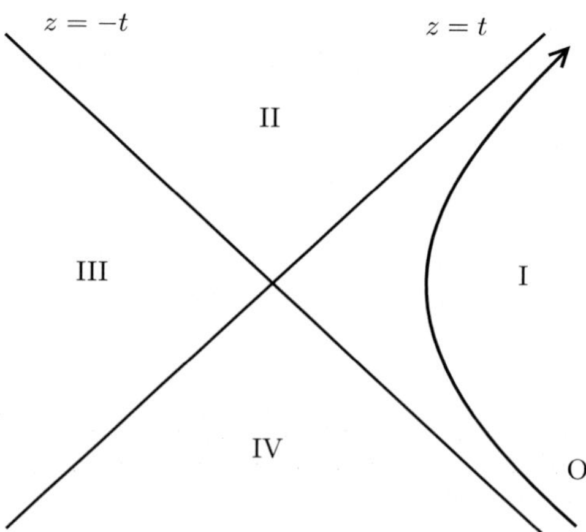

Fig. 15.2 Boulware's four regions of spacetime. The line $t = z$ is an event horizon, because the observer O can never receive any signal from regions II and III on the other side of it. Pictorially this is because the forward light cone of any potential signalling event in region II or III is entirely contained within those regions. For a similar reason, O can never signal to any event in regions III or IV, because the forward light cone of any point on the worldline of O is entirely contained within regions I and II

$$
\begin{aligned}
t &:= Z\sinh g\tau\,, \quad z := Z\cosh g\tau \quad &\text{(I)} \\
t &:= Z\cosh g\tau\,, \quad z := Z\sinh g\tau \quad &\text{(II)} \\
t &:= -Z\sinh g\tau\,,\ z := -Z\cosh g\tau \quad &\text{(III)} \\
t &:= -Z\cosh g\tau\,,\ z := -Z\sinh g\tau \quad &\text{(IV)}
\end{aligned}
\tag{15.11}
$$

The metric then takes the form given by the following line intervals in each of the four regions:

$$
\mathrm{d}s^2 = \varepsilon\left(-g^2 Z^2 \mathrm{d}\tau^2 + \mathrm{d}Z^2\right) + \mathrm{d}x^2 + \mathrm{d}y^2\,,
\tag{15.12}
$$

where

$$
\varepsilon = \begin{cases} 1 & \text{for } x \text{ in I and III}\,, \\ -1 & \text{for } x \text{ in II and IV}\,. \end{cases}
$$

Elsewhere [6, p. 183] Boulware describes this as the analytic continuation of the accelerated frame from region I to the other regions, or again [6, p. 171] as the natural extension of the coordinates (presumably the semi-Euclidean coordinates) to the other regions. But can we really speak of the (almost) complete coverage of the world by the τ, Z, x, y coordinates? (We have not specified what happens on the null lines.) As things stand, (15.11) just map the SE coordinates of the original spacetime four times, onto four different regions of Minkowski spacetime. How can this extend the τ, Z, x, y coordinates? For example, what are the τ, Z coordinates for a point in region II? Are they just the same as the τ, Z coordinates for some point in region I, and some point in each of the other two regions? In fact what we have here is an atlas for the Minkowski spacetime, with four charts. We shall discover the utility of this idea later, when we come to work out the electromagnetic potential due to an accelerating charge.

Presumably what Boulware really means is something like this. Under the coordinate transformation (15.9), we find that our original spacetime is just a quadrant of Minkowski spacetime, the one that we have called region I. We know that the metric of Minkowski spacetime on region I can be analytically extended (perhaps uniquely?) to the whole of Minkowski spacetime. In a sense we have imbedded our original spacetime in Minkowski spacetime. Note that we do not require the other coordinate transformations of (15.11) to make this claim. For what it is worth, these transformations (or rather their inverses) just show that one can map each of the other three regions of Minkowski spacetime to the original spacetime, except that the Minkowski metric then transforms to a different metric for the original spacetime in the case of regions II and IV [the case $\varepsilon = -1$ in (15.12)]. Alternatively (and this seems to be Boulware's intention), one can consider the inverses of the other coordinate transformations of (15.11) to be merely coordinate transformations restricted to each of the three other regions, whereupon the Minkowski metric takes on the forms (15.12).

Boulware's own summary of this in his introduction is [6, p. 171]:

> One may naturally extend these coordinates to regions II, III and IV by requiring that changes in the coordinate τ be the changes under Lorentz transformations just as in region I; the resulting metric in regions II and IV is independent of τ (invariant under Lorentz transformations), but the invariant hyperbolic cylinders are spacelike surfaces and τ is a space coordinate. The other coordinate Z, which is the space coordinate in region I, is the time coordinate in region II and the metric does depend on Z. There is no time-independent coordinate system for region II in which a uniformly accelerated charge in region I is at rest.

It is the last sentence that really spells out what Boulware needs here, although it is not at all proven by the above coordinate transformations. In fact, he does not prove it anywhere. We shall see why it is important for Boulware's arguments in Sect. 15.11.5.

Region I is the region of spacetime such that the uniformly accelerated observer at $Z = g^{-1}$ can both receive signals from any point and send signals to any point. Note that we single out the uniformly accelerated observer at this fixed Z position because she turns out to have proper acceleration g. It is quite obvious from the Minkowski diagram that the set of points from which the observer can receive signals is just $\{t < z\}$, which is the union of regions I and IV (see Fig. 15.3). Likewise, the set of points to which the observer can send signals is just $\{t > -z\}$, the union of regions I and II.

Boulware provides us with the following rather complex formula for the set of points from which the observer can receive signals:

$$t < \max\left\{g^{-1}\sinh g\tau - \left[(g^{-1}\cosh g\tau - z)^2 + \rho^2\right]^{1/2}\right\} = z\,. \tag{15.13}$$

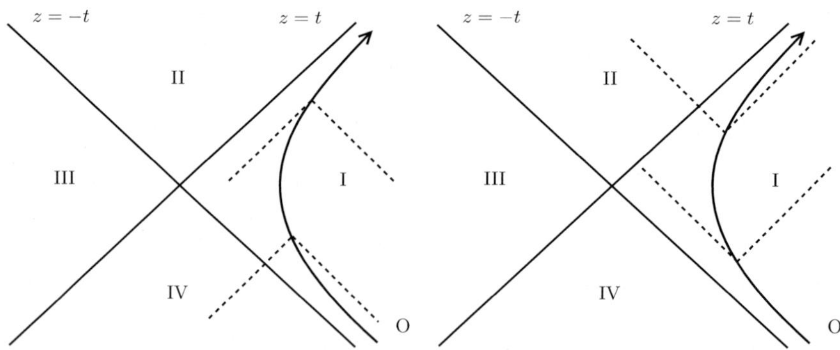

Fig. 15.3 Minkowski spacetime view of the eternally uniformly accelerating observer O. *Left*: The set of points from which O can receive signals is the union of all backward light cones of points on the worldline of O, i.e., the union of regions I and IV. *Right*: The set of points that O can signal to is the union of all forward light cones of points on the worldline of O, i.e., the union of regions I and II

This is proven as follows. At time τ, the observer is at $Z = g^{-1}$. In Minkowski coordinates, this worldline is

$$(g^{-1}\sinh g\tau, g^{-1}\cosh g\tau, 0, 0)\,.$$

Let (t,z,x,y) be a potential signalling point, whose aptitude we are to test, and immediately define $\rho^2 = x^2 + y^2$. For signalling to be possible, we must have $g^{-1}\sinh g\tau > t$ for some τ, and this will always be possible, because sinh is an increasing function. In fact, it happens for all τ with

$$\tau > g^{-1}\sinh^{-1}gt\,.$$

The other condition for (t,z,x,y) to be a signalling point is

$$(g^{-1}\sinh g\tau - t)^2 = (g^{-1}\cosh g\tau - z)^2 + \rho^2\,, \tag{15.14}$$

or

$$t = g^{-1}\sinh g\tau - \left[(g^{-1}\cosh g\tau - z)^2 + \rho^2\right]^{1/2}\,. \tag{15.15}$$

Now the function of τ on the right is the one appearing under the maximisation sign in (15.13). It is clear that, if maximisation of this function means anything, then t will have to be less than its maximum to have any hope of equalling it anywhere! If we write

$$f(\tau) := g^{-1}\sinh g\tau - \left[(g^{-1}\cosh g\tau - z)^2 + \rho^2\right]^{1/2}\,, \tag{15.16}$$

we find that

$$f'(\tau) = \cosh g\tau - \frac{(g^{-1}\cosh g\tau - z)\sinh g\tau}{\left[(g^{-1}\cosh g\tau - z)^2 + \rho^2\right]^{1/2}} \,.$$

This is only zero if

$$\cosh g\tau = \frac{(g^{-1}\cosh g\tau - z)\sinh g\tau}{\left[(g^{-1}\cosh g\tau - z)^2 + \rho^2\right]^{1/2}} \,,$$

and this can only happen if

$$(g^{-1}\cosh g\tau - z)^2 + \rho^2 \cosh^2 g\tau = 0 \,,$$

or

$$x = 0 = y \quad \text{and} \quad z = g^{-1}\cosh g\tau \,.$$

Since $\sinh \uparrow \cosh$, we conclude that $f'(\tau) > 0$ but that $f'(\tau) \to 0$ as $\tau \to \infty$. So $f(\tau)$ is an increasing function that levels out to a maximum.

What is this maximum value? As $\tau \to \infty$, we can neglect ρ^2 and

$$\begin{aligned} f(\tau) &\longrightarrow g^{-1}\sinh g\tau - (g^{-1}\cosh g\tau - z) \\ &= z + g^{-1}(\sinh g\tau - \cosh g\tau) \\ &\longrightarrow z \,, \end{aligned}$$

where we have used the fact that $g^{-1}\cosh g\tau > z$ when τ is large enough, to establish the right square root for the right-hand side of the first line.

We also note conversely, that if t is less than this maximum, then τ will eventually be big enough to receive a signal from the event (t,z,x,y). We have a similarly sophisticated analysis of the events that can receive signals from the observer, but it is not particularly illuminating, since the answer is so obvious anyway.

15.3 What the Uniformly Accelerated Observer Sees

We have to be a little careful when we talk about what the uniformly accelerated observer sees, since Boulware sometimes refers merely to what this observer observes relative to her coordinate system, and sometimes to light signals sent from events to this observer.

To begin with, we have the claim that a particle worldline which crosses the boundaries of region I, viz., $z = \pm t$, does so at the time $\tau = \pm\infty$ for the observer. This follows from (15.9), because $z = \pm t$ implies that

$$Z\cosh g\tau = \pm Z\sinh g\tau \,,$$

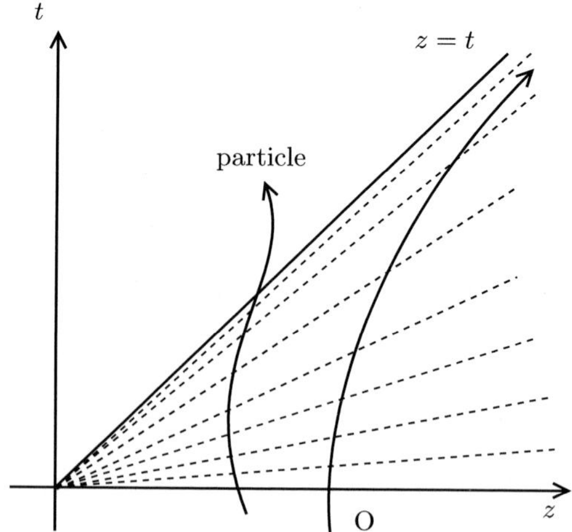

Fig. 15.4 View of the worldline of an eternally uniformly accelerating observer O relative to Minkowski coordinates (z,t) in a flat spacetime. The observer worldline approaches the null line $z = t$ asymptotically. A particle is moving around to the left of O. *Dotted lines* are hyperplanes of simultaneity which O borrows from instantaneously comoving inertial observers in order to set up semi-Euclidean coordinates. For a uniformly accelerating worldline like the worldline of O, these hyperplanes of simultaneity get steeper and steeper in just such a way that they all intersect at the origin (see Sect. 15.4). For each event at which the particle worldline crosses a HOS, the event on the worldline of O that O considers to be simultaneous gets further up the z axis at a very rapidly increasing rate, due to the fact that the worldline of O is curving down to stay below $z = t$

and hence,

$$\tanh g\tau = \pm 1 \ .$$

But this can only happen when $\tau = \pm\infty$. Strictly speaking, the coordinates τ, Z, x, y do not extend this far, but we can examine the limit as the particle worldline approaches these null lines. Figure 15.4 shows what is happening physically.

Next we have the claim that it takes an infinite time τ for a signal emitted from the particle worldline as it crosses the future boundary $z = t$ to catch up with the observer, and a signal must be emitted in the infinite past to meet the particle worldline as that worldline enters region I. Both things are obvious from the Minkowski diagram. For example, the first follows because our observer approaches the speed of light asymptotically and never actually crosses this null line $z = t$ (see Fig. 15.5 for a physical explanation). One can show that there is an infinite redshift for signals emitted from these null lines bounding region I (see Sect. 15.6).

Now Boulware plunges us straight into the paradox mentioned earlier when he describes the physical causes for all this. He says that the observer attributes the strange behaviour of light to the extremely strong gravitational field which produces a future event horizon at $Z = 0$, or $z = t$. But surely this is not an extremely strong gravitational field. It is a static and homogeneous gravitational field, as discussed above. But then how does the observer explain this event horizon and this infinite

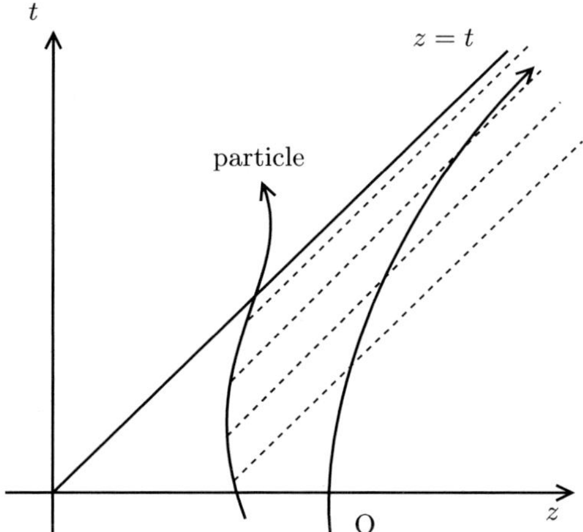

Fig. 15.5 View of the worldline of an eternally uniformly accelerating observer O relative to Minkowski coordinates (z,t) in a flat spacetime. The observer worldline approaches the null line $z = t$ asymptotically. A particle is moving around to the left of O. *Dotted lines* are signals sent from the particle toward the observer. It is obvious that they take longer and longer to catch up with O as the emission event approaches $z = t$

redshift if she considers herself to be supported in some SHGF? Viewed from the global inertial frame, the event horizon of the uniformly accelerating observer is due to the fact that her speed approaches the speed of light asymptotically as $\tau \rightarrow \pm\infty$. The redshift is just a Doppler shift, when viewed in the inertial frame, getting bigger without limit as the observer speed approaches the speed of light.

Indeed, these things seem obvious when we adopt the Minkowski frame and they are discussed in mathematical detail below. But we do have to ask what we would say if we were simply given the metric with interval

$$ds^2 = -(1+g\zeta)^2 d\tau^2 + d\zeta^2 + dx^2 + dy^2 , \tag{15.17}$$

and told to work out what it described. This is precisely the discussion of Chap. 2. As noted earlier, it turns out that g is the proper acceleration of a body sitting at $\zeta = 0$. It also turns out that any other body sitting at some fixed value of ζ (and fixed values of x and y) also has a uniform acceleration, although different for each value of ζ. So what is to stop us choosing one of the other uniformly accelerating points in the τ, ζ, x, y coordinates with fixed ζ and saying that it is this point which is in an SHGF of the appropriate value? The only answer we can give here is to say that the interval (15.17) picks out $\zeta = 0$ as the only such point for which τ is precisely the proper time of that point. This is obvious from the interval itself, when $d\zeta = 0 = dx = dy$, since we then have proper time $ds = (1+g\zeta)d\tau$, and this is only equal to $d\tau$ when $\zeta = 0$.

We note, however, that if we were not bothered about the proper time of our observer coinciding with the time coordinate, we could choose an observer with any fixed value of the spatial coordinate Z, no matter how small, and say that her uniform acceleration was the acceleration required to prevent her from falling freely relative to the gravitational effects modelled in this spacetime. This could give arbitrarily large 'gravitational accelerations', as we know. The real problem then is to establish that not all these different 'gravitational accelerations' can coexist. In other words, we are not in a situation with a varying gravitational field, which increases without limit as $Z \downarrow 0$. Our transition from a Minkowski spacetime via semi-Euclidean coordinate considerations would tend to confirm this, but how can we assert it purely on the basis of the interval (15.17) and general relativity?

A related and interesting question would be this: how would we model in general relativity a gravitational effect which was uniform in two spatial dimensions x and y but fell off as we moved away from $Z = 0$? We would presumably find a nonzero curvature for this metric, because there ought to be tidal effects. Just the simple fact that the spacetime is flat, hence involves no tidal effects, ought to tell us that the gravitational field in question is really homogeneous in all three spatial dimensions (in the Minkowski frame). Boulware seems to have missed this point. He imposed flatness of spacetime without realising that it was precisely this that made the gravitational field homogeneous. Indeed, he never says anything about why he imposed flatness.

But is there some sense in which one can say that it is the observer sitting at $\zeta = 0$, i.e., the one whose proper time is the time coordinate, whose uniform acceleration is equal to the 'gravitational acceleration' (by which we mean the acceleration required to prevent the observer from falling freely relative to the coordinates)? We can turn this into a related problem as follows. We have noted that the point $\zeta = 0$ has a specific property in the spacetime with interval (15.17), namely that, if a point follows the worldline $(\tau, 0, 0, 0)$, then the coordinate τ is actually its proper time. On the other hand, if a point follows the worldline $(\tau, \zeta_1, 0, 0)$, where $\zeta_1 \neq 0$, its proper time is not τ but $(1 + g\zeta_1)\tau$, which follows straight from the expression for the interval. Now since the proper time for ζ_1 is just a constant multiple of τ, we can change the time coordinate to $\tau_1 := (1 + g\zeta_1)\tau$, leaving the other coordinates as they were, and get the new metric defined by the interval

$$\mathrm{d}s^2 = -\left(\frac{1 + g\zeta}{1 + g\zeta_1}\right)^2 \mathrm{d}\tau_1^2 + \mathrm{d}\zeta^2 + \mathrm{d}x^2 + \mathrm{d}y^2 \,. \tag{15.18}$$

This can be rewritten by simple algebraic manipulation as

$$\mathrm{d}s^2 = -\left[1 + \frac{g(\zeta - \zeta_1)}{1 + g\zeta_1}\right]^2 \mathrm{d}\tau_1^2 + \mathrm{d}\zeta^2 + \mathrm{d}x^2 + \mathrm{d}y^2 \,. \tag{15.19}$$

We are now in trouble because we would appear to have the gravitational potential

$$\phi_1 := \frac{g(\zeta - \zeta_1)}{1 + g\zeta_1} \,, \tag{15.20}$$

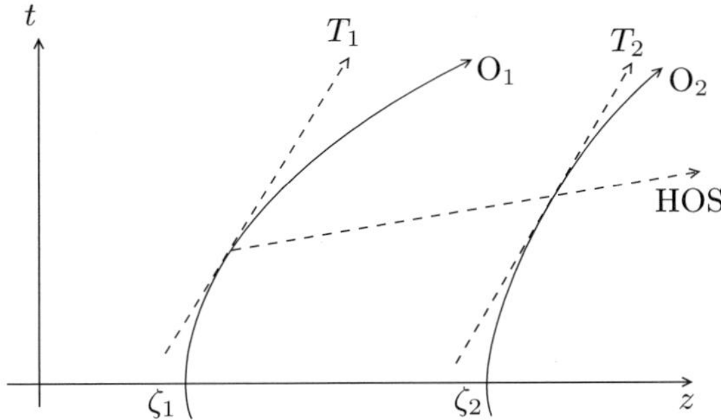

Fig. 15.6 Two observers O_1 and O_2 at fixed values ζ_1 and ζ_2 in a semi-Euclidean coordinate system, as viewed from a Minkowski frame. It turns out that both O_1 and O_2 are uniformly accelerating. The *dashed line* marked HOS is the hyperplane of simultaneity of O_1 at some event on its worldline, defined as the HOS of the ICIO at that event. It turns out that it is also a HOS for O_2 at the event where it intersects the worldline of O_2. The *dashed lines* marked T_1 and T_2 are time axes of the ICIOs for O_1 and O_2 at the two events. What we are saying is that they turn out to be parallel in Minkowski spacetime. These things are elucidated in Sect. 15.5

and the arguments outlined in Chap. 2 [see (2.52) on p. 18] would suggest taking the gradient of this to find the 'acceleration due to gravity', which delivers the value $g/(1+g\zeta_1)$. The worst of it is that this is precisely the uniform acceleration of the point moving with worldline $(\tau,\zeta_1,0,0)$.

Of course, this just confirms the discussion in Chap. 2, where it was concluded that GR is powerless to decide how much of an acceleration is due to a putative non-tidal effect and how much due to a choice of coordinates when the spacetime happens to be Minkowskian. Needless to say, one should not conclude that one can get a different physical situation just by changing coordinates. That is not at all the idea of covariance! There has to be some intrinsic way of deducing the physics, which does not depend on the coordinates.

It is useful at this point to examine the uniform acceleration case when we consider semi-Euclidean coordinates. If we consider any HOS of an observer following the worldline $(\tau,\zeta_1,0,0)$, we shall show that it is also a HOS of the observer following the worldline $(\tau,\zeta_2,0,0)$, for any different ζ_2 (see Fig. 15.6). Now if we were to set up the semi-Euclidean coordinates of either of these two observers, it turns out that the only difference would be that the two observers attribute different proper times to these spacelike hypersurfaces. So while observer 1 attributes proper time $(1+g\zeta_1)\tau$ to the HOS through her worldline at τ, observer 2 attributes proper time $(1+g\zeta_2)\tau$, but otherwise the coordinates agree on each HOS up to a change of origin. We shall now investigate all this algebraically.

15.4 Coordinate Singularity in the SE Metric

Let us begin by looking at the singularity in the semi-Euclidean metric, bearing in mind that it is not a true singularity, since it can be removed by changing coordinates (like the singularity at the black hole event horizon in the Schwarzschild metric when one uses the standard Schwarzschild coordinates). Changing notation and conventions from the Boulware paper, the transformation from the semi-Euclidean coordinates y^0 and y^1 of a uniformly accelerating observer located at $y^1 = 0$ is [see (2.14) and (2.15) on p. 9]

$$t = \frac{c}{g} \sinh \frac{gy^0}{c} + \frac{y^1}{c} \sinh \frac{gy^0}{c} , \tag{15.21}$$

$$x = \frac{c^2}{g} \left(\cosh \frac{gy^0}{c} - 1 \right) + y^1 \cosh \frac{gy^0}{c} , \tag{15.22}$$

ignoring the irrelevant coordinates. The coordinate y^0 here is just the proper time of the observer fixed at $y^1 = 0$, as can be seen directly from the metric, and the planes $y^0 = $ constant are the hyperplanes of simultaneity of the observer at $y^1 = 0$. So if we fix $y^0 = \kappa$ for some constant κ in the above transformation equations, allowing y^1 to vary, we pick out the relevant HOS in Minkowski spacetime. As we have suppressed two spatial dimensions, we shall consider the straight line in the (t,x) plane given parametrically by

$$x = \frac{c^2}{g} \left(\cosh \frac{g\kappa}{c} - 1 \right) + y^1 \cosh \frac{g\kappa}{c} , \tag{15.23}$$

$$t = \frac{c}{g} \sinh \frac{g\kappa}{c} + \frac{y^1}{c} \sinh \frac{g\kappa}{c} , \tag{15.24}$$

with y^1 as variable. Eliminating y^1, we obtain the HOS in Minkowski coordinates as

$$x - \frac{c^2}{g} \left(\cosh \frac{g\kappa}{c} - 1 \right) = c \coth \frac{g\kappa}{c} \left(t - \frac{c}{g} \sinh \frac{g\kappa}{c} \right) . \tag{15.25}$$

We do not have to do complicated algebra to find out that all these hyperplanes of simultaneity intersect on the $t = 0$ axis of Minkowski spacetime, since $t = 0$ in (15.25) implies that $x = -c^2/g$, regardless of the value of κ (see Fig. 15.7). In other words, if we looked at the HOS for $y^0 = \kappa_1$, we would get the Minkowski formula

$$x - \frac{c^2}{g} \left(\cosh \frac{g\kappa_1}{c} - 1 \right) = c \coth \frac{g\kappa_1}{c} \left(t - \frac{c}{g} \sinh \frac{g\kappa_1}{c} \right) ,$$

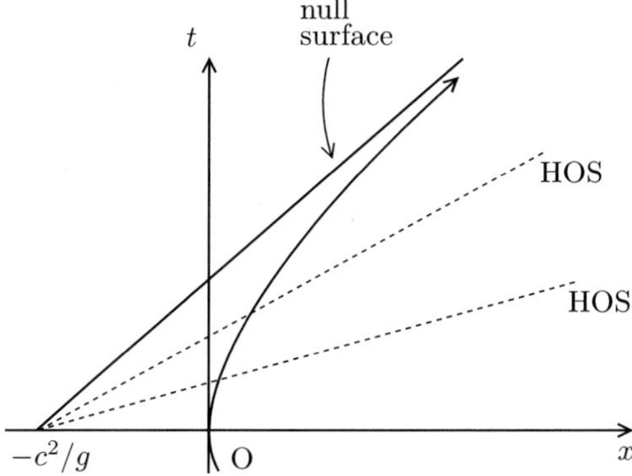

Fig. 15.7 The eternally uniformly accelerating observer O passes through the origin of the Minkowski frame. All hyperplanes of simultaneity of O intersect at $x = -c^2/g$ on the space axis. This is just Fig. 15.4 but with a shift in the origin of the space axis

and this too crosses the $t = 0$ axis at $x = -c^2/g$. This is precisely the point where the semi-Euclidean coordinate system breaks down, and it is also the singularity in the semi-Euclidean metric.

This is of course a potential problem for any semi-Euclidean coordinate system adapted to an accelerating worldline, since there is always a possibility that hyperplanes of simultaneity for different events on the worldline will intersect somewhere in spacetime, thereby making it impossible to extend this kind of adapted coordinate system beyond the intersection. As a relativist, the semi-Euclidean observer could of course use other coordinates and remove the singularity. However, as a real observer, either uniformly accelerating in a spacetime with no gravitational effects, not even non-tidal ones, or supported in a non-tidal gravitational field (an SHGF), she could never (to put it rather loosely) see beyond this set of events, i.e., receive messages from events prior to them. So there is a real physical significance to the singularity as far as this observer is concerned.

So what about the other uniformly accelerating observers that sit at fixed SE spatial coordinate positions in the frame of the first observer? What we need to do now is to consider the uniformly accelerating observer at another value of y^1 and show that her hyperplanes of simultaneity also intersect at the point $(0, -c^2/g)$ in the inertial coordinates. Since they are then merely the hyperplanes of simultaneity of the observer at $y^1 = 0$ (but she will attribute different times to the events in them, her own proper times), we shall find that the interval

$$\mathrm{d}s^2 = -\left[1 + \frac{g(\zeta - \zeta_1)}{1 + g\zeta_1}\right]^2 \mathrm{d}\tau_1^2 + \mathrm{d}\zeta^2 + \mathrm{d}x^2 + \mathrm{d}y^2 \tag{15.26}$$

(reverting briefly to Boulware's notation) is the interval for the metric of the semi-Euclidean observer at ζ_1.

We shall reach a conclusion that is perhaps obvious, and threatens any claim that the acceleration g is somehow intrinsic to the model we have made for this spacetime. For we may say, where is the surprise? If we begin with Minkowski spacetime, we can choose any uniformly accelerating observer and change to the semi-Euclidean coordinates of that observer. All these spacetimes are the same spacetime, with different coordinates, and a metric that merely looks different in each coordinate system. As noted in Chap. 2, general relativity is powerless to distinguish these physical situations.

15.5 Some Semi-Euclidean Geometry

We start by considering an eternally uniformly accelerating observer who remains at the origin of the semi-Euclidean system. Now we are interested in the worldline followed in Minkowski spacetime by some point at a fixed $y^1 = \kappa$ as y^0 varies, dropping the irrelevant coordinates y^2 and y^3 by setting them to zero.

As mentioned above, the transformation from semi-Euclidean to Minkowski coordinates is

$$x = \frac{c^2}{g}\left(\cosh\frac{gy^0}{c} - 1\right) + y^1 \cosh\frac{gy^0}{c}, \tag{15.27}$$

$$t = \frac{c}{g}\sinh\frac{gy^0}{c} + \frac{y^1}{c}\sinh\frac{gy^0}{c}. \tag{15.28}$$

Hence, the path of the point at a fixed $y^1 = \kappa$ is

$$\begin{cases} x(y^0) = \dfrac{c^2}{g}\left(\cosh\dfrac{gy^0}{c} - 1\right) + \kappa\cosh\dfrac{gy^0}{c}, \\[2ex] t(y^0) = \dfrac{c}{g}\sinh\dfrac{gy^0}{c} + \dfrac{\kappa}{c}\sinh\dfrac{gy^0}{c}, \end{cases} \tag{15.29}$$

parametrised by y^0. It is possible to solve the second of these for y^0 as a function of t and the constant κ, then express x as a function of t, and this gives

$$\frac{gy^0}{c} = \sinh^{-1}\frac{t}{c/g + \kappa/c},$$

whence

$$\cosh\frac{gy^0}{c} = \sqrt{1 + \frac{t^2}{(c/g + \kappa/c)^2}},$$

and

$$x(t) = c\sqrt{\left(\frac{c}{g} + \frac{\kappa}{c}\right)^2 + t^2} - \frac{c^2}{g}$$

$$= \frac{c^2}{g}\left[\sqrt{\left(1 + \frac{g\kappa}{c^2}\right)^2 + \frac{g^2 t^2}{c^2}} - 1\right]. \tag{15.30}$$

With $\kappa = 0$, one then has

$$x(t) = \frac{c^2}{g}\left(\sqrt{1 + \frac{g^2 t^2}{c^2}} - 1\right), \tag{15.31}$$

This is of course the equation for the path of the hyperbolically accelerating observer at the origin.

We can then work out the first and second derivatives of $x(t)$ with respect to t:

$$\frac{dx}{dt} = \frac{gt}{\sqrt{\left(1 + \frac{g\kappa}{c^2}\right)^2 + \frac{g^2 t^2}{c^2}}}, \tag{15.32}$$

$$\frac{d^2 x}{dt^2} = \frac{g\left(1 + \frac{g\kappa}{c^2}\right)^2}{\left[\left(1 + \frac{g\kappa}{c^2}\right)^2 + \frac{g^2 t^2}{c^2}\right]^{3/2}}. \tag{15.33}$$

We do not need these to work out the proper acceleration of the curve. However, we do need the proper time along the curve as a function of some parameter. A simple way is to take the parameter to be y^0, whereupon we find

$$\tau(y^0) = \frac{1}{c}\int_0^{y^0}\left[c^2\left(\frac{dt}{dy^0}\right)^2 - \left(\frac{dx}{dy^0}\right)^2\right]^{1/2}dy^0 = \left(1 + \frac{g\kappa}{c^2}\right)y^0, \tag{15.34}$$

using the expressions (15.29) directly to obtain the derivatives. Note the very simple result when $\kappa = 0$, viz., $\tau(y^0) = y^0$, which says that the proper time along the worldline of the observer at the origin is just the semi-Euclidean coordinate time.

As an aside, we have the 4-acceleration:

$$a^i = \left(c\frac{\mathrm{d}^2 t}{\mathrm{d}\tau^2}, \frac{\mathrm{d}^2 x}{\mathrm{d}\tau^2}, 0, 0 \right)$$

$$= \frac{1}{\left(1 + \frac{g\kappa}{c^2}\right)^2} \left(c\frac{\mathrm{d}^2 t}{\mathrm{d}t'^2}, \frac{\mathrm{d}^2 x}{\mathrm{d}t'^2}, 0, 0 \right)$$

$$= \frac{g}{\left(1 + \frac{g\kappa}{c^2}\right)} \left(\sinh\frac{gy^0}{c}, \cosh\frac{gy^0}{c}, 0, 0 \right), \tag{15.35}$$

which has pseudolength

$$a^2 = -\frac{g^2}{\left(1 + \frac{g\kappa}{c^2}\right)^2}, \tag{15.36}$$

a constant for given κ, but a function of κ going basically as an inverse square.

The following is a deliberately unsophisticated analysis in the hope that it will reveal the geometrical structure of the Minkowski spacetime when viewed by accelerating observers using semi-Euclidean adapted coordinate systems. We consider (15.30) as the Minkowski path of an observer fixed at $y^1 = \kappa > 0$ in the semi-Euclidean coordinates of the first observer. We call these observers 1 and 2. Since observer 2 has uniform acceleration, she could also set up semi-Euclidean coordinates of the kind set up by observer 1. With this in mind, let us show that the hyperplanes of simultaneity of observer 2 coincide with those of observer 1 (see Fig. 15.4 on p. 168). Note that, at Minkowski time $t = 0$, observer 2 is located at Minkowski coordinate $x = \kappa$. Since $\kappa > 0$, observer 2 has lower acceleration than observer 1. So observer 2 reaches a given Minkowski coordinate speed later with respect to Minkowski time than observer 1.

We saw above that the HOS of observer 1 which intersects the worldline of observer 1 at Minkowski time t_1 also passes through the Minkowski point $(0, -c^2/g)$ (see Fig. 15.7). Now the relevant Minkowski point on the worldline of observer 1 is clearly

$$\left[t_1, \frac{c^2}{g}\left(\sqrt{1 + \frac{g^2 t_1^2}{c^2}} - 1 \right) \right]. \tag{15.37}$$

The formula for the straight line in Minkowski spacetime representing this HOS of observer 1 is therefore

$$x(t) = \frac{c^2}{g}\left(\frac{t}{t_1}\sqrt{1 + \frac{g^2 t_1^2}{c^2}} - 1 \right). \tag{15.38}$$

It is easy to check that

$$x(0) = -\frac{c^2}{g}, \qquad x(t_1) = \frac{c^2}{g}\left(\sqrt{1 + \frac{g^2 t_1^2}{c^2}} - 1\right),$$

as required for the two known points on the HOS.

This straight line has a positive gradient and so will intersect the worldline of observer 2 at a higher value of the Minkowski time than t_1, which we shall call t_2. Where does this HOS of observer 1 intersect the worldline (15.30) of observer 2? We find t_2 from the relation

$$\frac{c^2}{g}\left[\sqrt{\left(1 + \frac{g\kappa}{c^2}\right)^2 + \frac{g^2 t_2^2}{c^2}} - 1\right] = \frac{c^2}{g}\left(\frac{t_2}{t_1}\sqrt{1 + \frac{g^2 t_1^2}{c^2}} - 1\right). \tag{15.39}$$

This in turn implies that

$$\sqrt{\left(1 + \frac{g\kappa}{c^2}\right)^2 + \frac{g^2 t_2^2}{c^2}} = \frac{t_2}{t_1}\sqrt{1 + \frac{g^2 t_1^2}{c^2}}, \tag{15.40}$$

and a little manipulation leads to the result

$$t_2 = t_1\left(1 + \frac{g\kappa}{c^2}\right). \tag{15.41}$$

So we are hoping that the straight line (15.38) is a HOS for observer 2 at Minkowski time t_2. Now this is true iff observer 2 has the same Minkowski coordinate speed at this time t_2 as observer 1 had at time t_1. We can see that such a thing is plausible because, although t_2 is later than t_1 for the Minkowski frame, observer 2 was accelerating more slowly.

In any case, we know the exact Minkowski coordinate speeds of observers 1 and 2 at Minkowski times t_1 and t_2, respectively, by applying (15.32). With the obvious notation,

$$\left.\frac{dx}{dt}\right|_1 = \frac{g t_1}{\sqrt{1 + \frac{g^2 t_1^2}{c^2}}}, \tag{15.42}$$

and

$$\left.\frac{dx}{dt}\right|_2 = \frac{gt_2}{\sqrt{\left(1+\frac{g\kappa}{c^2}\right)^2 + \frac{g^2 t_2^2}{c^2}}}$$

$$= \frac{gt_1\left(1+g\kappa/c^2\right)}{\sqrt{\left(1+\frac{g\kappa}{c^2}\right)^2 + \frac{g^2 t_1^2}{c^2}\left(1+\frac{g\kappa}{c^2}\right)^2}}$$

$$= \frac{gt_1}{\sqrt{1+\frac{g^2 t_1^2}{c^2}}} = \left.\frac{dx}{dt}\right|_1 ,$$

as required.

Therefore the HOS of observer 1 at t_1 is the HOS of observer 2 at t_2. Now observer 1 attributes the same value of y^0 to all events in this HOS, e.g., to the event where the worldline of observer 2 intersects this HOS. Indeed, both observers consider these two events, the intersection of this shared HOS with their two worldlines, to be simultaneous. This is perhaps the most remarkable thing about the whole affair: any two uniformly accelerated observers positioned in this way along the axis of acceleration always have the same notion of simultaneity. The proper time that observer 2 attributes to this HOS is $(1+g\kappa/c^2)y^0$, by (15.34). Using the transformation (15.28) from semi-Euclidean to Minkowski coordinates on p. 174,

$$t_1 = \frac{c}{g}\sinh\frac{gy_1^0}{c} , \tag{15.43}$$

so the proper time that observer 1 attributes to this HOS is

$$y_1^0 = \frac{c}{g}\sinh^{-1}\frac{gt_1}{c} . \tag{15.44}$$

Hence the proper time that observer 2 attributes to this HOS is

$$y_2^0 = \frac{c}{g}\left(1+\frac{g\kappa}{c^2}\right)\sinh^{-1}\frac{gt_1}{c} = \frac{c}{g}\left(1+\frac{g\kappa}{c^2}\right)\sinh^{-1}\frac{gt_2}{c\left(1+\frac{g\kappa}{c^2}\right)} . \tag{15.45}$$

If we now make the replacement

$$g_2 := \frac{g}{\left(1+\frac{g\kappa}{c^2}\right)} , \tag{15.46}$$

so that g_2 is just the uniform acceleration of observer 2, the relation (15.45) becomes

$$y_2^0 = \frac{c}{g_2}\sinh^{-1}\frac{g_2 t_2}{c} , \tag{15.47}$$

exactly the same relation as (15.43) for observer 1, but with the new value of g.

Now what semi-Euclidean coordinates would observer 2 set up? She uses the same hyperplanes of simultaneity, attributing her proper time, which differs by the factor $1 + g\kappa/c^2$, i.e., $\tau_2 = (1 + g\kappa/c^2)\tau_1$. What about the spatial coordinates in the HOS? The observers have the same speed for the shared HOS and so they agree on lengths. The only disagreement is on origin, since each takes her own position as origin in the HOS. However, observer 2 has a very simple motion when viewed by observer 1 in her semi-Euclidean system, because her position is quite simply fixed at κ. All we need to do to transform from the semi-Euclidean system of observer 1 to the semi-Euclidean system of observer 2 is to make the time dilation by the factor $1 + g\kappa/c^2$, as in (15.26) on p. 173, which leads to the interval

$$\mathrm{d}s^2 = -\left[1 + \frac{g(\zeta - \zeta_1)}{1 + g\zeta_1}\right]^2 \mathrm{d}\tau_1^2 + \mathrm{d}\zeta^2 + \mathrm{d}x^2 + \mathrm{d}y^2 \tag{15.48}$$

in Boulware's notation, then shift the spatial origin by

$$y^1 \longrightarrow y^1 - \kappa , \tag{15.49}$$

or

$$\zeta \longrightarrow \zeta' := \zeta - \zeta_1 , \tag{15.50}$$

in Boulware's notation, which leads to the interval

$$\mathrm{d}s^2 = -\left[1 + \frac{g\zeta'}{1 + g\zeta_1}\right]^2 \mathrm{d}\tau_1^2 + \mathrm{d}\zeta'^2 + \mathrm{d}x^2 + \mathrm{d}y^2 . \tag{15.51}$$

Since $g' := g/(1 + g\zeta_1)$ is the uniform acceleration of the new observer (observer 2 in the above analysis), the interval becomes

$$\mathrm{d}s^2 = -(1 + g'\zeta')^2 \mathrm{d}\tau_1^2 + \mathrm{d}\zeta'^2 + \mathrm{d}x^2 + \mathrm{d}y^2 . \tag{15.52}$$

We were therefore right to surmise that all these observers are on a par.

15.6 Redshift in a Uniformly Accelerating SE Frame

We consider the redshift of light emitted along the y^1 axis (the axis of acceleration) in the semi-Euclidean frame with metric given by the line element

$$\mathrm{d}s^2 = \left(1 + \frac{gy^1}{c^2}\right)^2 (\mathrm{d}y^0)^2 - (\mathrm{d}y^1)^2 - (\mathrm{d}y^2)^2 - (\mathrm{d}y^3)^2 . \tag{15.53}$$

Consider a null geodesic in the y^1 direction. A first integral of the geodesic equation is given by $\mathrm{d}s = 0$, i.e.,

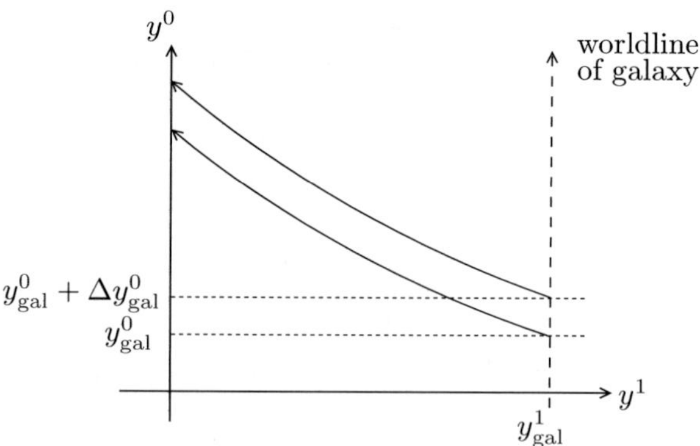

Fig. 15.8 Analysing redshift in the semi-Euclidean frame. A galaxy G on the y^1 axis at $y^1_{\text{gal}} > 0$ emits a signal towards the origin at semi-Euclidean time y^0_{gal}, and another shortly afterwards at time $y^0_{\text{gal}} + \Delta y^0_{\text{gal}}$. In fact, the null geodesics (*continuous curves*) followed by these signals on their way to the origin of the SE frame are identical, except that one of them is shifted uniformly up the y^0 axis, so that the SE time interval between emissions is exactly equal to the SE time interval between receptions [see (15.58) on p. 181]

$$\left(1 + \frac{gy^1}{c^2}\right) dy^0 = -dy^1 \, , \tag{15.54}$$

with the minus sign for a null geodesic coming in towards the origin. Let us adopt the kind of approach to redshift that one might in cosmology. Imagine a galaxy G on the y^1 axis at $y^1_{\text{gal}} > 0$, emitting a signal towards the origin at time y^0_{gal} (see Fig. 15.8). The signal is observed at the origin at time $y^0_{\text{obs}} > y^0_{\text{gal}}$.

The problem is to determine the time at which the signal is observed at the origin. Then consider another sent shortly afterwards and compare the proper time interval between the two emissions as measured in G with the proper time interval between the two receptions as measured by the observer at the origin. The reciprocal of the first proper time interval will be assimilated with the frequency of a light signal emitted from G, whilst the reciprocal of the second proper time interval will be assimilated with the frequency of the received light signal.

Now from (15.54), we have

$$\int_{y^0_{\text{gal}}}^{y^0_{\text{obs}}} dy^0 = -\int_{y^1_{\text{gal}}}^{y^1_{\text{obs}}} \frac{dy^1}{1 + gy^1/c^2} \, , \tag{15.55}$$

whence

$$y^0_{\text{obs}} - y^0_{\text{gal}} = -\left[\frac{c^2}{g}\ln\left(1 + \frac{gy^1}{c^2}\right)\right]^{y^1_{\text{obs}}}_{y^1_{\text{gal}}}$$

$$= \frac{c^2}{g}\ln\left(1 + \frac{gy^1_{\text{gal}}}{c^2}\right), \tag{15.56}$$

taking $y^1_{\text{obs}} = 0$, i.e., observer at the origin. So what happens to successive wave crests emitted by G at times y^0_{gal} and $y^0_{\text{gal}} + \Delta y^0_{\text{gal}}$? If G stays at y^1_{gal}, implying that it has a uniformly accelerated motion relative to a global Minkowski frame, as we saw in the last section, then we have the relation

$$y^0_{\text{obs}} + \Delta y^0_{\text{obs}} - (y^0_{\text{gal}} + \Delta y^0_{\text{gal}}) = \frac{c^2}{g}\ln\left(1 + \frac{gy^1_{\text{gal}}}{c^2}\right). \tag{15.57}$$

We thus find the simple relation

$$\Delta y^0_{\text{obs}} = \Delta y^0_{\text{gal}}, \tag{15.58}$$

which says that the coordinate time interval measured at reception is the same as the coordinate time interval measured at emission.

We must convert these to proper time intervals. Now the proper time interval between emissions in G is

$$\left(1 + \frac{gy^1_{\text{gal}}}{c^2}\right)\Delta y^0_{\text{gal}}, \tag{15.59}$$

whilst the proper time interval between receptions at the origin is just $\Delta y^0_{\text{obs}} = \Delta y^0_{\text{gal}}$. The reciprocals of these are the frequencies of emission and reception, respectively, and the redshift z is given by

$$1 + z = \frac{\text{frequency of emission}}{\text{frequency of reception}} = \left(1 + \frac{gy^1_{\text{gal}}}{c^2}\right)^{-1}. \tag{15.60}$$

Consequently, we have a blueshift ($z < 0$) for $y^1_{\text{gal}} > 0$ and a redshift ($z > 0$) for $y^1_{\text{gal}} < 0$.

Now all of this can be viewed by an observer using Minkowski coordinates, and we should have some explanation for this redshift effect in the ordinary framework of special relativity. Let us consider the claim that both the observer and the galaxy are uniformly accelerating. We set up semi-Euclidean coordinates for a uniform acceleration, as in Chap. 2, and start by considering an observer who remains at the origin and also the worldline followed in Minkowski spacetime by some point at a fixed y^1 as y^0 varies, dropping the irrelevant coordinates y^2 and y^3 by setting them to zero. We saw in the last section that such a worldline does indeed have a uniform acceleration in the sense that the pseudolength of the acceleration is constant. The worldline is given in Minkowski spacetime by:

$$x(t) = \frac{c^2}{g}\left[\sqrt{\left(1 + \frac{gy^1}{c^2}\right)^2 + \frac{g^2 t^2}{c^2}} - 1\right] , \tag{15.61}$$

$$\frac{dx}{dt} = \frac{gt}{\sqrt{\left(1 + \frac{gy^1}{c^2}\right)^2 + \frac{g^2 t^2}{c^2}}} , \tag{15.62}$$

$$\frac{d^2 x}{dt^2} = \frac{\left(1 + \frac{gy^1}{c^2}\right)^2}{\left[\left(1 + \frac{gy^1}{c^2}\right)^2 + \frac{g^2 t^2}{c^2}\right]^{3/2}} . \tag{15.63}$$

The proper time along the curve is

$$\tau(y^0) = \left(1 + \frac{gy^1}{c^2}\right) y^0 , \tag{15.64}$$

using y^0 to parametrise the worldline.

From (15.62), the functions $v_{\mathrm{obs}}(t)$ and $v_{\mathrm{gal}}(t)$ giving the coordinate speeds of the observer and the galaxy at Minkowski time t are

$$v_{\mathrm{obs}}(t) = \frac{gt}{\left(1 + \frac{g^2 t^2}{c^2}\right)^{1/2}} \tag{15.65}$$

and

$$v_{\mathrm{gal}}(t) = \frac{gt}{\left[\left(1 + \frac{gy^1_{\mathrm{gal}}}{c^2}\right)^2 + \frac{g^2 t^2}{c^2}\right]^{1/2}} . \tag{15.66}$$

Suppose G sends a signal from y^1_{gal} at t. By (15.61), the Minkowski spatial coordinate of the emission point is

$$x_{\mathrm{gal}}(t) = \frac{c^2}{g}\left[\sqrt{\left(1 + \frac{gy^1_{\mathrm{gal}}}{c^2}\right)^2 + \frac{g^2 t^2}{c^2}} - 1\right] . \tag{15.67}$$

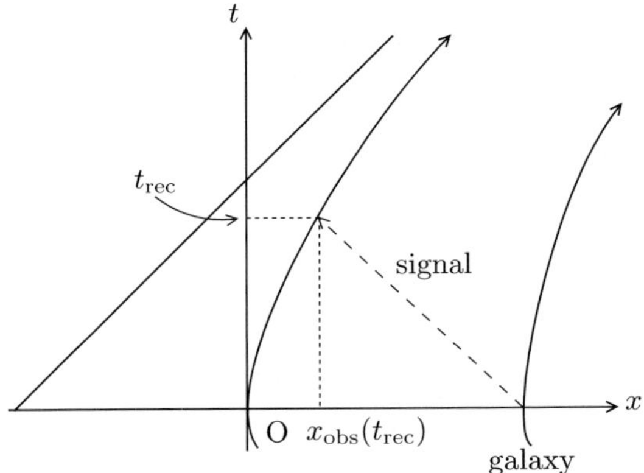

Fig. 15.9 Analysing redshift in the Minkowski frame. A galaxy at fixed SE coordinate y^1_{gal} sends a signal toward the SE observer O at Minkowski time $t = 0$ and it is received at a time t_{rec}. In this picture, we calculate the speed $v_{\mathrm{obs}}(t_{\mathrm{rec}})$ of the observer when she receives the signal, in order to calculate an ordinary Doppler shift

For a signal at $t = 0$, we have $x_{\mathrm{gal}}(0) = y^1_{\mathrm{gal}}$. At this time, $v_{\mathrm{gal}}(0) = 0$ and $v_{\mathrm{obs}}(0) = 0$, but the signal arrives at the observer later, at some time with Minkowski coordinate t_{rec}, when the observer has begun to move (see Fig. 15.9). Indeed, at the reception time, the observer has coordinate speed

$$v_{\mathrm{obs}}(t_{\mathrm{rec}}) = \frac{g t_{\mathrm{rec}}}{\left(1 + \frac{g^2 t_{\mathrm{rec}}^2}{c^2}\right)^{1/2}} . \tag{15.68}$$

At the reception time, the observer is at

$$x_{\mathrm{obs}}(t_{\mathrm{rec}}) = \frac{c^2}{g}\left(\sqrt{1 + \frac{g^2 t_{\mathrm{rec}}^2}{c^2}} - 1\right) . \tag{15.69}$$

The reception time is defined by the relation

$$y^1_{\mathrm{gal}} - x_{\mathrm{obs}}(t_{\mathrm{rec}}) = c t_{\mathrm{rec}} . \tag{15.70}$$

This becomes

$$y^1_{\mathrm{gal}} - \frac{c^2}{g}\left(\sqrt{1 + \frac{g^2 t_{\mathrm{rec}}^2}{c^2}} - 1\right) = c t_{\mathrm{rec}} . \tag{15.71}$$

When expanded out, it turns out that this is a linear equation in t_{rec} with solution

$$gt_{\mathrm{rec}} = \frac{\left(1 + \dfrac{g y^1_{\mathrm{gal}}}{c^2}\right)^2 - 1}{\dfrac{2}{c}\left(1 + \dfrac{g y^1_{\mathrm{gal}}}{c^2}\right)}\,. \tag{15.72}$$

We now calculate

$$v_{\mathrm{obs}}(t_{\mathrm{rec}}) = c\,\frac{\left(1 + \dfrac{g y^1_{\mathrm{gal}}}{c^2}\right)^2 - 1}{\left(1 + \dfrac{g y^1_{\mathrm{gal}}}{c^2}\right)^2 + 1}\,. \tag{15.73}$$

Therefore, although the source had no motion relative to the observer when the signal was emitted, the observer had begun to move with this speed towards the signal by the time the signal was received. What Doppler shift will this cause? The formula for an ordinary Doppler shift in Minkowski spacetime is

$$1 + z = \frac{(1 - v/c)^{1/2}}{(1 + v/c)^{1/2}}\,, \quad \text{with } v = v_{\mathrm{obs}}(t_{\mathrm{rec}})\,. \tag{15.74}$$

Carrying out this calculation, we find

$$1 + z = \left(1 + \frac{g y^1_{\mathrm{gal}}}{c^2}\right)^{-1}, \tag{15.75}$$

precisely as in (15.60).

Although the spacetime has no curvature, there is redshift. In the semi-Euclidean view, it seems rather like the kind of redshift discussed in cosmology, but in the Minkowski view, it is just an ordinary Doppler shift due to relative motion. The same situation arises in the Milne universe. In the cosmological view, where the metric attributes a hyperbolic geometry to the spatial hypersurfaces, we have the same kind of cosmological redshift as for any other hyperbolic Robertson–Walker universe, but in the Minkowski view (and the Milne universe is the only RW universe that is actually a flat spacetime), the redshift is just the ordinary Doppler effect due to relative motion of source and observer.

So we can explain infinite redshifts in two different ways:

- Viewed from a global inertial frame, they are basically due to the fact that the origin of the SE frame asymptotically approaches the speed of light. Consider a series of emitters $G_1, G_2, \ldots, G_n$, etc., at fixed values $y^1_1, y^1_2, \ldots y^1_n$, etc., of the SE coordinate, approaching the forbidden point where the SE metric is singular (see Fig. 15.10). According to (15.60) above, the redshift observed by the uniformly accelerating observer at the origin of the SE frame depends only on the fixed value y^1_n. The point is that, following the sequence of galaxies toward the

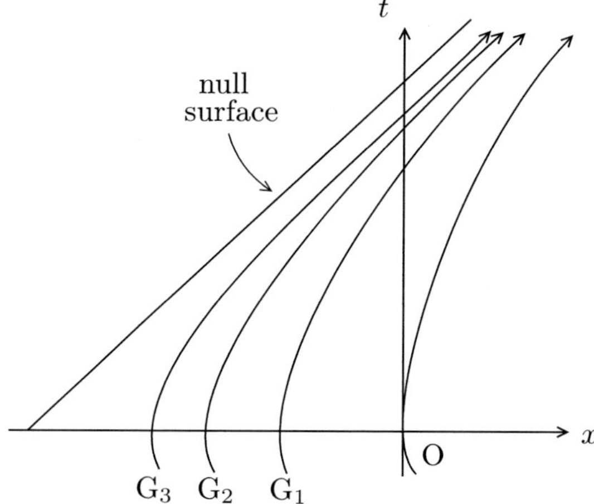

Fig. 15.10 $G_1, G_2, \ldots, G_n, \ldots$ is a series of galaxies at fixed values $y_1^1, y_2^1, \ldots y_n^1, \ldots$ of the SE coordinate, ever closer to the forbidden point where the metric is singular. Each therefore has a uniform acceleration. They emit signals to the observer O at the origin of the SE frame. It turns out that the difference of speed between G_n and O at the event when G_n emits its signal and the event when O receives it is constant for a given emitter G_n, i.e., independent of when the emission event occurs, but that this difference of speed increases without limit as n increases, i.e., as G_n approaches the singularity

singularity, the difference of speed between G_n and O at the event when G_n emits its signal and the event when O receives it is constant for a given emitter G_n, i.e., independent of when the emission event occurs, but that this difference of speed increases without limit as n increases, i.e., as G_n approaches the singularity.

- Viewed from the SE frame itself, one might well like to imagine a series of galaxies sitting fixed at values of y_{gal}^1 that approach the fateful $y^1 = -c^2/g$, with these galaxies sending light signals to the observer at the origin of the SE frame (see Fig. 15.11), but in order to sit fixed at these values they have to have a series of (uniform) 4-accelerations that tend to infinity, as can be seen from (15.36) on p. 176. And it is the fact that the factor $(1 + gy_{\text{gal}}^1/c^2)$ goes to zero that makes the proper time interval (15.59) between emissions go to zero and leads to ever larger redshifts. But something has to provide these galaxies with these accelerations. It is pure assumption to suppose that they arise merely because each galaxy is held up against a gravitational field. In fact, there is no explanation for how they manage to move like this. There is no explanation for the way any of the other points at fixed values of y^1 should be able to move like this. One solution to this paradox might just be to point out that nobody ever said that this spacetime had to be populated by objects like this.

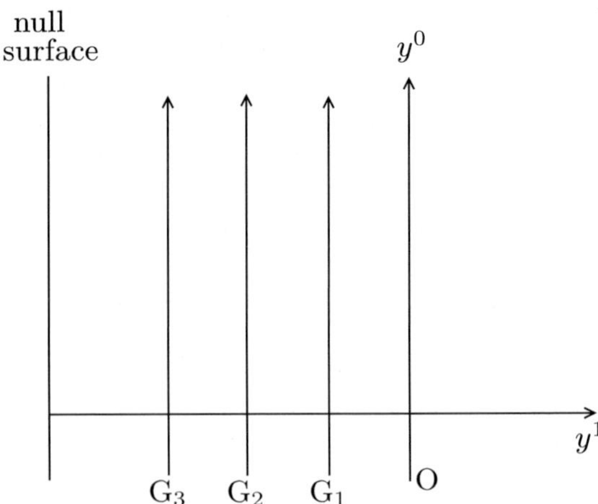

Fig. 15.11 $G_1, G_2, \ldots, G_n, \ldots$ is a series of galaxies at fixed values $y_1^1, y_2^1, \ldots y_n^1, \ldots$ of the SE coordinate, ever closer to the forbidden point where the metric is singular. Each therefore has a uniform acceleration. The null surface indicated is the event horizon for the observer O, intersecting the y^1 axis at the singularity. The galaxies emit signals to the observer O at the origin of the SE frame with a redshift that is constant for any given galaxy, but increases without limit as n increases, i.e., as G_n approaches the singularity

15.7 Interpreting Semi-Euclidean Coordinates

The last two sections are a reaction to Boulware's claim that there is a strong gravitational field causing redshift near the boundaries of his region I. Of course, the real question here is: how do we interpret a set of coordinates and a metric? If we are given a metric of semi-Euclidean form in which the origin of the coordinate system has uniform acceleration g, then we are dealing with a static and truly homogeneous gravitational field (an SHGF), but on what grounds could we conclude that the 'acceleration due to gravity' is everywhere equal to g? This is certainly one possible interpretation, but as we have seen, it is not the only one.

There remains a clear ambiguity in the application of GR due to the fact that this spacetime is flat. There can be no intrinsic way to extract the specific value of g that turns up in the coordinates we have chosen. Indeed, we know that, starting from the flat spacetime and choosing any uniformly accelerating observer, with any value of g, we can always find some coordinates for the spacetime in which g turns up in the metric. At the same time, we would like to say that the flatness of the metric rules out the possibility of a strong gravitational field in some zones and a weaker one in others, because such a field would involve tidal forces that would show up in a nonzero curvature. Whatever the value of g we eventually accept, we would like therefore to insist that it be the same everywhere in the spacetime. We might argue, referring to the discussion in Sect. 2.4, that insofar as we can apply the usual approximations that link Einstein's equation with Poisson's equation for a gravitational potential, these approximations lead to a gravitational potential of the type gz, which implies a constant 'acceleration due to gravity' of g everywhere.

In this view, it is not so worrying that we can merely change coordinates and obtain a different value of g, since we should be able to explain this as viewing the original situation from an accelerating frame. How much of the acceleration is considered to be gravitational and how much due to frame motion is an inevitable ambiguity in a theory which was partly inspired by the idea that any inertial acceleration can be confused with a gravitational acceleration. In this line of argument, it is only when we can examine the source of the gravity that we can make this breakdown. On the other hand, looking back at Sect. 2.4, we did in fact fail to identify an energy–momentum distribution that would generate these metrics via Einstein's equation (apart from the zero distribution).

So one might imagine the following line of argument that opposes the simple idea, expressed two paragraphs ago, that the value of g should be the same everywhere in spacetime. As an observer in this world, it would be natural to consider the views of observers placed at constant proper distances. Each of these would come up with a different value of g, and without other landmarks in the spacetime, such as a source of gravity, nobody would know whose value of g to accept, and this despite the fact that they would all measure zero curvature. Democratically, one would have no way of distinguishing one particular value. Although this may seem a sorry state of affairs, it does look as though one could argue this way in the very ideal case of the SHGF.

In the present view it is nevertheless a poor strategy to accept the idea that the gravitational acceleration may be stronger in some places than others, whatever that may mean. To begin with, recall from Sect. 2.4 that, in order to see a gravitational potential from the standpoint of general relativity, in which gravity is not modelled that way, an observer has to pretend that slightly non-inertial coordinates are in fact inertial, in the sense that they are things she can understand in that way, as though she were doing Newtonian gravitation. This ought to throw some doubt on the democratic process. The present view is that it is better to consider that the SHGF is a good approximation to some real spacetime with well-defined gravitational sources, relative to which distances can be assessed. One then simply remembers that the fact that a four-acceleration a^μ is required to hold an object at some fixed coordinate position does not mean a priori that a^μ is a measure of the gravitational field. The curvature gives an indication of the gravitational field when there are tidal effects, but when the gravitational field is non-tidal, we have to look at the sources.

This view perhaps looks more convincing if we reconsider the semi-Euclidean coordinates set up by an observer with uniform acceleration in a Minkowski spacetime without gravitational effects. Such an observer adopts the planes of simultaneity of instantaneously comoving inertial observers in order to attribute her own instantaneous proper time to spacelike hypersurfaces and the space coordinates of those inertial observers. This leads to a remarkable situation which is unique to the hyperbolic motion of a uniformly accelerating observer. When we consider the motion of a point with a fixed value of the spatial coordinate in the semi-Euclidean frame, it also turns out to be uniformly accelerating, and its worldline could be the worldline of another uniformly accelerating observer, but with a different value of the acceleration (and a different Minkowski spatial coordinate at Minkowski time zero). Furthermore, such an observer, located at this specific point relative to the first, would always agree with the first about the simultaneity of pairs of events. In other words, their planes of simultaneity coincide, although they attribute different times to them (their own proper times). Moreover, if the new observer set up her own semi-Euclidean coordinate system, with her own acceleration, it would be related to the first semi-Euclidean coordinate system by a simple transformation, merely scaling the time coordinate by a constant factor and shifting the spatial origin.

In this context, it would be easy to look at the worldlines of points at different values of the space coordinate ζ along the axis of acceleration in the coordinates with non-Minkowski metric and point out that each has a different uniform acceleration, claiming as a consequence that there is a non-uniform gravitational field which turns out to be infinite at certain values of ζ. The trouble with this is that the 4-accelerations of such points are irrelevant. Nobody ever said that such points represented actual motions, let alone the motions of bodies supported in a gravitational field. What the coordinates do relative to the Minkowski frame is irrelevant, because they are just coordinates, and one can do anything with coordinates. A better policy in this non-gravitational Minkowski spacetime is thus to refer to a global inertial frame to understand what is happening. Then in the SHGF case, referring in the same way to the happily available global inertial frame, one should select one uniformly accelerating observer as supported relative to whatever source one can observe, in the sense that her four-acceleration corresponds to the strength of the gravitational field.

In the fully ideal case where there is no source and the SHGF permeates the whole of spacetime, the theory of GR would appear to be mute. Such gravitational fields without tidal effects have zero Riemann tensor and that is all we can ascertain. In fact it is in this sense that the theory implements the well-advertised idea that a gravitational effect and an inertial acceleration cannot be distinguished, at least when the gravitational effect is static and spatially homogeneous. But this does not mean that different observers held fixed at different spatial coordinates should be allowed to disagree about the gravitational potential in the Newtonian approximation.

15.8 Accelerations

We return to Boulware's paper to consider what he has to say about this coordinate system in his region I (the quadrant of spacetime in which the hyperbolic motion takes place). We also return to Boulware's notation. As a consequence of having required the curvature to be zero, the metric with interval

$$\mathrm{d}s^2 = -g^2 Z^2 \mathrm{d}\tau^2 + \mathrm{d}Z^2 + \mathrm{d}x^2 + \mathrm{d}y^2 \tag{15.76}$$

is simply a coordinate transformation of the Minkowski metric. It is obviously invariant under τ translation. In terms of the Minkowski coordinates (t, z), this corresponds to Lorentz invariance of the Minkowski metric. Boulware explains this claim as follows. Since

$$z = Z\cosh g\tau \,, \quad t = Z\sinh g\tau \,, \tag{15.77}$$

we have

$$z \pm t = Z\mathrm{e}^{\pm g\tau} \,.$$

Hence, under the translation $\tau \to \tau + \alpha$,

$$z \pm t \longrightarrow (z \pm t)\mathrm{e}^{\pm g\alpha} \,.$$

This is a neat result for those who recognise this as the essence of the Lorentz transformation. One can also go straight from (15.77) and apply the addition laws for hyperbolic functions. More will be said about this later in the context of Parrott's criticism of Boulware's paper (see Sect. 16.2).

A particle at rest at $Z = Z_0$ in sector I of the (Z, τ) coordinate system would have the Minkowski trajectory

$$z = Z_0\cosh g\tau \,, \quad t = Z_0\sinh g\tau \,, \tag{15.78}$$

with 4-velocity

$$u^\nu := \frac{\mathrm{d}x^\nu}{\mathrm{d}\lambda} = gZ_0(\cosh g\tau, \sinh g\tau)\frac{\mathrm{d}\tau}{\mathrm{d}\lambda} \,, \tag{15.79}$$

where λ is the proper time of the particle, to be determined by normalisation of the 4-velocity:

$$u^2 = -1 \quad \Longrightarrow \quad 1 = \left(\frac{\mathrm{d}\tau}{\mathrm{d}\lambda}\right)^2 g^2 Z_0^2 \,,$$

whence

$$\frac{\mathrm{d}\tau}{\mathrm{d}\lambda} = \frac{1}{gZ_0} \,. \tag{15.80}$$

We now have

$$a^\mu := \frac{\mathrm{d}u^\mu}{\mathrm{d}\lambda} = \frac{1}{Z_0}(\sinh g\tau, \cosh g\tau) \,, \tag{15.81}$$

so that the square of the 4-acceleration is $1/Z_0^2$, as we well know.

At this point Boulware says something very revealing about his view of what is happening:

> The (essentially unique) coordinate frame describing a static gravitational field indeed has bodies at rest in it undergoing constant acceleration in an inertial frame. However, bodies located at different points undergo different accelerations: it is not possible to find a single static coordinate system in which bodies at rest at different points undergo the same proper acceleration because two bodies experiencing the same proper acceleration do not maintain the same proper distance.

What one must object to here is the suggestion that there are bodies, because a priori we merely have trajectories defined by coordinates, and we are not claiming that there are bodies following these motions. There is a certain risk of believing that these bodies might somehow be moving solely under the effects of being supported against free fall due to a gravitational field. In the present view, they should not in fact all be considered as supported by a force that only has to counter the gravitational effect. Of course, they are merely coordinate points.

The last claim in the quote is also false as stated, viz., it is not possible to find a single static coordinate system in which bodies at rest at different points undergo the same proper acceleration because two bodies experiencing the same proper acceleration do not maintain the same proper distance. At the end of this section, we shall show that we can find such a static coordinate system, but that the spacetime is not flat. But first, remaining in the Minkowski spacetime, let us analyse the subclaim that two bodies experiencing the same proper acceleration do not maintain the same proper distance, which is essentially the beginning of Bell's famous paper [18]. We follow Boulware's analysis.

The instantaneously comoving spatial frame for one of the uniformly accelerating observers at a fixed value of the coordinate Z, which Boulware calls the rest frame, is just $\tau = $ constant. We understand this from our analysis of the semi-Euclidean frame, at least in the case of the observer who sets up this frame, because she uses the HOS in the instantaneously comoving frame and attributes her proper time to all events in it. This then becomes the time coordinate for the semi-Euclidean system. But we have also shown that all the uniformly accelerating observers associated with the first one share the same planes of simultaneity. Boulware says that the distance to a point at Z', τ is just $|Z - Z'|$, which is a rather vague way of expressing things. In fact, from the metric with interval (15.76), any two events with the same value of τ have a spacelike separation of magnitude $|Z - Z'|$. Indeed, this is just the proper distance between two events that are simultaneous in the instantaneous rest frame of the observer at Z.

We did not need to labour this point, but it serves to show the pitfalls of an over-mathematical approach. We consider two bodies starting out at rest and initially separated by l_0 as measured in their common rest frame (see Figs. 15.12 and 15.13). In Minkowski coordinates, the first has trajectory

$$z_1 = \frac{1}{g}\cosh g\tau\,, \qquad t_1 = \frac{1}{g}\sinh g\tau\,, \tag{15.82}$$

which we can see from (15.78) gives just the value $Z = 1/g$ of the observer who would set up the coordinates (τ, Z, x, y) with interval (15.76). Indeed, we know that

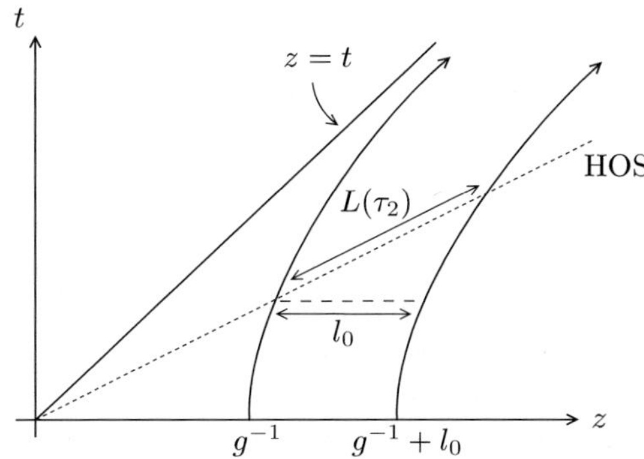

Fig. 15.12 Particle 1 sits at the fixed value $1/g$ of the SE spatial coordinate, so follows the usual hyperbolic worldline in Minkowski spacetime. Particle 2 has exactly the same Minkowski coordinate acceleration at any given Minkowski time, so moves along at a fixed Minkowski distance l_0 ahead of particle 1, as judged in the Minkowski hyperplanes of simultaneity. As judged in the SE hyperplanes of simultaneity of particle 1, the second particle gets further and further ahead without limit

$$Z_1 = (z_1^2 - t_1^2)^{1/2} = g^{-1} \, ,$$

as Boulware states. This is independent of the proper time of the particle, one of the reasons for setting up such semi-Euclidean coordinates.

The second body has trajectory

$$z_2 = \frac{1}{g} \cosh g\tau_2 \pm l_0 \, , \qquad t_2 = \frac{1}{g} \sinh g\tau_2 \, , \tag{15.83}$$

with proper time τ_2. Now this has semi-Euclidean spatial coordinate (in the system of an observer moving with the first body)

$$Z_2 = (z_2^2 - t_2^2)^{1/2} = (g^{-2} + l_0^2 \pm 2g^{-1} l_0 \cosh g\tau_2)^{-1/2} \, . \tag{15.84}$$

This depends on the proper time of the particle, precisely because it is not a fixed spatial coordinate in the semi-Euclidean system. The proper distance $L(\tau_2)$ between the particles as measured by an observer co-accelerating with particle 1 is $|Z_2 - Z_1|$, so that

$$L(\tau_2) = \pm \big[(g^{-2} + l_0^2 \pm 2g^{-1} l_0 \cosh g\tau_2)^{1/2} - g^{-1} \big] \, . \tag{15.85}$$

As mentioned on several occasions, this assumes that an observer co-accelerating with particle 1 would use instantaneously comoving inertial frames to gauge distances. We have indicated by the argument of L that it depends on the proper time of the second particle on its worldline.

If particle 1 chases particle 2, the case $(+)$ illustrated in Fig. 15.12, it never gains in the view of the observer co-accelerating with particle 1. In fact the lead always increases. We have to show that

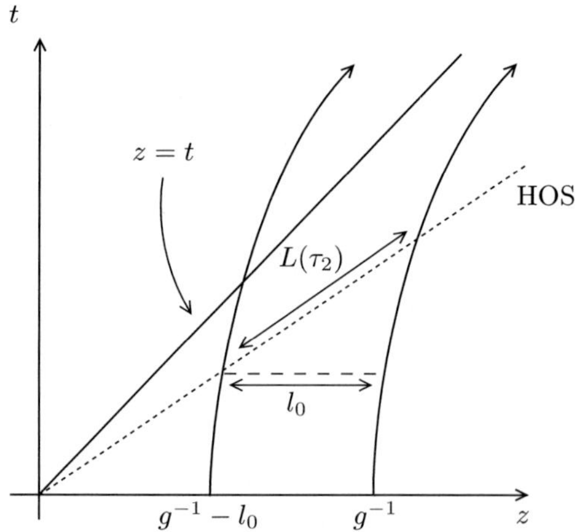

Fig. 15.13 Particle 1 sits at the fixed value $1/g$ of the SE spatial coordinate, so follows the usual hyperbolic worldline in Minkowski spacetime. Particle 2 has exactly the same Minkowski coordinate acceleration at any given Minkowski time, so moves along at a fixed Minkowski distance l_0 behind particle 1, as judged in the Minkowski hyperplanes of simultaneity. As judged in the SE hyperplanes of simultaneity of particle 1, the second particle gets further and further behind. Although it crosses the event horizon $z = t$ of particle 1, and can then no longer signal to particle 1, the proper time of particle 1 has gone to infinity before this happens, where 'before' refers to the SE time of particle 1. This is because the hyperplanes of simultaneity of particle 1 (like the one marked HOS here) get steeper and steeper, and their slope tends to the slope of $z = t$

$$L(\tau_2) = (g^{-2} + l_0^2 + 2g^{-1}l_0 \cosh g\tau_2)^{1/2} - g^{-1} \tag{15.86}$$

increases with τ_2, and this is obvious if we assume $\tau_2 > 0$, since $\cosh \tau_2$ is an increasing function of τ_2 in this region.

As an aside, this is therefore why the string would break in Bell's article [18]. Bell considered two rockets accelerating along an axis in flat spacetime, one placed in front of the other on that axis, and with exactly the same acceleration. The rockets are joined by a fragile thread that is already stretched to its limit when the acceleration begins. The point here is that the thread must break.

If particle 1 is leading particle 2, which is the case $(-)$ illustrated in Fig. 15.13, its lead as measured by its co-accelerating observer increases, but only until the square root vanishes, at which point $L_{\max} = 1/g > l_0$. Note that one must assume $1/g > l_0$, which just says that one cannot choose l_0 so that particle 2 is outside the horizon of particle 1. We now have

$$L(\tau_2) = g^{-1} - (g^{-2} + l_0^2 - 2g^{-1}l_0 \cosh g\tau_2)^{1/2} , \tag{15.87}$$

and we can see that, as $\tau_2 > 0$ increases, so does $\cosh \tau_2$, and we are subtracting less and less from $1/g$ in the formula for L, but only until

$$g^{-2} + l_0^2 = 2g^{-1}l_0 \cosh g\tau_2 . \tag{15.88}$$

Thereafter we have a problem with our formula, because we can no longer take the square root.

Fortunately, no problem arises physically, because the proper time of particle 1 goes to infinity as we approach the problem situation specified in (15.88). Let us prove this. In the semi-Euclidean system of particle 1, the proper time τ attributed by particle 1 to an event with Minkowski coordinates t, z is obtained from

$$
\begin{aligned}
\tau &= \frac{1}{g}\tanh^{-1}\frac{t}{z} \\
&= \frac{1}{g}\tanh^{-1}\frac{g^{-1}\sinh g\tau_2}{g^{-1}\cosh g\tau_2 - l_0} \\
&= \frac{1}{g}\tanh^{-1}\frac{2\sinh g\tau_2}{g^{-1}l_0^{-1} - gl_0} \, ,
\end{aligned}
$$

when t and z satisfy (15.88). But with this condition (15.88), we also have

$$
\sinh g\tau_2 = (\cosh^2 g\tau_2 - 1)^{1/2} = \frac{1}{2}\left(\frac{1}{l_0 g} - gl_0\right) \, .
$$

We find that

$$
\tau = \frac{1}{g}\tanh^{-1}1 = \infty \, .
$$

It is then clear that the proper separation L between the two particles, as viewed by particle 1, reaches its maximum value of $1/g$. (This may seem rather small, but when we reinstate the speed of light, it turns out to be c^2/g, which is generally a large number.)

Finally we note that the maximal proper separation is reached when particle 2 crosses the null line $z = t$ between Boulware's regions I and II. We have to find out where particle 2 is located in the Minkowski coordinate system when (15.88) holds. Firstly,

$$
z_2 = -l_0 + g^{-1}\cosh g\tau_2 = \frac{1}{2g}\left(\frac{1}{gl_0} - gl_0\right) \, ,
$$

and secondly,

$$
t_2 = g^{-1}\sinh g\tau_2 = g^{-1}(\cosh^2 g\tau_2 - 1)^{1/2} = \frac{1}{2g}\left(\frac{1}{l_0 g} - gl_0\right) \, ,
$$

which is essentially the same calculation as we carried out above. Indeed, we showed above that the hyperbolic tangent of z/t was evaluated when $z/t = 1$, and this is the same result.

Note that, in the above calculations, we have been using the semi-Euclidean coordinates to understand what an observer co-accelerating with particle 1 will observe, since we assume that such an observer will use the instantaneously comoving inertial frame to coordinatise.

Static Coordinate Systems in Which Bodies at Rest at Different Points Undergo the Same Proper Acceleration

It is claimed that such systems cannot exist because two bodies experiencing the same proper acceleration do not maintain the same proper distance. However, there are revealing cases where this is not true. We have shown above that, in Minkowski spacetime, two bodies experiencing the same proper acceleration do not maintain the same proper distance as measured in the instantaneously comoving inertial frame of one of the particles. The point is that proper distance is a relative concept, depending on the choice of spacelike hypersurfaces in the coordinate system.

Let us approach the problem from a general perspective. We consider metrics with intervals of the form

$$\mathrm{d}s^2 = -\alpha^2(X)\mathrm{d}T^2 + \beta^2(X)\mathrm{d}X^2 + \mathrm{d}Y^2 + \mathrm{d}Z^2 \ . \tag{15.89}$$

This is a static metric. It should be compared with (15.1) on p. 160. We saw there that we can replace $\beta(X)$ by unity at the expense of changing the coordinate X, and that the requirement of zero curvature forces α to be a linear function of X. However, let us keep the interval as (15.89) for the moment.

Consider two worldlines with fixed values of X, viz.,

$$X_1(T) := (T, \chi_1, 0, 0) \quad \text{and} \quad X_2(T) := (T, \chi_2, 0, 0) \ ,$$

where χ_1, χ_2 are fixed. The 4-velocities of these worldlines are $u_{1,2}$ given by

$$u^\mu_{1,2} = \frac{\mathrm{d}X^\mu_{1,2}}{\mathrm{d}\tau_{1,2}} = \frac{\mathrm{d}X^\mu_{1,2}}{\mathrm{d}T}\frac{\mathrm{d}T}{\mathrm{d}\tau_{1,2}} = (1,0,0,0)\frac{\mathrm{d}T}{\mathrm{d}\tau_{1,2}} \ , \tag{15.90}$$

where $\tau_{1,2}$ is the proper time along each worldline. Since each 4-velocity has pseudolength minus 1, we deduce that

$$-1 = u^2_{1,2} = -\alpha^2(\chi_{1,2})\left(\frac{\mathrm{d}T}{\mathrm{d}\tau_{1,2}}\right)^2 \ ,$$

whence

$$\frac{\mathrm{d}\tau_{1,2}}{\mathrm{d}T} = \alpha(\chi_{1,2}) \ . \tag{15.91}$$

These are constant and we may even deduce therefore that

$$\tau_{1,2} = \alpha(\chi_{1,2})T \ , \tag{15.92}$$

measuring the proper time from $T = 0$ for each particle. This is obvious from the above interval (15.89). Hence

$$u_1 = (1/\alpha_1, 0, 0, 0) \ , \qquad u_2 = (1/\alpha_2, 0, 0, 0) \ , \tag{15.93}$$

where $\alpha_{1,2} := \alpha(\chi_{1,2})$. These are constant, but the accelerations are not!

We must now find certain connection coefficients in order to work out

$$a^i = \frac{\mathrm{d}u^i}{\mathrm{d}\tau} + \Gamma^i_{jk}u^j u^k = \Gamma^i_{00}/\alpha^2 . \tag{15.94}$$

With the usual formula for the Levi-Civita connection coefficients, we soon find

$$\Gamma^0_{00} = 0 = \Gamma^2_{00} = \Gamma^3_{00} , \quad \Gamma^1_{00} = \frac{\alpha\alpha'}{\beta^2} , \tag{15.95}$$

where α' denotes the derivative of $\alpha(X)$ with respect to X. The 4-accelerations of our two worldlines are

$$a_{1,2} = \left(0, \frac{\alpha'_{1,2}}{\alpha_{1,2}\beta^2_{1,2}}, 0, 0 \right) . \tag{15.96}$$

These are constant with squares

$$a^2_{1,2} = \frac{\alpha'^2_{1,2}}{\alpha^2_{1,2}\beta^4_{1,2}} . \tag{15.97}$$

Let us now see whether a^2 can be independent of the values χ_1 and χ_2 that we have chosen for X, so that all such stationary points relative to the spatial coordinates of our coordinate system can have the same proper acceleration. This gives the condition

$$\alpha' = \kappa\alpha\beta^2 , \tag{15.98}$$

where κ^2 is just the constant value of a^2. This has the solution

$$\alpha = \mathrm{e}^{\kappa \int \beta^2} . \tag{15.99}$$

If we make the simple choice $\beta = 1$, and we know that we can arrange for this at the outset by adjusting the coordinate X, then the condition is

$$\alpha' = \kappa\alpha , \tag{15.100}$$

and this equation has the solution

$$\alpha = \mathrm{e}^{\kappa X} . \tag{15.101}$$

We are proposing the interval

$$\mathrm{d}s^2 = -\mathrm{e}^{2\kappa X}\mathrm{d}T^2 + \mathrm{d}X^2 + \mathrm{d}Y^2 + \mathrm{d}Z^2 . \tag{15.102}$$

This is one of the metrics mentioned in Parrott's paper [7] and will be discussed later (see Sect. 16.5). For the moment, note that, for an interval of the form

$$ds^2 = c(X)^2 dT^2 - dX^2 - dY^2 - dZ^2 \,, \tag{15.103}$$

we have

$$a = \frac{c'}{c} \partial_X = \kappa \partial_X \quad \Longrightarrow \quad a^2 = -\kappa^2 \,, \tag{15.104}$$

in the present case, where $c(X) = e^{\kappa X}$.

So here is a static metric and particles moving with fixed spatial coordinates all have the same uniform acceleration. However, it is not a flat spacetime, because the coefficient of the time interval is not linear in the coordinate X [see (15.5) on p. 160]. Let us note the three following points:

- In his statement above, claiming that it is not possible to find a single static coordinate system in which bodies at rest at different points undergo the same proper acceleration, because two bodies experiencing the same proper acceleration do not maintain the same proper distance, Boulware merely omitted to mention that he was insisting on the spacetime being flat.
- At the same time, this suggests that he was not aware of the link. Even in his mathematical derivation of the metric for a static gravitational field, he assumes that the spacetime is flat without comment or justification.
- He is clearly concerned by the fact that the fixed spatial points in Rindler coordinates do not actually have the same 4-acceleration, because intuitively one would expect this model to describe a spatiotemporally constant gravitational field (precisely because the spacetime is flat). We note that he describes the gravitational field as static, but never mentions that it may also be homogeneous.

15.9 Fields of a Uniformly Accelerated Charge

Boulware's derivation of these fields is the most complete, although much abbreviated for the requirements of his publication. The aim here is to spell out the details, since some important features are missing in [6]. Boulware's notation is used throughout. Note that the semi-Euclidean coordinates are referred to as the Rindler coordinates here.

15.9.1 Obtaining the Vector Potential

We begin with the Lienard–Wiechert potential of a particle of charge e moving on the worldline $x(\lambda)$, where λ is its proper time:

$$A^\mu(x) = \frac{e}{2\pi} \int_{-\infty}^{\infty} d\lambda' \frac{dx^\mu}{d\lambda}(\lambda') \theta\big(t - x^0(\lambda')\big) \delta\big([x - x(\lambda')]^2\big) \,. \tag{15.105}$$

In the case of the uniformly accelerated charge, we have

$$x(\lambda) = 0 = y(\lambda)\,, \quad t = g^{-1}\sinh g\lambda\,, \quad z(\lambda) = g^{-1}\cosh g\lambda\,, \qquad (15.106)$$

whence

$$\frac{dx^{\mu}}{d\lambda} = (\cosh g\lambda, 0, 0, \sinh g\lambda)\,. \qquad (15.107)$$

We now calculate the vector potential. It turns out to be convenient to calculate the Rindler components, and to do so, we must consider the three different regions (see Fig. 15.2 on p. 164 for the definitions):

- $z > |t|$ (region I),
- $t > |z|$ (region II),
- $t + z < 0$ (regions III and IV).

We can deal with the last region immediately. Since it is outside the future light cone of all points on the particle worldline, the vector potential is zero there. By the usual transformation relations and the form of the Minkowski metric,

$$A_{\tau} = \frac{\partial t}{\partial \tau}A_t + \frac{\partial z}{\partial \tau}A_z = -\frac{\partial t}{\partial \tau}A^t + \frac{\partial z}{\partial \tau}A^z\,. \qquad (15.108)$$

For the purposes of the calculation, we define $\rho^2 := x^2 + y^2$.

We begin with region I. The argument of the delta function in (15.105) is the square of the displacement from the field point (t, x, y, z) to the charge for the relevant value of λ. In the Rindler coordinates for region I, the field point is

$$(Z\sinh g\tau, x, y, Z\cosh g\tau)\,.$$

Note that these are just the Minkowski coordinates expressed in terms of the Rindler coordinates, for it is much easier to work out this quantity using the Minkowski metric. The charge is at

$$(g^{-1}\sinh g\tau, x, y, g^{-1}\cosh g\tau)\,.$$

The quantity we require is therefore

$$\begin{aligned}
\left[x - x(\lambda)\right]^2 &= -(Z\sinh g\tau - g^{-1}\sinh g\tau)^2 + \rho^2 + (Z\cosh g\tau - g^{-1}\cosh g\tau)^2 \\
&= Z^2 + \rho^2 + g^{-2} + 2Zg^{-1}(\sinh g\tau \sinh g\lambda - \cosh g\tau \cosh g\lambda)^2 \\
&= Z^2 + \rho^2 + g^{-2} - 2Zg^{-1}\cosh g(\tau - \lambda)\,.
\end{aligned} \qquad (15.109)$$

Let us consider this same factor for region II. This is where we understand the utility of the relations (15.11) on p. 164, because we can now express the Minkowski coordinates of a field point in region II in the form

$$(Z\cosh g\tau, x, y, Z\sinh g\tau)\,.$$

The argument of the delta function in this region then becomes

$$
\begin{aligned}
\left[x - x(\lambda)\right]^2 &= -(Z\cosh g\tau - g^{-1}\sinh g\tau)^2 + \rho^2 + (Z\sinh g\tau - g^{-1}\cosh g\tau)^2 \\
&= -Z^2 + \rho^2 + g^{-2} + 2Zg^{-1}(\cosh g\tau \sinh g\lambda - \sinh g\tau \cosh g\lambda)^2 \\
&= -Z^2 + \rho^2 + g^{-2} - 2Zg^{-1}\sinh g(\tau - \lambda)\,.
\end{aligned}
\tag{15.110}
$$

Let us now think about the terms coming from

$$
-\frac{\partial t}{\partial \tau}\frac{\mathrm{d}t}{\mathrm{d}\lambda} + \frac{\partial z}{\partial \tau}\frac{\mathrm{d}z}{\mathrm{d}\lambda}\,.
$$

It is because this works out to a neat expression that we considered the Rindler components of the potential, rather than the Minkowski components, which we subsequently derive from them. From (15.107), this gives

$$
-\frac{\partial t}{\partial \tau}\cosh g\lambda + \frac{\partial z}{\partial \tau}\sinh g\lambda\,.
$$

Now going back to region I, this gives

$$
\begin{aligned}
-\frac{\partial t}{\partial \tau}\cosh g\lambda + \frac{\partial z}{\partial \tau}\sinh g\lambda &= -gZ\cosh g\tau \cosh g\lambda + gZ\sinh g\tau \sinh g\lambda \\
&= -gZ\cosh g(\tau - \lambda)\,.
\end{aligned}
$$

In region II, we obtain

$$
\begin{aligned}
-\frac{\partial t}{\partial \tau}\cosh g\lambda + \frac{\partial z}{\partial \tau}\sinh g\lambda &= -gZ\sinh g\tau \cosh g\lambda + gZ\cosh g\tau \sinh g\lambda \\
&= -gZ\sinh g(\tau - \lambda)\,.
\end{aligned}
$$

It only remains to sort out the step function in (15.105).

Now the delta function in (15.105) ensures that the only relevant λ in the integration range will be such that $x^\mu(\lambda)$ lies on the intersection of the forward or backward light cone of the given field point x^μ and the hyperbolic worldline. The purpose of the step function is then to ensure that only that λ for which $x^\mu(\lambda)$ is on the backward light cone of the field point will contribute. Now for a field point in region II, this extra condition will be unnecessary because, for any field point in this region, we know that the forward light cone of the field point never intersects the hyperbolic worldline (see Fig. 15.14). We automatically have $t > x^0(\lambda)$ for any value of λ that makes $x^\mu - x^\mu(\lambda)$ null.

For a field point in region I, there will be an event on the hyperbolic worldline that lies on the future light cone of the field point, as well as an event on the hyperbolic worldline that lies on the past light cone of the field point. However, if the field point is (τ, x, y, Z) in Rindler coordinates, we need only require $\tau > \lambda$, because the value λ of the proper time when the charge could have signalled to it must be less than τ, while the value λ' such that $x^\mu(\lambda')$ lies on the future light cone of (τ, x, y, Z)

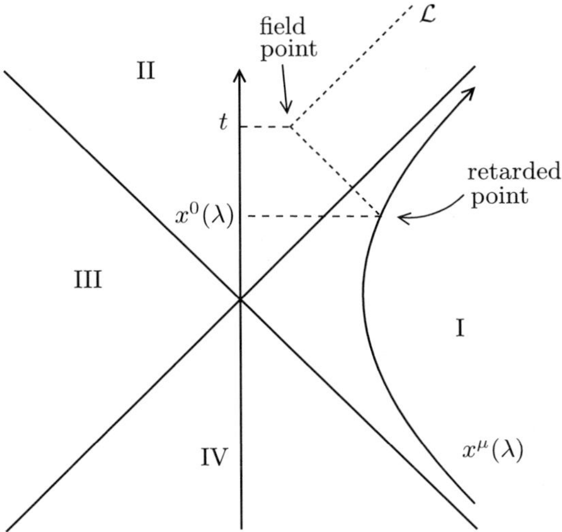

Fig. 15.14 Dealing with the step function. First case: the field point (t,x,y,z) is in region II so its future light cone $\mathcal{L}$ never intersects the hyperbolic worldline. Requiring $x^\mu - x^\mu(\lambda)$ to be null thus entirely determines λ and we automatically have $t > x^0(\lambda)$ for that value of λ. We can simply drop the step function which stipulates this from (15.105)

must be greater than τ. This is best seen from a spacetime diagram (see Fig. 15.15). From the field point, we draw the Rindler plane of simultaneity through the charge worldline, which it intersects at proper time τ (definition of Rindler time τ). This τ is after the appropriate retarded time λ and before the advanced time λ'.

We thus arrive at the expressions

$$A_\tau = -\frac{egZ}{2\pi} \int_{-\infty}^{\infty} d\lambda \, \cosh g(\tau - \lambda)\theta(\tau - \lambda) \tag{15.111}$$

$$\delta\left(g^{-2} + \rho^2 + Z^2 - 2Zg^{-1}\cosh g(\tau - \lambda)\right),$$

in region I,

$$A_\tau = -\frac{egZ}{2\pi} \int_{-\infty}^{\infty} d\lambda \, \sinh g(\tau - \lambda)\delta\left(g^{-2} + \rho^2 - Z^2 - 2Zg^{-1}\sinh g(\tau - \lambda)\right), \tag{15.112}$$

in region II, and

$$A_\tau = 0, \qquad \text{in regions III and IV}. \tag{15.113}$$

We now have to evaluate the integral in each region.

In region I, the step function instructs us to integrate only up to $\lambda = \tau$. Now the delta function has the form $\delta\left(f(\lambda)\right)$, where

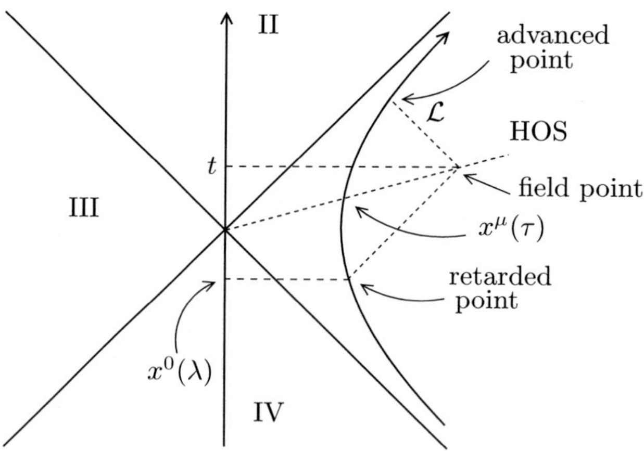

Fig. 15.15 Dealing with the step function. Second case: the field point (t,x,y,z) is in region I so both its future and its past light cone intersect the hyperbolic worldline. The future light cone is denoted by $\mathcal{L}$. Requiring $x^\mu - x^\mu(\lambda)$ to be null thus allows two possible values of λ. If the field point has Rindler coordinates (τ,x,y,Z) then τ will be less than one of the possible values of λ, the one such that $x^\mu(\lambda)$ is on the future light cone of the field point, and greater than the other value of λ, the one such that $x^\mu(\lambda)$ is on the past light cone of the field point. In order to pick up only the latter value of λ in the integral of (15.105), it suffices to stipulate that $\tau > \lambda$ by inserting the step function $\theta(\tau - \lambda)$

$$f(\lambda) = g^{-2} + \rho^2 + Z^2 - 2Zg^{-1}\cosh g(\tau - \lambda) \, ,$$

and this instructs us to divide the rest of the integrand by $|f'(\lambda)|$ and then evaluate everything at the zeros of f. We note that the step function rules out one of the two zeros of f, which is an even function of $\tau - \lambda$. Now the condition $f = 0$ reads

$$\cosh g(\tau - \lambda) = \frac{g^{-2} + \rho^2 + Z^2}{2Zg^{-1}} \, . \tag{15.114}$$

Furthermore,

$$f'(\lambda) = 2Z \sinh g(\tau - \lambda) \, .$$

Equation (15.114) implies that

$$2Z \sinh g(\tau - \lambda) = 2Z \left[\cosh^2 g(\tau - \lambda) - 1 \right]^{1/2}$$

$$= 2Z \left[\frac{(g^{-2} + \rho^2 + Z^2)^2}{4Z^2 g^{-2}} - 1 \right]^{1/2}$$

$$= g \left[(g^{-2} + \rho^2 + Z^2)^2 - 4Z^2 g^{-2} \right]^{1/2} \, .$$

Hence, in region I,

$$A_\tau = -\frac{eg}{4\pi} \frac{g^{-2} + \rho^2 + Z^2}{\left[(g^{-2} + \rho^2 + Z^2)^2 - 4Z^2 g^{-2}\right]^{1/2}} \, . \tag{15.115}$$

If we had not had the step function, the answer would have been precisely twice this.

But now when we turn to region II, we no longer have a step function. What happens here? This time the condition $f = 0$ in the argument of the delta function reads

$$\sinh g(\tau - \lambda) = \frac{g^{-2} + \rho^2 - Z^2}{2Zg^{-1}} \, . \tag{15.116}$$

This is an odd function of $\tau - \lambda$ and it has only one zero. Furthermore,

$$f'(\lambda) = 2Z \cosh g(\tau - \lambda) \, .$$

Equation (15.116) implies that

$$2Z \cosh g(\tau - \lambda) = 2Z \left[\sinh^2 g(\tau - \lambda) + 1\right]^{1/2}$$

$$= 2Z \left[\frac{(g^{-2} + \rho^2 - Z^2)^2}{4Z^2 g^{-2}} + 1\right]^{1/2}$$

$$= g\left[(g^{-2} + \rho^2 - Z^2)^2 + 4Z^2 g^{-2}\right]^{1/2} \, .$$

Hence, in region II,

$$A_\tau = -\frac{eg}{4\pi} \frac{g^{-2} + \rho^2 - Z^2}{\left[(g^{-2} + \rho^2 - Z^2)^2 + 4Z^2 g^{-2}\right]^{1/2}} \, . \tag{15.117}$$

This is the first stage of the calculation.

Next we turn to the calculation of the other nonzero component of the vector potential in the Rindler coordinates, viz., A_Z, noting in passing that A_x and A_y are the same in both Minkowski and Rindler coordinates, and that they are zero according to (15.105), because of the factor $dx^\mu/d\lambda$ in the integrand. Now

$$A_Z = \frac{\partial t}{\partial Z} A_t + \frac{\partial z}{\partial Z} A_z = -\frac{\partial t}{\partial Z} A^t + \frac{\partial z}{\partial Z} A^z \, . \tag{15.118}$$

The only thing that changes in the integrand of (15.105) is the subcalculation with the components of the 4-velocity.

In region I,

$$A_Z = \frac{e}{2\pi} \int_{-\infty}^{\infty} d\lambda \left[-\frac{\partial t}{\partial Z} \cosh g\lambda + \frac{\partial z}{\partial Z} \sinh g\lambda\right]$$

$$\theta(\tau - \lambda) \delta\left(g^{-2} + \rho^2 + Z^2 - 2Zg^{-1} \cosh g(\tau - \lambda)\right) \, .$$

The only change compared with what we had for A_τ is the part in square brackets, which now contributes

$$-\sinh g\tau \cosh g\lambda + \cosh g\tau \sinh g\lambda = -\sinh g(\tau - \lambda)\,.$$

This now simply cancels with the term $|f'(\lambda)|$ from the delta function, which requires us to divide by $2Z \sinh g(\tau - \lambda)$. Once again, there are two zeros of the function f in the argument of the delta function, and the step function in the integrand rules one of them out. The final result in region I is

$$A_Z = -\frac{e}{4\pi Z}\,. \tag{15.119}$$

Finally, for A_Z in region II, we have

$$A_Z = \frac{e}{2\pi} \int_{-\infty}^{\infty} d\lambda \left[-\frac{\partial t}{\partial Z} \cosh g\lambda + \frac{\partial z}{\partial Z} \sinh g\lambda \right]$$

$$\delta\left(g^{-2} + \rho^2 - Z^2 - 2Zg^{-1} \sinh g(\tau - \lambda)\right),$$

where this time the part in square brackets contributes

$$-\cosh g\tau \cosh g\lambda + \sinh g\tau \sinh g\lambda = -\cosh g(\tau - \lambda)\,.$$

This cancels with the term $|f'(\lambda)|$ from the delta function, which requires us to divide by $2Z \cosh g(\tau - \lambda)$. This time there is only one zero of the function f in the argument of the delta function and no step function in the integrand. The final result in region II is

$$A_Z = -\frac{e}{4\pi Z}\,. \tag{15.120}$$

So much for the Rindler components of the vector potential. But what we want are the Minkowski components.

We have

$$A^t = \frac{\partial t}{\partial \tau} A^\tau + \frac{\partial t}{\partial Z} A^Z$$

$$= -\frac{1}{\varepsilon g^2 Z^2} \frac{\partial t}{\partial \tau} A_\tau + \frac{1}{\varepsilon} \frac{\partial t}{\partial Z} A_Z$$

$$= -\frac{z}{\varepsilon g Z^2} A_\tau + \frac{t}{\varepsilon Z} A_Z\,, \tag{15.121}$$

and this holds in both regions I and II, using the formulas (15.11) and (15.12) on p. 164, with $\varepsilon = +1$ in region I and $\varepsilon = -1$ in region II.

In region I, this gives

$$A^t = \frac{z}{gZ^2} \frac{eg}{4\pi} \frac{g^{-2}+\rho^2+Z^2}{\left[(g^{-2}+\rho^2+Z^2)^2 - 4Z^2g^{-2}\right]^{1/2}} - \frac{t}{Z} \frac{e}{4\pi Z}$$

$$= \frac{e}{4\pi Z^2} \left\{ \frac{z(g^{-2}+\rho^2+Z^2)}{\left[(g^{-2}+\rho^2+Z^2)^2 - 4Z^2g^{-2}\right]^{1/2}} - t \right\}$$

$$= \frac{e}{4\pi Z^2} \left\{ \frac{zg(g^{-2}+\rho^2+Z^2)}{2R} - t \right\} , \tag{15.122}$$

where we have defined a convenient quantity

$$R := \frac{g}{2}\left[(g^{-2}+\rho^2+Z^2)^2 - 4Z^2g^{-2}\right]^{1/2} , \tag{15.123}$$

and, in terms of Minkowski coordinates,

$$Z^2 = z^2 - t^2 . \tag{15.124}$$

It is the last innocuous equation which helps to find a uniform formula for region II, because, if we look at (15.11) on p. 164, we find that the Z in region II satisfies $Z^2 = t^2 - z^2$. In fact, Boulware uses another symbol for it, viz., $\tilde{Z}$.

What we have in region II is

$$A^t = \frac{z}{g\tilde{Z}^2} A_\tau - \frac{t}{\tilde{Z}} A_Z$$

$$= -\frac{z}{g\tilde{Z}^2} \frac{eg}{4\pi} \frac{g^{-2}+\rho^2-\tilde{Z}^2}{\left[(g^{-2}+\rho^2-\tilde{Z}^2)^2 + 4\tilde{Z}^2g^{-2}\right]^{1/2}} + \frac{t}{\tilde{Z}} \frac{e}{4\pi \tilde{Z}}$$

$$= -\frac{e}{4\pi \tilde{Z}^2} \left\{ \frac{z(g^{-2}+\rho^2-\tilde{Z}^2)}{\left[(g^{-2}+\rho^2-\tilde{Z}^2)^2 + 4\tilde{Z}^2g^{-2}\right]^{1/2}} - t \right\}$$

$$= \frac{e}{4\pi Z^2} \left\{ \frac{zg(g^{-2}+\rho^2+Z^2)}{2R} - t \right\} , \tag{15.125}$$

where we have put $\tilde{Z}^2 = -Z^2$ at the end, because it does have the same expression in Minkowski coordinates, viz., $z^2 - t^2$.

We can summarise the t component of the potential in all four regions of spacetime by

$$A^t = \frac{e}{4\pi Z^2} \left[\frac{zg(g^{-2}+\rho^2+Z^2)}{2R} - t \right] \theta(t+z) , \tag{15.126}$$

but we should remember that the step function does not tell us the correct values on the null surface $t+z = 0$. Its purpose here is just to rule out signals into regions III and IV. We investigate this in detail below.

By similar arguments, we get the other Minkowski components of the potential as

$$A^z = \frac{e}{4\pi Z^2}\left[\frac{tg(g^{-2}+\rho^2+Z^2)}{2R} - z\right]\theta(t+z)\,, \qquad (15.127)$$

together with the obvious $A^x = 0 = A^y$.

15.9.2 Obtaining the Electromagnetic Fields

We shall denote the fields corresponding to the above potential with a tilde, because we shall subsequently adjust them on the null hypersurface $z+t = 0$ where they are not correct. When we calculate partial derivatives of the potential in (15.126) and (15.127), we just forget the step function, then reinsert it afterwards. This amounts to getting the right functions in the interiors of the regions on either side of the null hypersurface, and then making a correction to the fields (rather than the potentials) in a final stage.

We begin with

$$\tilde{F}^{tz} = E^z = -\frac{\partial A^z}{\partial t} - \frac{\partial A^t}{\partial z}\,. \qquad (15.128)$$

Inside regions I and II,

$$\tilde{F}^{tz} = -\frac{e}{4\pi}\left\{\frac{\partial}{\partial t}\left[\frac{tg(g^{-2}+\rho^2+Z^2)}{2RZ^2} - z\right] + \frac{\partial}{\partial z}\left[\frac{zg(g^{-2}+\rho^2+Z^2)}{2RZ^2} - t\right]\right\}$$

$$= -\frac{e}{4\pi}\left[\frac{g(g^{-2}+\rho^2+Z^2)}{RZ^2} + \frac{g}{2}\left(t\frac{\partial}{\partial t} + z\frac{\partial}{\partial z}\right)\left(\frac{g^{-2}+\rho^2+Z^2}{RZ^2}\right)\right]\,.$$

We note that the thing to be differentiated in the last expression only depends on t and z through Z^2, so we begin by differentiating the relevant term with respect to Z^2, viz.,

$$\frac{\partial}{\partial(Z^2)}\left(\frac{g^{-2}+\rho^2+Z^2}{RZ^2}\right) = \frac{1}{RZ^2} - \frac{g^{-2}+\rho^2+Z^2}{RZ^4} - \frac{g^{-2}+\rho^2+Z^2}{R^2Z^2}\frac{\partial R}{\partial(Z^2)}\,,$$

where

$$\frac{\partial R}{\partial(Z^2)} = \frac{g^2}{4}\frac{1}{2R}\frac{\partial}{\partial(Z^2)}\left[(g^{-2}+\rho^2+Z^2)^2 - 4Z^2g^{-2}\right]$$

$$= \frac{g^2}{4R}(\rho^2+Z^2-g^{-2})\,.$$

We also have

$$\left(t\frac{\partial}{\partial t} + z\frac{\partial}{\partial z}\right)Z^2 = -2t^2 + 2z^2 = 2Z^2\,.$$

Hence, by the chain rule,

$$\frac{g}{2}\left(t\frac{\partial}{\partial t}+z\frac{\partial}{\partial z}\right)\left(\frac{g^{-2}+\rho^2+Z^2}{RZ^2}\right)=gZ^2\frac{\partial}{\partial(Z^2)}\left(\frac{g^{-2}+\rho^2+Z^2}{RZ^2}\right).$$

It is not a good idea to work this term out fully. If we put it into the relevant expression for the field, a term cancels and this helps the algebra. So we now have

$$\begin{aligned}
\tilde{F}^{tz} &= -\frac{e}{4\pi}\left[\frac{g(g^{-2}+\rho^2+Z^2)}{RZ^2}+gZ^2\frac{\partial}{\partial(Z^2)}\left(\frac{g^{-2}+\rho^2+Z^2}{RZ^2}\right)\right]\\
&= -\frac{eg}{4\pi R}\left[1-\frac{g^{-2}+\rho^2+Z^2}{R}\frac{\partial R}{\partial(Z^2)}\right]\\
&= -\frac{eg}{4\pi R}\frac{g^{-2}+\rho^2-Z^2}{2R^2}\\
&= \frac{eg}{4\pi}\frac{Z^2-\rho^2-g^{-2}}{2R^3}.
\end{aligned}$$

Having done this calculation and noted that the fields are zero in regions III and IV, we now reinsert the step function to get

$$\tilde{F}^{tz}=\frac{eg}{4\pi}\frac{Z^2-\rho^2-g^{-2}}{2R^3}\theta(t+z)\,. \tag{15.129}$$

This is not valid on the null hypersurface $t+z=0$, as we shall see.

The formula for $\tilde{F}^{tx}=E^x$ is simpler, partly because A^x is zero. We have, in regions I and II,

$$\begin{aligned}
E^x &= -\frac{\partial A^t}{\partial x}\\
&= -\frac{e}{4\pi Z^2}z\frac{\partial}{\partial\rho^2}\left\{\frac{g^{-2}+\rho^2+Z^2}{\left[(g^{-2}+\rho^2+Z^2)^2-4Z^2g^{-2}\right]^{1/2}}\right\}\frac{\partial(\rho^2)}{\partial x}\\
&= -\frac{ez}{4\pi Z^2}\frac{2x}{(2g^{-1}R)^3}\left[4g^{-2}R^2-(g^{-2}+\rho^2+Z^2)^2\right]\\
&= \frac{ezxg}{4\pi R^3}\,.
\end{aligned}$$

Similarly, in regions I and II,

$$E^y=\frac{ezyg}{4\pi R^3}\,.$$

The neat formula for all of spacetime, which is actually wrong on the null hypersurface $z+t=0$, is

$$\tilde{F}^{t\rho} = E^{\rho} = \frac{egz\rho}{4\pi R^3}\,\theta(t+z)\,,\tag{15.130}$$

with the obvious notation for the two components here.

We now seek the magnetic field. To begin with,

$$B^z = \tilde{F}^{xy} = \frac{\partial A^y}{\partial x} - \frac{\partial A^x}{\partial y} = 0\,,\tag{15.131}$$

because $A^x = 0 = A^y$. However,

$$B^x = \tilde{F}^{yz} = \frac{\partial A^z}{\partial y} - \frac{\partial A^y}{\partial z} = \frac{\partial A^z}{\partial y}\tag{15.132}$$

and

$$B^y = \tilde{F}^{zx} = \frac{\partial A^x}{\partial z} - \frac{\partial A^z}{\partial x} = -\frac{\partial A^z}{\partial x}\,,\tag{15.133}$$

whence we may deduce that, for the unit vector $\hat{\mathbf{e}}_z$ in the z direction,

$$\hat{\mathbf{e}}_z \times \mathbf{B} = \left(\frac{\partial A^z}{\partial x},\frac{\partial A^z}{\partial y},0\right) = \nabla_{\rho}A^z\,,\tag{15.134}$$

with the obvious notation. (Several sign conventions are required along the way, but we have been lucky.)

Now if we look at (15.126) and (15.127) for A^t and A^z, respectively, we find that they differ only by exchanging the explicit mention of z and t in the part in square brackets. We have calculated $\partial A^t/\partial x$ and $\partial A^t/\partial y$ above, and we can copy these deductions to obtain $\partial A^z/\partial x$ and $\partial A^z/\partial y$. We also note that $\tilde{F}^{z\rho} = -\hat{\mathbf{e}}_z \times \mathbf{B}$, so finally,

$$\tilde{F}^{z\rho} = -\hat{\mathbf{e}}_z \times \mathbf{B} = -\nabla_{\rho}A^z = \frac{egt\rho}{4\pi R^3}\,\theta(t+z)\,.\tag{15.135}$$

Once again, this is incorrect on the null hypersurface $z+t = 0$.

15.9.3 Electromagnetic Fields on the Null Surface $z + t = 0$

The fields we have found satisfy Maxwell's equations when $z+t > 0$, and trivially when $z+t < 0$, where they are identically zero. Let us now show that they do not satisfy Maxwell's equations on the null surface $z+t = 0$. We would expect this, as we have not used the formula (15.105) carefully in this region. Indeed, we just stuck in the step function $\theta(t+z)$ for convenience. In fact, we shall show that

$$\partial_{\nu}\tilde{F}^{x\nu} = 0 = \partial_{\nu}\tilde{F}^{y\nu}\,,\tag{15.136}$$

$$\partial_\nu \tilde{F}^{t\nu} = -\frac{4eg^2}{4\pi(1+g^2\rho^2)^2}\delta(z+t) = -\partial_\nu \tilde{F}^{z\nu} . \tag{15.137}$$

All these quantities should be zero on the null hypersurface in question, because the source is zero there.

Since we are hoping to fix up this problem, we also have to look at the other Maxwell equations, which normally follow because the electromagnetic field tensor comes from a potential. This is an important point, because we will soon add a field which cancels out the unwanted term in (15.137) and itself satisfies the Maxwell equations arising because it stems from a potential. The superposition of the two fields would not then satisfy the source-free Maxwell equations unless $\tilde{F}$ did. It is also interesting to see why the source-free Maxwell equations get through despite our careless attitude with the step function above. This issue is neglected by Boulware, presumably through want of space.

Consider first the equation $\nabla \cdot \mathbf{B} = 0$. By (15.131), (15.132) and (15.133),

$$B^x = \frac{\partial A^z}{\partial y} , \qquad B^y = -\frac{\partial A^z}{\partial x} , \qquad B^z = 0 . \tag{15.138}$$

It is now immediately obvious that $\nabla \cdot \mathbf{B} = 0$, just by the standard calculation when $\mathbf{B}$ comes from a potential. What is crucial here is that we can leave the step function $\theta(z+t)$ in the expressions for the potential as they are unaffected by the differentiations. It does not matter whether the potential was the right one on the null hypersurface, since we only require $\nabla \cdot \mathbf{B} = 0$. Looking at the calculation of E^z, we see that it was more convenient to drop the step function and then reinsert it. It should be stressed that the expressions for the potentials in (15.126) and (15.127) are already incorrect on the null hypersurface, but this does not matter in the above check of this Maxwell equation.

Now consider the other source-free Maxwell equation, viz.,

$$\nabla \times \mathbf{E} + \frac{\partial \mathbf{B}}{\partial t} = 0 . \tag{15.139}$$

We said that

$$\mathbf{E} = \frac{eg}{4\pi R^3}\left(xz, yz, \frac{1}{2}(Z^2 - g^{-2} - \rho^2)\right)\theta(t+z) \tag{15.140}$$

and

$$\mathbf{B} = \frac{eg}{4\pi R^3}(-yt, xt, 0)\theta(t+z) . \tag{15.141}$$

We must be very careful in taking derivatives because R depends on all the Minkowski coordinates. For reference,

$$R := \frac{g}{2}\left[(g^{-2} + \rho^2 + Z^2)^2 - 4Z^2 g^{-2}\right]^{1/2} , \tag{15.142}$$

where $Z^2 = z^2 - t^2$ and $\rho^2 = x^2 + y^2$.

We begin with the easiest component of (15.139), which is the third. We have to show that

$$\frac{\partial E^y}{\partial x} = \frac{\partial E^x}{\partial y} \, . \qquad (15.143)$$

Now

$$\frac{\partial E^y}{\partial x} = \frac{egyz}{4\pi}\theta(t+z)\frac{\partial}{\partial x}\frac{1}{R^3}$$

and

$$\frac{\partial E^x}{\partial y} = \frac{egxz}{4\pi}\theta(t+z)\frac{\partial}{\partial y}\frac{1}{R^3} \, .$$

We get equality because

$$\frac{\partial}{\partial x}\frac{1}{R^3} = \frac{\partial}{\partial(\rho^2)}\left(\frac{1}{R^3}\right)\frac{\partial(\rho^2)}{\partial x} = 2x\frac{\partial}{\partial(\rho^2)}\left(\frac{1}{R^3}\right)$$

and

$$\frac{\partial}{\partial y}\frac{1}{R^3} = \frac{\partial}{\partial(\rho^2)}\left(\frac{1}{R^3}\right)\frac{\partial(\rho^2)}{\partial y} = 2y\frac{\partial}{\partial(\rho^2)}\left(\frac{1}{R^3}\right) \, .$$

So the third component of (15.139) is satisfied.

Let us just check the first component now, since it is bound to be very similar to the second, and it is already enough work. The first component of (15.139) reads

$$\frac{\partial E^z}{\partial y} - \frac{\partial E^y}{\partial z} = -\frac{\partial B^x}{\partial t} \, , \qquad (15.144)$$

and this in turn reads

$$\frac{\partial}{\partial y}\left[\frac{1}{2}\frac{1}{R^3}(Z^2 - g^{-2} - \rho^2)\theta(t+z)\right] - \frac{\partial}{\partial z}\left[\frac{yz}{R^3}\theta(t+z)\right] = \frac{\partial}{\partial t}\left[\frac{yt}{R^3}\theta(t+z)\right] \, . \qquad (15.145)$$

Exactly two of the terms lead to delta functions. We have

$$-\frac{yz}{R^3}\delta(t+z) \quad \text{on the left} \, ,$$

$$\frac{yt}{R^3}\delta(t+z) \quad \text{on the right} \, ,$$

and these are equal, because we can put $t = -z$ in the prefactors. Without this cancellation, all would have been lost.

The rest of (15.145) leaves us with

$$\frac{1}{2}\theta(t+z)\left[-\frac{2y}{R^3}+(Z^2-g^{-2}-\rho^2)\frac{\partial}{\partial y}\frac{1}{R^3}\right]-y\theta(t+z)\left[\frac{1}{R^3}+z\frac{\partial}{\partial z}\frac{1}{R^3}\right]$$

$$=y\theta(t+z)\left[\frac{1}{R^3}+t\frac{\partial}{\partial t}\frac{1}{R^3}\right]. \tag{15.146}$$

Cancelling the step function and rearranging this, we now have to check that

$$\frac{1}{2}(Z^2-g^{-2}-\rho^2)\frac{\partial}{\partial y}\frac{1}{R^3}-zy\frac{\partial}{\partial z}\frac{1}{R^3}=\frac{3y}{R^3}+ty\frac{\partial}{\partial t}\frac{1}{R^3}. \tag{15.147}$$

Now $1/R^3$ depends on t or z only through Z^2 so

$$\frac{\partial}{\partial t}\frac{1}{R^3}=\frac{\partial}{\partial(Z^2)}\left(\frac{1}{R^3}\right)\frac{\partial(Z^2)}{\partial t}=-2t\frac{\partial}{\partial(Z^2)}\left(\frac{1}{R^3}\right)$$

and

$$\frac{\partial}{\partial z}\frac{1}{R^3}=\frac{\partial}{\partial(Z^2)}\left(\frac{1}{R^3}\right)\frac{\partial(Z^2)}{\partial z}=2z\frac{\partial}{\partial(Z^2)}\left(\frac{1}{R^3}\right).$$

Our equation becomes

$$\frac{1}{2}(Z^2-g^{-2}-\rho^2)\frac{\partial}{\partial y}\frac{1}{R^3}=\frac{3y}{R^3}+2yZ^2\frac{\partial}{\partial(Z^2)}\frac{1}{R^3}. \tag{15.148}$$

Likewise, $1/R^3$ depends on y only through ρ^2, and hence

$$\frac{\partial}{\partial y}\frac{1}{R^3}=\frac{\partial}{\partial(\rho^2)}\left(\frac{1}{R^3}\right)\frac{\partial(\rho^2)}{\partial y}=2y\frac{\partial}{\partial(\rho^2)}\left(\frac{1}{R^3}\right).$$

Cancelling a factor of y on both sides, the equation is now

$$(Z^2-g^{-2}-\rho^2)\frac{\partial}{\partial(\rho^2)}\frac{1}{R^3}=\frac{3}{R^3}+2Z^2\frac{\partial}{\partial(Z^2)}\frac{1}{R^3}. \tag{15.149}$$

Finally, we feed in

$$\frac{\partial}{\partial(\rho^2)}\frac{1}{R^3}=-\frac{(g/2)^2}{R^5}3(g^{-2}+\rho^2+Z^2) \tag{15.150}$$

and

$$\frac{\partial}{\partial(Z^2)}\frac{1}{R^3}=-\frac{(g/2)^2}{R^5}3(-g^{-2}+\rho^2+Z^2), \tag{15.151}$$

and then all that remains is to do a little algebra. The two sides of (15.139) do turn out to be equal.

This is crucial to the reasoning in Boulware's paper, although it seems to have been overlooked. Shortly we will show that the other Maxwell equations are not satisfied on the null cone $z + t = 0$ by showing (15.137) on p. 207. We will then add another field, which is only nonzero on this null cone, but which already satisfies the source-free Maxwell equations. If the superposed field is to satisfy the source-free Maxwell equations, then the first field must too, as indeed it does.

Let us now turn to the other Maxwell equations. To begin with, we shall show that some of these other equations are actually satisfied, viz.,

$$0 = \partial_\nu \tilde{F}^{x\nu} = \partial_t \tilde{F}^{xt} + \partial_x \tilde{F}^{xx} + \partial_y \tilde{F}^{xy} + \partial_z \tilde{F}^{xz} . \tag{15.152}$$

Now $\tilde{F}^{xx} = 0$ by antisymmetry of $\tilde{F}$ and $\tilde{F}^{xy}$ is proportional to B^z, so we have to show that

$$0 = \partial_t \tilde{F}^{xt} + \partial_z \tilde{F}^{xz} . \tag{15.153}$$

This requires

$$0 = \frac{\partial}{\partial t} \left[\frac{xgz}{R^3} \theta(t+z) \right] + \frac{\partial}{\partial z} \left[\frac{xgt}{R^3} \theta(t+z) \right] . \tag{15.154}$$

On the right-hand side we obtain two delta functions, viz.,

$$\frac{xgz}{R^3} \delta(t+z) + \frac{xgt}{R^3} \delta(t+z) .$$

These cancel because we can replace t by $-z$ in one of the prefactors. We have to show that

$$0 = \frac{\partial}{\partial t} \left(\frac{xgz}{R^3} \right) + \frac{\partial}{\partial z} \left(\frac{xgt}{R^3} \right) , \tag{15.155}$$

where we have cancelled the step function on both sides. This in turn is equivalent to

$$0 = z \frac{\partial}{\partial t} \left(\frac{1}{R^3} \right) + t \frac{\partial}{\partial z} \left(\frac{1}{R^3} \right) . \tag{15.156}$$

Now we have the usual trick, using the fact that $1/R^3$ depends only on z and t through Z^2, and the equation follows.

We now consider one of the equations which actually deviates from Maxwell's equations, viz., the equation referring to

$$\partial_\nu \tilde{F}^{t\nu} = \partial_t \tilde{F}^{tt} + \partial_x \tilde{F}^{tx} + \partial_y \tilde{F}^{ty} + \partial_z \tilde{F}^{tz}$$

$$= \frac{eg}{4\pi}\left\{ \frac{\partial}{\partial x}\left[\frac{xz}{R^3}\theta(t+z)\right] + \frac{\partial}{\partial y}\left[\frac{yz}{R^3}\theta(t+z)\right] \right.$$

$$\left. + \frac{\partial}{\partial z}\left[\frac{Z^2 - g^{-2} - \rho^2}{2R^3}\theta(t+z)\right] \right\}$$

$$= \frac{eg\theta(t+z)}{4\pi}\left[\frac{3z}{R^3} + xz\frac{\partial}{\partial x}\frac{1}{R^3} + yz\frac{\partial}{\partial y}\frac{1}{R^3} + \frac{1}{2}(Z^2 - g^{-2} - \rho^2)\frac{\partial}{\partial z}\frac{1}{R^3}\right]$$

$$+ \frac{eg}{4\pi}\frac{Z^2 - g^{-2} - \rho^2}{2R^3}\delta(t+z) .$$

With the same kind of algebra as we used to show the similar relation (15.147), we can prove that the expression in square brackets here is zero. This means that we do not have $\partial_\nu \tilde{F}^{t\nu} = 0$, but instead

$$\partial_\nu \tilde{F}^{t\nu} = \frac{eg}{4\pi}\frac{Z^2 - g^{-2} - \rho^2}{2R^3}\delta(t+z) . \tag{15.157}$$

This is just what we would expect: things go wrong by a delta function, because we were careless about a step function. The last expression simplifies when we put $t = -z$, i.e., $Z^2 = 0$, in the prefactor, giving

$$\partial_\nu \tilde{F}^{t\nu} = -\frac{e}{4\pi}\frac{4g^2}{(1+g^2\rho^2)^2}\delta(t+z) . \tag{15.158}$$

To finish off this section, we should show that we have

$$\partial_\nu \tilde{F}^{t\nu} = -\partial_\nu \tilde{F}^{z\nu} . \tag{15.159}$$

Now this calculation begins with

$$\partial_\nu \tilde{F}^{z\nu} = \partial_t \tilde{F}^{zt} + \partial_x \tilde{F}^{zx} + \partial_y \tilde{F}^{zy} + \partial_z \tilde{F}^{zz}$$

$$= \frac{eg}{4\pi}\left\{ -\frac{\partial}{\partial t}\left[\frac{Z^2 - g^{-2} - \rho^2}{2R^3}\theta(t+z)\right] + \frac{\partial}{\partial x}\left[\frac{xt}{R^3}\theta(t+z)\right] \right.$$

$$\left. + \frac{\partial}{\partial y}\left[\frac{yt}{R^3}\theta(t+z)\right] \right\} ,$$

and continues along exactly the same lines.

15.9.4 Fixing up the Fields on the Null Surface

The aim here is to find another field $\Delta F^{\mu\nu}$ which can be added to $\tilde{F}^{\mu\nu}$ to obtain the correct retarded field everywhere. It will be a delta function (distribution) with support on the null surface $z+t=0$ so that we keep $\tilde{F}$ elsewhere, since we know that $\tilde{F}$ was the correct field off the null surface. Since $\tilde{F}$ satisfies the source-free Maxwell equations, the same must be true of ΔF. Finally, we shall require

$$\partial_\nu \Delta F^{t\nu} = -\partial_\nu \Delta F^{z\nu} = \frac{e}{4\pi}\frac{4g^2}{(1+g^2\rho^2)^2}\delta(t+z)\,. \tag{15.160}$$

Note that, if we find such a field ΔF, then $\tilde{F}+\Delta F$ must be the retarded field due to our charge. The logic is this: we have a field which satisfies Maxwell's equations everywhere for the charge distribution represented by our uniformly accelerated charge. This is because the last adjustment only affects the result outside the charge distribution, i.e., off the worldline of the charge.

Now Boulware points out that the charge worldline is invariant under Lorentz transformations along the z axis. Indeed, it has the formula $z^2 - t^2 = 1$ in any inertial frame. This implies that the resultant field must be Lorentz invariant too. We shall look at what this means in a moment (but see also p. 73). Now the argument is that the only field ΔF which is invariant and restricted to the null surface $z+t=0$ is of the form

$$\Delta F^{tz} = 0\,, \qquad \Delta F^{t\rho} = -\Delta F^{z\rho} = \rho A(\rho)\delta(z+t)\,. \tag{15.161}$$

Unfortunately, there is no proof of this claim in Boulware's paper (sufficiency is straightforward, but necessity does not appear to be). Worse, we have not shown that $\tilde{F}$ is Lorentz invariant. If it were not, and we added a Lorentz invariant field, we would get a field that was not Lorentz invariant and that would not do.

In the present circumstances it seems unreasonable to spend too much time on this, since it would appear to involve a good deal of calculation. It matters little how we are guided to the solution, provided that it satisfies all the conditions listed above. However, let us just investigate the idea that a field of the form (15.161) is Lorentz invariant. Then we will have the following logic: since we know that $\tilde{F}+\Delta F$ is Lorentz invariant, and we will show that ΔF is Lorentz invariant, we will be able to deduce that $\tilde{F}$ was actually Lorentz invariant.

We have a field of the form

$$\Delta F = \begin{pmatrix} 0 & \Delta F^{tx} & \Delta F^{ty} & 0 \\ -\Delta F^{tx} & 0 & 0 & -\Delta F^{zx} \\ -\Delta F^{ty} & 0 & 0 & -\Delta F^{zy} \\ 0 & \Delta F^{zx} & \Delta F^{zy} & 0 \end{pmatrix}\,, \tag{15.162}$$

where

$$\Delta F^{t\rho} = -\Delta F^{z\rho} = \rho A(\rho)\delta(z+t)\,. \tag{15.163}$$

Consider a Lorentz transformation

$$t' = \gamma(t - uz) , \qquad z' = \gamma(z - ut) .$$
(15.164)

It is easy to see that

$$z' + t' = \gamma(z - ut + t - uz) = \gamma(1 - u)(z + t) ,$$
(15.165)

whence

$$\delta(z' + t') = \delta\big(\gamma(1 - u)(z + t)\big) = \frac{1}{\gamma(1 - u)}\delta(z + t) .$$
(15.166)

We also note that

$$\rho A(\rho) = \rho' A(\rho') ,$$
(15.167)

since ρ means (x, y), and this is just (x', y'), which is just what we would mean by ρ'.

Let us now Lorentz transform the above field:

$$\Delta F'^{ij} = L^i_{\ k} L^j_{\ l} \Delta F^{kl} , \quad \text{or} \quad \Delta F' = L(\Delta F)L^{\mathrm{T}} ,$$
(15.168)

whence

$$\Delta F' = \begin{pmatrix} \gamma & 0 & 0 & -\gamma u \\ 0 & 1 & 0 & 0 \\ 0 & 0 & 1 & 0 \\ -\gamma u & 0 & 0 & \gamma \end{pmatrix} (\Delta F) \begin{pmatrix} \gamma & 0 & 0 & -\gamma u \\ 0 & 1 & 0 & 0 \\ 0 & 0 & 1 & 0 \\ -\gamma u & 0 & 0 & \gamma \end{pmatrix}$$

$$= \gamma \begin{pmatrix} 0 & K & L & 0 \\ -K & 0 & 0 & M \\ -L & 0 & 0 & N \\ 0 & -M & -N & 0 \end{pmatrix} ,$$

where we have used the shorthand

$$K := \Delta F^{tx} - u\Delta F^{zx} , \quad L := \Delta F^{ty} - u\Delta F^{zy} ,$$

$$M := u\Delta F^{tx} - \Delta F^{zx} , \quad N := u\Delta F^{ty} - \Delta F^{zy} ,$$

in order to fit the matrix onto the line.

Consider now

$$\Delta F'^{tx} = \gamma(\Delta F^{tx} - u\Delta F^{zx})$$
$$= \gamma\big[xA(\rho)\delta(z+t) + uxA(\rho)\delta(z+t)\big]$$
$$= \gamma^2(1-u)\delta(z'+t')(1+u)xA(\rho)$$
$$= x'A(\rho')\delta(z'+t') \, .$$

By symmetry,

$$\Delta F'^{ty} = y'A(\rho')\delta(z'+t') \, .$$

The expressions for these primed components are identical to the expressions for the corresponding unprimed components, if we replace primed coordinates by unprimed coordinates. This is what we mean when we say that the field ΔF is invariant under such Lorentz transformations. We can also check that this works for the other primed components, e.g.,

$$\Delta F'^{zx} = \gamma(-u\Delta F^{tx} + \Delta F^{zx})$$
$$= \gamma\big[-uxA(\rho)\delta(z+t) - xA(\rho)\delta(z+t)\big]$$
$$= -\gamma^2(1-u)\delta(z'+t')(1+u)xA(\rho)$$
$$= -x'A(\rho')\delta(z'+t') \, ,$$

which should be compared with

$$\Delta F^{zx} = -xA(\rho)\delta(z+t) \, .$$

This calculation has been done to show what is meant when we say that the field ΔF is invariant under Lorentz transformations in the z direction. We mean that, when the components are Lorentz transformed and expressed as functions of the new coordinates, the formulas are identical in form to those for the original components as functions of the original coordinates. This is indeed what would be expected for the total retarded field when it is engendered by a charge whose worldline is Lorentz invariant.

This deduction will only allow us to conclude that the fields $\tilde{F}$ are invariant under these Lorentz transformations. In fact, we do not need this. We have to show that the proposed field ΔF satisfies the source-free Maxwell equations and that we can arrange for (15.160) by suitable choice of $A(\rho)$ in (15.162) and (15.163). This ensures that the superposition $\tilde{F} + \Delta F$ satisfies Maxwell's equations, and is therefore the retarded field (assumed unique) due to the uniformly accelerating charge.

We consider now the quantities $\partial_\nu \Delta F^{\mu\nu}$ for $\mu = t, x, y, z$. Firstly,

$$\partial_\nu \Delta F^{x\nu} = \partial_t \Delta F^{xt} + \partial_y \Delta F^{xy} + \partial_z \Delta F^{xz}$$
$$= -xA(\rho)\left(\frac{\partial}{\partial t} - \frac{\partial}{\partial z}\right)\delta(z+t) = 0 \, . \tag{15.169}$$

Likewise $\partial_\nu \Delta F^{y\nu} = 0$. Now

$$\partial_\nu \Delta F^{t\nu} = \partial_x \Delta F^{tx} + \partial_y \Delta F^{ty} \,, \tag{15.170}$$

and also

$$\partial_\nu \Delta F^{z\nu} = \partial_x \Delta F^{zx} + \partial_y \Delta F^{zy} = -\partial_\nu \Delta F^{t\nu} \,. \tag{15.171}$$

When we analyse $\partial_\nu \Delta F^{t\nu}$, we obtain

$$\partial_\nu \Delta F^{t\nu} = \left\{ \frac{\partial}{\partial x}\left[xA(\rho) \right] + \frac{\partial}{\partial y}\left[yA(\rho) \right] \right\} \delta(z+t)$$

$$= \left[2A(\rho) + \rho \frac{\mathrm{d}A}{\mathrm{d}\rho} \right] \delta(z+t)$$

$$= \frac{1}{\rho}\frac{\mathrm{d}}{\mathrm{d}\rho}\left[\rho^2 A(\rho) \right] \delta(z+t) \,. \tag{15.172}$$

Comparing with (15.158), viz.,

$$\partial_\nu \tilde{F}^{t\nu} = -\frac{e}{4\pi}\frac{4g^2}{(1+g^2\rho^2)^2}\delta(t+z) \,, \tag{15.173}$$

we see that $F := \tilde{F} + \Delta F$ will satisfy Maxwell's source equations everywhere provided that we arrange for

$$\frac{1}{\rho}\frac{\mathrm{d}}{\mathrm{d}\rho}\left[\rho^2 A(\rho) \right] = \frac{e}{4\pi}\frac{4g^2}{(1+g^2\rho^2)^2} \,. \tag{15.174}$$

Let us therefore solve this equation for $A(\rho)$. We have

$$\rho^2 A(\rho) = \int \mathrm{d}\rho \, \frac{e}{4\pi}\frac{4g^2\rho}{(1+g^2\rho^2)^2}$$

$$= \frac{eg^2}{\pi}\left[\kappa - (1+g^2\rho^2)^{-1} \right]\frac{1}{2g^2} \,,$$

where κ is a constant. If we choose $\kappa = 1$, we obtain

$$\rho^2 A(\rho) = \frac{e}{2\pi}\frac{g^2\rho^2}{1+g^2\rho^2} \,.$$

This choice of ρ is the only one ensuring that A will be finite at $\rho = 0$. Finally,

$$A(\rho) = \frac{eg^2}{2\pi(1+g^2\rho^2)} \,. \tag{15.175}$$

We still have to check that the field ΔF satisfies the source-free Maxwell equations

$$\nabla \cdot \mathbf{B} = 0 , \qquad \nabla \times \mathbf{E} + \frac{\partial \mathbf{B}}{\partial t} = 0 .$$

This is because we know that $\tilde{F}$ does, and hence the superposition $\tilde{F} + \Delta F$ will satisfy these equations iff ΔF does.

The best approach is just to demonstrate that ΔF derives from a potential, since we know this automatically leads to the above two Maxwell equations. Indeed, consider

$$\Delta A_\mu := (0, -x, -y, 0) A(\rho) \theta(t+z) . \tag{15.176}$$

This leads to a field $\Delta F_{\mu\nu} = \partial_\mu \Delta A_\nu - \partial_\nu \Delta A_\mu$, which we calculate to be

$$\Delta F_{\text{covariant}} = \begin{pmatrix} 0 & -x & -y & 0 \\ x & 0 & 0 & x \\ y & 0 & 0 & y \\ 0 & -x & -y & 0 \end{pmatrix} A(\rho) \delta(z+t) , \tag{15.177}$$

to be compared with (15.162). Note that (15.162) gives the contravariant field tensor components, so we have to lower the indices with the Minkowksi metric.

Final Retarded Field

We record here the field $F := \tilde{F} + \Delta F$, which we claim to be the retarded field due to the uniformly accelerated charge:

$$F^{tz} = \frac{e}{4\pi} g \frac{Z^2 - g^{-2} - \rho^2}{2R^3} \theta(z+t) , \tag{15.178}$$

$$F^{tx} = \frac{e}{4\pi} x \left[\frac{gz}{R^3} \theta(z+t) + \frac{2g^2}{1 + g^2\rho^2} \delta(z+t) \right] , \tag{15.179}$$

$$F^{ty} = \frac{e}{4\pi} y \left[\frac{gz}{R^3} \theta(z+t) + \frac{2g^2}{1 + g^2\rho^2} \delta(z+t) \right] , \tag{15.180}$$

$$F^{zx} = \frac{e}{4\pi} x \left[\frac{gt}{R^3} \theta(z+t) - \frac{2g^2}{1 + g^2\rho^2} \delta(z+t) \right] , \tag{15.181}$$

$$F^{zy} = \frac{e}{4\pi} y \left[\frac{gt}{R^3} \theta(z+t) - \frac{2g^2}{1 + g^2\rho^2} \delta(z+t) \right] . \tag{15.182}$$

It seems likely that these could be derived directly from the potential (15.105) if we were more careful about the step function in the integrand.

15.10 Origin of the Delta Function in the Field

Boulware has discovered an interesting way to see why the delta function terms arise. He considers a charge that was initially free, i.e., it has not always undergone uniform acceleration. (This is reminiscent of the derivation of the Lorentz–Dirac equation.) We assume that the charge was at rest at $z = 1/g$ for $t < 0$ and subsequently underwent uniform acceleration (see Fig. 15.16). We do not concern ourselves about the transition between these two motions, a possible source of problems.

If we consider a field point outside the future light cone of the event at which the acceleration began, it will only be affected by signals from the part of the worldline where the charge was at rest, so the field at such a field point will be the Coulomb field. If we consider a field point inside the future light cone of the event at which the acceleration began, and consider the backward light cone of this point, we find that it contains some inertial points and some accelerating points on the worldline. However, the only point on the worldline that can signal to our chosen field point will be a point within the future light cone of the point where the acceleration began, and so the field there will be the field of a uniformly accelerated charge which we have just calculated.

If we therefore define the quantity

$$r := \left[\rho^2 + (z - g^{-1})^2\right]^{1/2} , \tag{15.183}$$

which is the spatial separation of the field point from the spatial point at which the acceleration began, as measured in the rest frame at $t = 0$, the field in the scenario we are describing here must be the superposition:

$$F^{tx} = \frac{e}{4\pi} \frac{x}{r^3} \theta(r - t) + \frac{e}{4\pi} \frac{xgz}{R^3} \theta(t - r) , \tag{15.184}$$

$$F^{ty} = \frac{e}{4\pi} \frac{y}{r^3} \theta(r - t) + \frac{e}{4\pi} \frac{ygz}{R^3} \theta(t - r) , \tag{15.185}$$

$$F^{tz} = \frac{e}{4\pi} \frac{z - 1/g}{r^3} \theta(r - t) + \frac{e}{4\pi} \frac{g(Z^2 - g^{-2} - \rho^2)}{2R^3} \theta(t - r) , \tag{15.186}$$

$$F^{zx} = \frac{e}{4\pi} \frac{xgt}{R^3} \theta(t - r) , \qquad F^{zy} = \frac{e}{4\pi} \frac{ygt}{R^3} \theta(t - r) . \tag{15.187}$$

We now carry out a Lorentz transformation. Boulware is vague about this, but we shall be very explicit, even though the result appears slightly contorted. We eventually accord with Boulware on the final result.

The approach here is illustrated in Figs. 15.16 and 15.17. Figure 15.16 shows the past and future light cones at the origin of an inertial frame, the spatial axis of

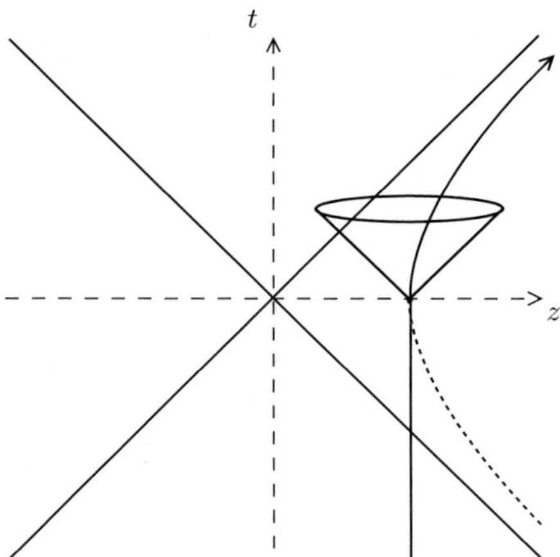

Fig. 15.16 The charge has been sitting at $z = 1/g$ for all time, but begins to uniformly accelerate at Minkowski time $t = 0$. The EM fields within the future light cone of the event where acceleration begins are precisely those of the eternally uniformly accelerating charge, while those outside the future light cone are precisely the Coulomb field of a motionless charge at $z = 1/g$

acceleration as a dashed line, the hyperbolic trajectory that a perpetually uniformly accelerating charge would have followed as a dotted line, and the vertical line representing the newly imagined worldline of a charge that is stationary in this inertial frame up until time $t = 0$ whereafter it coincides with the hyperbolic trajectory. Figure 15.17 is the same except that it is viewed from an inertial frame (primed coordinates) moving to the right with respect to the first, so that the straight part of the new worldline is slanting to the left as t' increases, and meets the hyperbolic trajectory when $t' < 0$.

So we take Fig. 15.17 to represent the same physical situation as Fig. 15.16, but viewed from a primed coordinate system given by

$$z' = z\cosh\alpha - t\sinh\alpha , \qquad t' = t\cosh\alpha - z\sinh\alpha . \tag{15.188}$$

In this system, the old time axis is given by $z = 0$ and it leans over to the left. The old space axis is given by $t = 0$ and it moves off at an angle below the new space axis. In the primed picture, the charge worldline for positive unprimed time looks the same because it has the same formula, viz., $z'^2 - t'^2 = g^{-2}$, as it did in the unprimed picture, viz., $z^2 - t^2 = g^{-2}$. However, in the primed picture, the old space axis intersects it below the new space axis, in the zone of negative primed time. This intersection is the event specifying the beginning of the acceleration. As α is increased in our Lorentz transformation, i.e., as we consider bigger boosts to view the system from a faster-moving frame, the event specifying the beginning of acceleration moves further back in the time of that frame. This is a key point.

The part of the charge worldline in the negative unprimed time zone, where the charge motion is inertial, is still a straight line in the primed view, but leaning over to

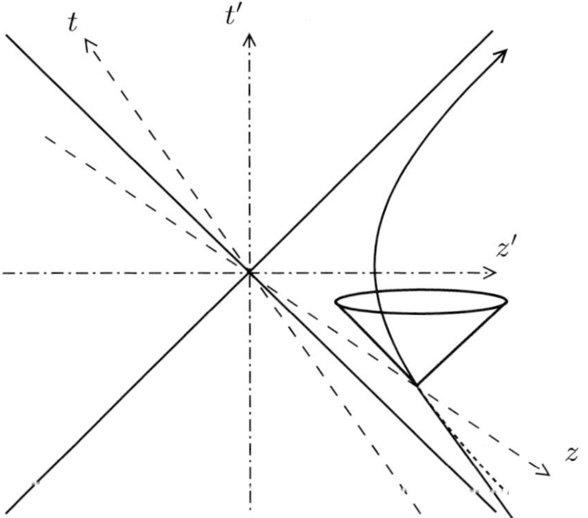

Fig. 15.17 View of the situation in Fig. 15.16 from an inertial frame moving to the right with respect to the inertial frame in the first figure. The EM fields within the future light cone of the event where acceleration begins are precisely those of the eternally uniformly accelerating charge, while those outside the future light cone are precisely the Coulomb field of a motionless charge at $z = 1/g$. As the speed of this new frame increases, the event where acceleration begins slips back into the past of the primed frame and the future light cone of this event occupies more and more of the region $z' + t' > 0$

the left, and in fact, parallel to the old time axis. The bigger the value of α, the more it leans over towards the null line $z + t = 0$ (which is the same line as $z' + t' = 0$). As $\alpha \to \infty$, this inertial part of the charge motion begins to approach the null line and the point where it ends moves down to more and more negative values of the primed time. Eventually, in the limit, we retrieve our picture of a charge decelerating uniformly from the speed of light at high z values, moving in to $z = g^{-1}$, and then accelerating uniformly back up to the speed of light. At this point, we drop all the primes in our transformed fields and compare them with (15.178)–(15.182) to see where the delta-function term came from.

The fields in (15.184)–(15.187) are the unprimed fields, giving an electromagnetic field tensor

$$
F = \begin{pmatrix}
0 & F^{tx} & F^{ty} & F^{tz} \\
-F^{tx} & 0 & 0 & F^{xz} \\
-F^{ty} & 0 & 0 & F^{yz} \\
-F^{tz} & -F^{xz} & -F^{yz} & 0
\end{pmatrix}.
\tag{15.189}
$$

The primed version of this, although the arguments of functions have yet to be reexpressed in terms of the primed coordinates, is then

$$F' = \begin{pmatrix} C(\alpha) & 0 & 0 & -S(\alpha) \\ 0 & 1 & 0 & 0 \\ 0 & 0 & 1 & 0 \\ -S(\alpha) & 0 & 0 & C(\alpha) \end{pmatrix} \begin{pmatrix} 0 & F^{tx} & F^{ty} & F^{tz} \\ -F^{tx} & 0 & 0 & F^{xz} \\ -F^{ty} & 0 & 0 & F^{yz} \\ -F^{tz} & -F^{xz} & -F^{yz} & 0 \end{pmatrix}$$

$$\begin{pmatrix} C(\alpha) & 0 & 0 & -S(\alpha) \\ 0 & 1 & 0 & 0 \\ 0 & 0 & 1 & 0 \\ -S(\alpha) & 0 & 0 & C(\alpha) \end{pmatrix} , \tag{15.190}$$

where we have used the shorthand $C(\alpha) := \cosh\alpha$, $S(\alpha) := \sinh\alpha$. Carrying out the matrix multiplication, we have

$$F' = \begin{pmatrix} 0 & F'^{tx} & F'^{ty} & F'^{tz} \\ -F'^{tx} & 0 & 0 & F'^{xz} \\ -F'^{ty} & 0 & 0 & F'^{yz} \\ -F'^{tz} & -F'^{xz} & -F'^{yz} & 0 \end{pmatrix} , \tag{15.191}$$

where

$$F'^{tx} = F^{tx}C(\alpha) + F^{xz}S(\alpha) , \quad F'^{ty} = F^{ty}C(\alpha) + F^{yz}S(\alpha) , \quad F'^{tz} = F^{tz} , \tag{15.192}$$

$$F'^{xz} = F^{tx}S(\alpha) + F^{xz}C(\alpha) , \quad F'^{yz} = F^{ty}S(\alpha) + F^{yz}C(\alpha) . \tag{15.193}$$

Now

$$\begin{aligned} F'^{tx} &= F^{tx}\cosh\alpha + F^{xz}\sinh\alpha \\ &= \frac{e}{4\pi}\left[\frac{x}{r^3}\theta(r-t) + \frac{xgz}{R^3}\theta(t-r)\right]\cosh\alpha - \frac{e}{4\pi}\frac{xgt}{R^3}\theta(t-r)\sinh\alpha \\ &= \frac{e}{4\pi}\frac{x}{r^3}\theta(r-t)\cosh\alpha + \frac{e}{4\pi}\frac{xg}{R^3}\theta(t-r)(z\cosh\alpha - t\sinh\alpha) \\ &= \frac{e}{4\pi}\left[\frac{x}{r^3}\theta(r-t)\cosh\alpha + \frac{xgz'}{R^3}\theta(t-r)\right] , \end{aligned}$$

where we have used the fact that

$$z' = z\cosh\alpha - t\sinh\alpha .$$

We have to express everything in terms of the primed coordinates, so we need the inverse Lorentz transformation, viz.,

$$z = z'\cosh\alpha + t'\sinh\alpha , \qquad t = t'\cosh\alpha + z'\sinh\alpha . \tag{15.194}$$

Of course, $x = x'$, $y = y'$, and $\rho^2 = x'^2 + y'^2$. We now have

$$r = \left[\rho^2 + (z' \cosh\alpha + t' \sinh\alpha - g^{-1})^2\right]^{1/2} . \tag{15.195}$$

Hence,

$$F'^{tx} = \frac{e}{4\pi} \frac{x' \cosh\alpha}{r^3} \theta(r - t' \cosh\alpha - z' \sinh\alpha)$$
$$+ \frac{e}{4\pi} \frac{x' g z'}{R^3} \theta(t' \cosh\alpha + z' \sinh\alpha - r) ,$$

where we note that R is unchanged in form when expressed in terms of the primed coordinates (i.e., it is invariant under the Lorentz transformation in the z direction), because it is a function of ρ^2 and Z^2 alone, and

$$\rho'^2 = x'^2 + y'^2 = x^2 + y^2 ,$$

$$Z'^2 = z'^2 - t'^2 = z^2 - t^2 .$$

At some risk of confusion, we now drop all the primes in our expression. The reason for this is simply that we hope to compare the result of the Lorentz transformation (at present primed) with the fields (15.178)–(15.182). The result is finally

$$F^{tx} = \frac{e}{4\pi} \frac{x \cosh\alpha}{r^3} \theta(r - t \cosh\alpha - z \sinh\alpha)$$
$$+ \frac{e}{4\pi} \frac{x g z}{R^3} \theta(t \cosh\alpha + z \sinh\alpha - r) , \tag{15.196}$$

where

$$r := \left[\rho^2 + (z \cosh\alpha + t \sinh\alpha - g^{-1})^2\right]^{1/2} . \tag{15.197}$$

In this way, we also establish the following:

$$F^{ty} = \frac{e}{4\pi} \frac{y \cosh\alpha}{r^3} \theta(r - t \cosh\alpha - z \sinh\alpha)$$
$$+ \frac{e}{4\pi} \frac{y g z}{R^3} \theta(t \cosh\alpha + z \sinh\alpha - r) , \tag{15.198}$$

$$F^{tz} = \frac{e}{4\pi} \frac{z \cosh\alpha + t \sinh\alpha - g^{-1}}{r^3} \theta(r - t \cosh\alpha - z \sinh\alpha)$$
$$+ \frac{e}{4\pi} \frac{g(Z^2 - g^{-2} - \rho^2)}{2R^3} \theta(t \cosh\alpha + z \sinh\alpha - r) , \tag{15.199}$$

$$F^{zx} = -\frac{e}{4\pi}\frac{x\sinh\alpha}{r^3}\theta(r-t\cosh\alpha-z\sinh\alpha)$$

$$+\frac{e}{4\pi}\frac{xgt}{R^3}\theta(t\cosh\alpha+z\sinh\alpha-r)\,,\qquad(15.200)$$

$$F^{zy} = -\frac{e}{4\pi}\frac{y\sinh\alpha}{r^3}\theta(r-t\cosh\alpha-z\sinh\alpha)$$

$$+\frac{e}{4\pi}\frac{ygt}{R^3}\theta(t\cosh\alpha+z\sinh\alpha-r)\,.\qquad(15.201)$$

Note the difference with Boulware's stated result in the last two, where we have a minus sign before the first term.

As mentioned above, in the limit as $\alpha \to \infty$, the time at which the uniform acceleration began goes to $-\infty$ and the initial speed of approach tends to c. Furthermore, any point with $z+t > 0$ lies inside the forward light cone of the point at which the acceleration started. This is totally obvious from the spacetime diagram in Fig. 15.17, and nothing is gained by formulating algebraically as Boulware does. Likewise, any point with $z+t < 0$ lies outside the light cone of the point at which the acceleration started.

We note that the value of the field inside this light cone does not depend on α. This can be seen directly from (15.196)–(15.201), by looking at the coefficients of the terms in $\theta(t\cosh\alpha + z\sinh\alpha - r)$ in these formulas. Therefore, for any point with $z+t > 0$, the field eventually (in the limit as $\alpha \to \infty$) becomes just the field of the uniformly accelerated charge. This too can be seen directly from the formulas, comparing the coefficients of the terms in $\theta(t+z)$ in (15.178)–(15.182) with the coefficients of the terms in $\theta(t\cosh\alpha + z\sinh\alpha - r)$ in (15.196)–(15.201). (We now see why it was useful to drop the primes earlier, as they might have confused the comparison we are making now.)

In the region $z+t < 0$, we have

$$r = \left[\rho^2 + (z\cosh\alpha + t\sinh\alpha - g^{-1})^2\right]^{1/2}$$

$$= \left\{\rho^2 + \left[(z+t)\sinh\alpha + z(\cosh\alpha - \sinh\alpha) - g^{-1}\right]^2\right\}^{1/2}$$

$$\sim |z+t|\sinh\alpha + g^{-1}$$

$$\sim \frac{1}{2}|z+t|e^\alpha + g^{-1}$$

$$\sim -\frac{1}{2}(z+t)e^\alpha + g^{-1}\,.$$

Now the field in this region is the limit of the terms in $\theta(r - t\cosh\alpha - z\sinh\alpha)$ in (15.196)–(15.201). But each of these has a $\sinh\alpha$ or a $\cosh\alpha$ divided by r^3, so they go as $e^{-2\alpha}$, which means that they go to zero, as $\alpha \to \infty$. In the limit there is zero field in this zone, just as predicted by (15.178)–(15.182).

It remains to analyse what happens on the null line $z + t = 0$ itself in the limit as $\alpha \to \infty$. We now have

$$
\begin{aligned}
r &= \left[\rho^2 + (e^{-\alpha} - g^{-1})^2\right]^{1/2} \\
&\sim (\rho^2 + g^{-2})^{1/2} \\
&= g^{-1}(1 + \rho^2 g^2)^{1/2} \, .
\end{aligned}
\tag{15.202}
$$

Since r is now finite as $\alpha \to \infty$, and since the parts of the fields (15.196)–(15.201) which are relevant when $z + t = 0$ are the terms in $\theta(r - t \cosh\alpha - z \sinh\alpha)$, which all contain factors of $\sinh\alpha$ or $\cosh\alpha$, these fields tend to infinity on this null line. We expect this from the delta functions in (15.178)–(15.182). In order to show that we obtain these delta functions in the limit $\alpha \to \infty$ when $z + t = 0$, we can integrate the field components over t from $-\infty$ up to the light cone of the event when the acceleration began and then take the limit as $\alpha \to \infty$.

We begin by integrating (15.178)–(15.182) in this way to see what we should expect. To begin with, the formula (15.178) for F^{tz} has no delta function, so we expect zero. Looking at the delta function terms in (15.179) and (15.181), we also have

$$
\int_{-\infty}^{\infty} (F^{tx} + F^{zx}) \, dt = 0 \, ,
\tag{15.203}
$$

$$
\int_{-\infty}^{\infty} (F^{tx} - F^{zx}) \, dt = \frac{e}{4\pi} \int_{-\infty}^{\infty} \left[\frac{4g^2 x}{1 + g^2 \rho^2} \delta(z + t) \right] dt = \frac{e}{4\pi} \frac{4g^2 x}{1 + g^2 \rho^2} \, .
\tag{15.204}
$$

If we use (15.196) and (15.200) to calculate

$$
\int_{-\infty}^{\infty} (F^{tx} + F^{zx}) \theta(r - t \cosh\alpha - z \sinh\alpha) \, dt \, ,
$$

we obtain

$$
\frac{e}{4\pi} \int_{-\infty}^{r = t \cosh\alpha + z \sinh\alpha} \frac{x(\cosh\alpha - \sinh\alpha)}{r^3} \, dt \, .
$$

Now this contains the factor $\cosh\alpha - \sinh\alpha = e^{-\alpha} \to 0$ as $\alpha \to \infty$. This confirms that the result (15.200) is right, despite the sign difference with Boulware.

If we now look at (15.199), we have

$$\int_{-\infty}^{\infty} F'^{tz}\theta(r - t\cosh\alpha - z\sinh\alpha)\,dt$$

$$= \frac{e}{4\pi}\int_{-\infty}^{r=t\cosh\alpha+z\sinh\alpha} \frac{z\cosh\alpha + t\sinh\alpha - g^{-1}}{r^3}\,dt$$

$$= \frac{e}{4\pi}\int_{-\infty}^{r=t\cosh\alpha+z\sinh\alpha} \frac{z\cosh\alpha + t\sinh\alpha - g^{-1}}{\left[\rho^2 + (z\cosh\alpha + t\sinh\alpha - g^{-1})^2\right]^{3/2}}\,dt$$

$$= -\frac{e}{4\pi}\frac{1}{\sinh\alpha}\left[\left[\rho^2 + (z\cosh\alpha + t\sinh\alpha - g^{-1})^2\right]^{-1/2}\right]_{-\infty}^{r=t\cosh\alpha+z\sinh\alpha}.$$

The lower bound of the integral ($t = -\infty$) gives zero. What about the upper bound? We have $r = t\cosh\alpha + z\sinh\alpha$ iff

$$\rho^2 + (z\cosh\alpha + t\sinh\alpha - g^{-1})^2 = (t\cosh\alpha + z\sinh\alpha)^2,$$

where ρ and z are fixed and we eliminate t. This gives a quadratic in t, viz.,

$$t^2 + 2g^{-1}(z\cosh\alpha + t\sinh\alpha) - \rho^2 - z^2 - g^{-2} = 0. \tag{15.205}$$

What happens to the value of t as $\alpha \to \infty$? We have to be very careful with the limits here! Solving the quadratic and taking the correct root, we have

$$t = \left[(g^{-1}\cosh\alpha - z)^2 + \rho^2\right]^{1/2} - g^{-1}\sinh\alpha. \tag{15.206}$$

From the Minkowski diagram of Fig. 15.17, we can see that the future light cone of the event where the acceleration began moves down to touch $z = -t$ in the limit as $\alpha \to \infty$. But if we put $z = -t$ in (15.205) and then take the limit $\alpha \to \infty$, we obtain the absurd requirement $\rho^2 + g^{-2} = 0$. Furthermore, $z = -t$ implies that

$$t\cosh\alpha + z\sinh\alpha = te^{-\alpha} \to 0,$$

which is equally absurd. In fact (15.206) tells us that

$$t = -z + \varepsilon(\alpha), \quad \text{where } \varepsilon(\alpha) \to 0 \quad \text{as } \alpha \to \infty. \tag{15.207}$$

Now we have

$$t\cosh\alpha + z\sinh\alpha = \varepsilon(\alpha)\cosh\alpha - z(\cosh\alpha - \sinh\alpha)$$

$$= \varepsilon(\alpha)\cosh\alpha - ze^{-\alpha},$$

and this does not have to tend to zero as $\alpha \to \infty$. Indeed it must tend to the same limit as the expression for r, viz.,

$$\left[\rho^2 + (z\cosh\alpha + t\sinh\alpha - g^{-1})^2\right]^{1/2},$$

when $t = -z + \varepsilon(\alpha)$.

Now we have to find

$$-\frac{e}{4\pi}\frac{1}{\sinh\alpha}\frac{1}{r}\bigg|_{t=-z+\varepsilon(\alpha)}.$$

Without going into the details here, let us surmise that, as $\alpha \to \infty$ and $t \downarrow -z$, we do have the finite limit

$$r \longrightarrow g^{-1}(1 + \rho^2 g^2)^{1/2}$$

which we estimated in (15.202). This should really be checked very carefully by estimating $\varepsilon(\alpha)$. In fact, the result for r is not quite the same, as we shall see in (15.210) on p. 227. The result there is a finite limit anyway, so we immediately deduce that

$$\lim_{\alpha\to\infty} -\frac{e}{4\pi}\frac{1}{\sinh\alpha}\frac{1}{r}\bigg|_{t=-z+\varepsilon(\alpha)} = 0 \,,$$

because of the factor of $1/\sinh\alpha$. This agrees with what was expected from the formula (15.178) for F^{tz}, which has no delta function term.

Let us now find $\varepsilon(\alpha)$. We have from (15.206),

$$t = \left[(g^{-1}\cosh\alpha - z)^2 + \rho^2\right]^{1/2} - g^{-1}\sinh\alpha$$

$$= \left[g^{-2}\cosh^2\alpha - 2g^{-1}z\cosh\alpha + z^2 + \rho^2\right]^{1/2} - g^{-1}\sinh\alpha$$

$$= g^{-1}\cosh\alpha\left[1 - \frac{2z}{g^{-1}\cosh\alpha} + \frac{z^2 + \rho^2}{g^{-2}\cosh^2\alpha}\right]^{1/2} - g^{-1}\sinh\alpha \,.$$

We now expand

$$\left[1 - 2z\delta + (z^2 + \rho^2)\delta^2\right]^{1/2}$$

to second order in $\delta = g/\cosh\alpha$, which is small. We have the Taylor expansion

$$\left[1 - 2z\delta + (z^2 + \rho^2)\delta^2\right]^{1/2} \approx 1 - z\delta + \frac{1}{2}(z^2 + \rho^2)\delta^2$$

$$+ \frac{1}{2!}\left(-\frac{1}{4}\right)\left[-2z\delta + (z^2 + \rho^2)\delta^2\right]^2$$

$$\approx 1 - z\delta + \frac{1}{2}(z^2 + \rho^2)\delta^2 - \frac{1}{2}z^2\delta^2$$

$$= 1 - z\delta + \frac{1}{2}\rho^2\delta^2 \,.$$

We now return to

$$t \approx g^{-1}\cosh\alpha\left(1 - z\delta + \frac{1}{2}\rho^2\delta^2\right) - g^{-1}\sinh\alpha$$

$$= g^{-1}(\cosh\alpha - \sinh\alpha) - z + \frac{\rho^2 g}{2\cosh\alpha}$$

$$\approx -z + (g^{-1} + \rho^2 g)e^{-\alpha} \, . \tag{15.208}$$

As predicted, t has the form $t = -z + \varepsilon(\alpha)$, as proposed in (15.207), with

$$\varepsilon(\alpha) = (g^{-1} + \rho^2 g)e^{-\alpha} \, , \tag{15.209}$$

which does indeed tend to zero as $\alpha \to \infty$.

We can now just check that the two expressions

$$t\cosh\alpha + z\sinh\alpha \quad \text{and} \quad \left[\rho^2 + (z\cosh\alpha + t\sinh\alpha - g^{-1})^2\right]^{1/2}$$

tend to the same limit as $\alpha \to \infty$, when t has the form given in (15.207). To begin with, note that

$$t\cosh\alpha + z\sinh\alpha \approx -ze^{-\alpha} + (g^{-1} + \rho^2 g)e^{-\alpha}\cosh\alpha$$

$$\approx \frac{1}{2}(g^{-1} + \rho^2 g) \, .$$

We note that this squares to

$$\frac{1}{4}(g^{-1} + \rho^2 g)^2 = \frac{1}{4}(g^{-2} + 2\rho^2 + \rho^4 g^2) \, .$$

The calculation for the other expression is slightly more long-winded. We calculate the square:

$$\rho^2 + (z\cosh\alpha + t\sinh\alpha - g^{-1})^2 \approx \rho^2 + \left[ze^{-\alpha} + (g^{-1} + \rho^2 g)e^{-\alpha}\sinh\alpha - g^{-1}\right]^2$$

$$\approx \rho^2 + \left(-\frac{1}{2}g^{-1} + \frac{1}{2}\rho^2 g\right)^2$$

$$= \frac{1}{4}(g^{-2} + 2\rho^2 + \rho^4 g^2) \, ,$$

as required.

We can now check the integral for F^{tz}. We obtained

$$-\frac{e}{4\pi}\frac{1}{\sinh\alpha}\frac{1}{r}\bigg|_{t=-z+\varepsilon(\alpha)} \, .$$

We have shown above that, as $\alpha \to \infty$ and $t \downarrow -z$, we do have the finite limit

$$r \longrightarrow \frac{1}{2} g^{-1}(1 + \rho^2 g^2) \,, \tag{15.210}$$

which is not quite the same as (15.202) on p. 223, due to a factor of 1/2 and a power of 1/2. Hence,

$$-\frac{e}{4\pi} \frac{1}{\sinh \alpha} \frac{1}{r} \bigg|_{t=-z+\varepsilon(\alpha)} = -\frac{e}{4\pi} \frac{1}{\sinh \alpha} \frac{2g}{1+\rho^2 g^2} \longrightarrow 0 \,.$$

We only required the $1/r$ factor to have a finite limit, since we then divide by $\sinh \alpha$.

Why are the limits (15.202) and (15.210) for r different? In (15.202), we were just examining r as a function of t, x, y, z for different values of α, and noting that, whenever $z + t = 0$, we get the finite limit (15.202) as $\alpha \to \infty$. In (15.210), we have examined the condition $r = t \cosh \alpha + z \sinh \alpha$ as a condition on t for given x, y, z and α, so that z is not a priori equal to $-t$, although it is in the limit $\alpha \to \infty$, and we have then evaluated the limit of the associated value of r as $\alpha \to \infty$.

If we use (15.196) and (15.200) to calculate

$$\int_{-\infty}^{\infty} (F^{tx} - F^{zx}) \theta(r - t \cosh \alpha - z \sinh \alpha) \mathrm{d}t \,,$$

we obtain

$$\frac{e}{4\pi} \int_{-\infty}^{r=t \cosh \alpha + z \sinh \alpha} \frac{x(\cosh \alpha + \sinh \alpha)}{r^3} \mathrm{d}t \,.$$

We must now carry out the integral

$$\frac{ex}{4\pi} e^{\alpha} \int_{-\infty}^{r=t \cosh \alpha + z \sinh \alpha} \frac{\mathrm{d}t}{\left[\rho^2 + (z \cosh \alpha + t \sinh \alpha - g^{-1})^2\right]^{3/2}} \,. \tag{15.211}$$

We make the substitution

$$T := \frac{1}{\rho}(z \cosh \alpha + t \sinh \alpha - g^{-1}) \,,$$

so that

$$\mathrm{d}T = \frac{1}{\rho} \sinh \alpha \, \mathrm{d}t \,,$$

and the integral becomes

$$\frac{ex}{4\pi} \frac{e^{\alpha}}{\rho^2 \sinh \alpha} \int_{-\infty}^{T_{\mathrm{upper}}} \frac{\mathrm{d}T}{(1 + T^2)^{3/2}} \,, \tag{15.212}$$

where T_{upper} is the value of T when $r = t \cosh \alpha + z \sinh \alpha$.

The whole problem now is to establish the value of T_{upper}, because the integral is easy:

$$\int \frac{dT}{(1+T^2)^{3/2}} = \text{constant} + \frac{T}{(1+T^2)^{1/2}} \ .$$

It follows that our integral is

$$\frac{ex}{4\pi}\frac{e^\alpha}{\rho^2 \sinh\alpha}\left[1 + \frac{T_{\text{upper}}}{(1+T_{\text{upper}}^2)^{1/2}}\right] \ . \tag{15.213}$$

We note that $e^\alpha/\sinh\alpha \to 2$ as $\alpha \to \infty$. We must therefore calculate the limit of

$$\left[1 + \frac{T_{\text{upper}}}{(1+T_{\text{upper}}^2)^{1/2}}\right]$$

as $\alpha \to \infty$. We only need the highest order terms, i.e., powers of e^α, if there are any, or constant terms, or powers of $e^{-\alpha}$, if there are no higher order terms.

Now we have established that $t_{\text{upper}} = -z + \varepsilon(\alpha)$, where $\varepsilon(\alpha)$ is given by (15.209), viz.,

$$\varepsilon(\alpha) = (g^{-1} + \rho^2 g)e^{-\alpha} \ . \tag{15.214}$$

Furthermore,

$$T_{\text{upper}} = \frac{1}{\rho}(z\cosh\alpha + t_{\text{upper}}\sinh\alpha - g^{-1}) \ . \tag{15.215}$$

This means that the terms of order e^α in T_{upper} cancel, leaving

$$T_{\text{upper}} = \frac{1}{\rho}\left[ze^{-\alpha} + (g^{-1} + \rho^2 g)e^{-\alpha}\sinh\alpha - g^{-1}\right] \ . \tag{15.216}$$

We can drop the terms in $e^{-\alpha}$ or smaller. Hence,

$$\lim_{\alpha\to\infty} T_{\text{upper}} = \frac{\rho^2 g^2 - 1}{2\rho g} \ . \tag{15.217}$$

With a little manipulation, we now find

$$1 + \frac{T_{\text{upper}}}{(1+T_{\text{upper}}^2)^{1/2}} = \frac{2\rho^2 g^2}{1+\rho^2 g^2} \ , \tag{15.218}$$

in the limit as $\alpha \to \infty$, and finally, the integral we started with becomes

$$\int_{-\infty}^{\infty}(F^{tx} - F^{zx})\theta(r - t\cosh\alpha - z\sinh\alpha)dt = \frac{ex}{4\pi}\frac{4g^2}{1+\rho^2 g^2} \ . \tag{15.219}$$

This agrees entirely with (15.204), which was found by looking at the delta function terms in (15.179) and (15.181).

What we have shown is that the delta function terms in the retarded fields (15.178)–(15.182) can be thought of as arising in the above limit from the Coulomb field of the charge before it began its acceleration, but viewed from a fast-moving frame. In a way, this feature indicates how idealistic this situation is, not just because a uniform acceleration is already idealistic, but because of the symmetry in time and the fact that it started infinitely long ago and will continue forever.

15.11 Conclusions Regarding the Fields

The aim here is to outline and criticise Boulware's conclusions regarding the electromagnetic fields produced by the uniformly accelerating charge.

15.11.1 Fields in Region I

Region I is the region $z > |t| \geq 0$. The fields here are just the retarded fields $\tilde{F}$ which we first worked out in (15.129), (15.130), and (15.135), i.e., the non-delta-function pieces of (15.178)–(15.182). Not surprisingly these are Lorentz invariant, since the charge worldline is itself Lorentz invariant. This invariance can be seen at a glance from the formulas for these fields in region I:

$$F^{tz} = \frac{e}{4\pi} g \frac{Z^2 - g^{-2} - \rho^2}{2R^3} , \tag{15.220}$$

$$F^{t\rho} = \frac{e}{4\pi} \rho \frac{gz}{R^3} , \tag{15.221}$$

$$F^{z\rho} = \frac{e}{4\pi} \rho \frac{gt}{R^3} , \tag{15.222}$$

in conjunction with the component transformation formulas (15.192) and (15.193) on p. 220, viz.,

$$F'^{tx} = F^{tx}C(\alpha) + F^{xz}S(\alpha) , \quad F'^{ty} = F^{ty}C(\alpha) + F^{yz}S(\alpha) , \quad F'^{tz} = F^{tz} , \tag{15.223}$$

$$F'^{xz} = F^{tx}S(\alpha) + F^{xz}C(\alpha) , \quad F'^{yz} = F^{ty}S(\alpha) + F^{yz}C(\alpha) , \tag{15.224}$$

where the Lorentz transformation is

$$z' = z\cosh\alpha - t\sinh\alpha\,, \qquad t' = t\cosh\alpha - z\sinh\alpha\,. \tag{15.225}$$

Note that this differs from Boulware's transformation by $\alpha \to -\alpha$, which is just a different choice of α, and irrelevant to the result. We note that

$$Z^2 = z^2 - t^2 = z'^2 - t'^2 = Z'^2 \quad\text{and}\quad \rho^2 = x^2 + y^2 = x'^2 + y'^2 = \rho'^2\,.$$

Hence, both $Z^2 - g^{-2} - \rho^2$ and R^3 are invariant. We have

$$F^{t'z'}(z',t') = F^{tz}(z,t) = \frac{e}{4\pi}g\frac{Z'^2 - g^{-2} - \rho'^2}{2R'^3}\,,$$

$$\begin{aligned}
F^{t'\rho'}(z',t') &= F^{t\rho}(z,t)\cosh\alpha + F^{\rho z}(z,t)\sinh\alpha \\
&= \frac{e}{4\pi}\rho\,\frac{g(z\cosh\alpha - t\sinh\alpha)}{R^3} \\
&= \frac{e}{4\pi}\rho'\frac{gz'}{R'^3}\,,
\end{aligned}$$

$$\begin{aligned}
F^{z'\rho'}(z',t') &= F^{z\rho}(z,t)\cosh\alpha + F^{\rho t}(z,t)\sinh\alpha \\
&= \frac{e}{4\pi}\rho\,\frac{g(t\cosh\alpha - z\sinh\alpha)}{R^3} \\
&= \frac{e}{4\pi}\rho'\frac{gt'}{R'^3}\,.
\end{aligned}$$

Each expression is identical to the expression for the unprimed component, with primed coordinates replacing unprimed coordinates in functional arguments. This is what we mean by invariance.

There is another obvious, and totally unsurprising invariance, namely, time-reversal invariance. It is easy to check that the fields are invariant under

$$t \longrightarrow -t\,, \qquad F^{tv} \longrightarrow F^{tv}\,, \qquad F^{z\rho} \longrightarrow -F^{z\rho}\,. \tag{15.226}$$

Note that the magnetic fields change sign, but not the electric fields. Our whole scenario is symmetric under reflection in the space axis of the spacetime diagram. Note, however, that these are the retarded fields, referring to a direction in time. In fact, it is a general result that retarded fields are transformed into advanced fields under the above transformation. Since the retarded fields turn out to be invariant, this shows that the retarded fields just happen to be equal to the advanced fields in this precise case. Once again, this points to the idealistic nature of our scenario.

Rindler Components of Fields in Region I

This is perhaps the best moment to show two key results. In the Rindler (SE) frame:

- the electric field components are independent of the Rindler time coordinate,
- the magnetic field components are zero.

This can be shown by transforming the fields (15.220)–(15.222) in the Minkowski frame, but since we have the vector potential in the Rindler frame from (15.115) on p. 201 and (15.119) on p. 202, this provides a simpler proof. We thus begin with

$$A_\tau = -\frac{eg}{4\pi}\,\frac{g^{-2}+\rho^2+Z^2}{\left[(g^{-2}+\rho^2+Z^2)^2-4Z^2g^{-2}\right]^{1/2}}\,, \qquad (15.227)$$

$$A_Z = -\frac{e}{4\pi Z}\,, \qquad A_x = 0 = A_y\,. \qquad (15.228)$$

The electromagnetic field tensor is

$$F_{\mu\nu} = A_{\nu,\mu} - A_{\mu,\nu} = \begin{pmatrix} 0 & E_1 & E_2 & E_3 \\ -E_1 & 0 & -B_3 & B_2 \\ -E_2 & B_3 & 0 & -B_1 \\ -E_3 & -B_2 & B_1 & 0 \end{pmatrix}\,, \qquad (15.229)$$

which defines the 'electric' and 'magnetic' fields in any coordinates and explains how to get them in terms of the components of the 4-potential. Hence,

$$\mathbf{E}_{\mathrm{SE}} = \begin{pmatrix} F_{01} \\ F_{02} \\ F_{03} \end{pmatrix} = \begin{pmatrix} A_{1,0}-A_{0,1} \\ A_{2,0}-A_{0,2} \\ A_{3,0}-A_{0,3} \end{pmatrix} = \begin{pmatrix} -\dfrac{\partial A_\tau}{\partial Z}+\dfrac{\partial A_Z}{\partial \tau} \\[2mm] -\dfrac{\partial A_\tau}{\partial x}+\dfrac{\partial A_x}{\partial \tau} \\[2mm] -\dfrac{\partial A_\tau}{\partial y}+\dfrac{\partial A_y}{\partial \tau} \end{pmatrix}\,, \qquad (15.230)$$

$$\mathbf{B}_{\mathrm{SE}} = \begin{pmatrix} F_{32} \\ F_{13} \\ F_{21} \end{pmatrix} = \begin{pmatrix} A_{2,3}-A_{3,2} \\ A_{3,1}-A_{1,3} \\ A_{1,2}-A_{2,1} \end{pmatrix} = \begin{pmatrix} \dfrac{\partial A_x}{\partial y}-\dfrac{\partial A_y}{\partial x} \\[2mm] \dfrac{\partial A_y}{\partial Z}-\dfrac{\partial A_Z}{\partial y} \\[2mm] \dfrac{\partial A_Z}{\partial x}-\dfrac{\partial A_x}{\partial Z} \end{pmatrix}\,. \qquad (15.231)$$

Since A_Z is independent of x and y, we immediately have

$$\mathbf{B}_{\mathrm{SE}} = 0\,. \qquad (15.232)$$

We also have

$$\mathbf{E}_{\mathrm{SE}} = - \begin{pmatrix} \dfrac{\partial A_\tau}{\partial Z} \\[2mm] \dfrac{\partial A_\tau}{\partial x} \\[2mm] \dfrac{\partial A_\tau}{\partial y} \end{pmatrix} . \tag{15.233}$$

There is no need to calculate the rather complicated partial derivatives of A_τ from (15.227) to see that the result will not depend on τ, because A_τ does not.

Apparently, these results are often taken to mean that there is no radiation for the SE observer. That is not the view adopted in this book. This is discussed further in Sects. 15.12.1 and 16.8. Some also find that $\mathbf{E}_{\mathrm{SE}}$ looks like a Coulomb field, which is rather surprising. It certainly does not have the same form relative to SE coordinates as the usual Coulomb field relative to inertial coordinates in Minkowski spacetime. In fact, one only has to begin to carry out the partial derivatives $\partial A_\tau/\partial Z$, $\partial A_\tau/\partial x$, and $\partial A_\tau/\partial y$ to discover that these field components look rather complicated functions of the coordinates in this frame, just as one would expect.

15.11.2 Fields Along Forward Light Cone of Point on Worldline

The field at the point (ρ,z,t) was found as the retarded field of some point along the worldline of the charge. For any (ρ,z,t), we can always Lorentz transform to a frame in which the charge was at rest at the relevant retarded point. As we have seen, by the Lorentz invariance of all this theory, we will then just be considering the case where the retarded point happens to be $z(t) = g^{-1}$ on the worldline. [Recall that the worldline is given by $z(t)^2 - t^2 = g^{-2}$.] We can introduce a new variable θ which is an angle in the spacelike hypersurface $t = 0$ in this frame, namely, the angle between the axis of acceleration and the line joining the field point (ρ,z) to the retarded point $(0,0,g^{-1})$, as shown in Fig. 15.18. If we take r to be the radius at the field point of the light cone centered on the retarded point (where the charge is at rest for this frame), we then have

$$r = t , \qquad z = g^{-1} + r\cos\theta , \qquad \rho = \hat{\rho}r\sin\theta , \tag{15.234}$$

where $\hat{\rho}$ is a unit vector in the direction of ρ. We can calculate the value of the Rindler coordinate Z for the field point from

$$Z^2 = z^2 - t^2 = g^{-2} + 2g^{-1}r\cos\theta - r^2\sin^2\theta . \tag{15.235}$$

We also have the value of the variable R defined earlier, viz.,

$$R = \frac{g}{2}\left[(g^{-2} + \rho^2 + Z^2)^2 - 4g^{-2}Z^2\right]^{1/2} = r , \tag{15.236}$$

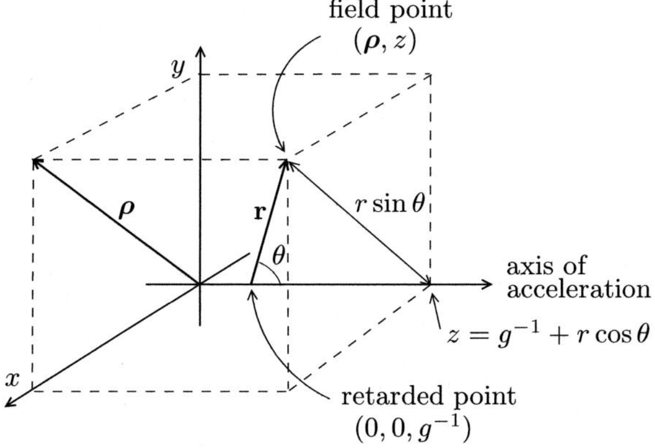

Fig. 15.18 New variables for analysing the fields along the forward light cone of a point on the charge worldline

after a little algebra. Note that R is just a function of the field point, which happens to equal r. This gives us an interpretation of R as the radius at the field point of the light cone centered on the retarded point, when the frame is such that the charge is at rest.

The reason for introducing all these notions is that we can now rewrite the field along the forward light cone of the point at which the charge is at rest. We start with (15.129), (15.130), and (15.135), viz.,

$$F'^{tz} = \frac{e}{4\pi} \frac{Z^2 - \rho^2 - g^{-2}}{2R^3} \theta(t+z) , \tag{15.237}$$

$$F'^{t\rho} = E^\rho = \frac{egz\rho}{4\pi R^3} \theta(t+z) , \tag{15.238}$$

and

$$F^{z\rho} = \frac{egt\rho}{4\pi R^3} \theta(t+z) . \tag{15.239}$$

In region I, we always have $\theta(t+z) = 1$. Substituting in the expressions for z, t, Z, and R, we come up with

$$F'^{tz} = \frac{e}{4\pi} \left(\frac{\cos \theta}{r^2} - \frac{g}{r} \sin^2 \theta \right) , \tag{15.240}$$

$$F'^{t\rho} = \frac{e}{4\pi} \hat{\rho} \left(\frac{\sin \theta}{r^2} + \frac{g}{r} \sin \theta \cos \theta \right) , \tag{15.241}$$

and

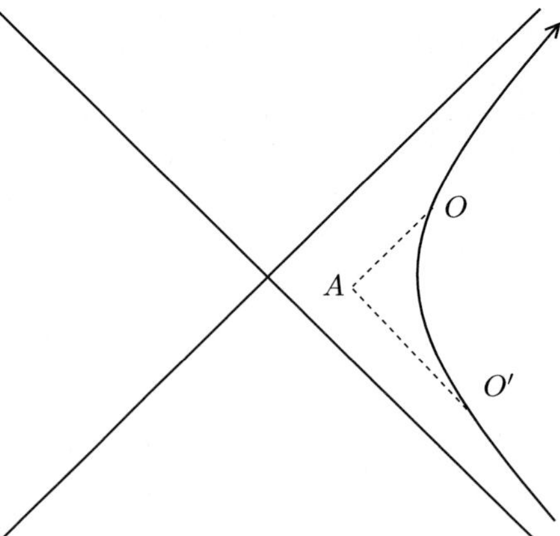

Fig. 15.19 The fields at A can be considered as the retarded field due to the passage of the charge at O' or the advanced field due to the passage of the charge at O

$$F^{z\rho} = \frac{e}{4\pi}\hat{\rho}\frac{g\sin\theta}{r}. \tag{15.242}$$

We can interpret these formulas as follows, bearing in mind that we have chosen a frame in which the charge was at rest at the relevant retarded point. (This is like putting $\mathbf{v} = 0$ in the formulas for the electric and magnetic fields of an instantaneously stationary accelerating charge.) The terms proportional to $1/r^2$ are the Coulomb field of the charge and the terms proportional to $g\sin\theta/r$ are the radiation field of a charge with acceleration g.

15.11.3 Equivalence of Advanced and Retarded Fields

Given a field point, let us call the retarded point O. Then the field along the backward light cone of O is precisely the same as the fields given on the forward light cone in the last section, except that the magnetic fields are in the opposite direction. What we mean here is that, if we consider a point (ρ, z, t) on the forward light cone and the corresponding point (same values of ρ and z) on the backward light cone, viz., $(\rho, z, -t)$, then we can see from (15.237), (15.238), and (15.239) that the fields are the same at both points except that (15.239) gives minus the value for $F^{z\rho}$ because it is proportional to t. But this feature, i.e., the same electric fields and reversed magnetic fields, is the signature of the advanced field. We conclude that the field along the backward light cone is indeed the advanced field due to O. We could work this out directly from the advanced potentials, etc.

Let A be a point on the backward light cone of O (see Fig. 15.19). It is also on the forward light cone of some O' on the worldline and the fields there are therefore the retarded fields due to the passage of the charge at O'. This proves that the advanced field in region I is the retarded field, something we already concluded at the end of the section before last. The field at any point in region I may be interpreted either as the Coulomb field plus outgoing radiation field of the charge at the retarded time, or as the Coulomb field plus ingoing radiation field of the charge at the advanced time.

It is interesting to note the connection with Dirac's work, already mentioned briefly in Chap. 11. Dirac redefines the field of radiation $F_{\text{rad}}^{\mu\nu}$ produced by the electron as

$$F_{\text{rad}}^{\mu\nu} = F_{\text{ret}}^{\mu\nu} - F_{\text{adv}}^{\mu\nu} \ . \tag{15.243}$$

He then obtains his equation of motion of the electron in an external field $F_{\text{in}}^{\mu\nu}$, taking its own field into account. To do so, he examines the flow of energy and momentum out of a thin tube enclosing the electron worldline. If ε is the spatial radius of the tube, he absorbs a term going as ε^{-1} into the electron mass and obtains the usual Lorentz-type equation of motion. However, there is now an effective electromagnetic field $f^{\mu\nu}$ on the force side of the equation:

$$m\dot{v}^\mu = ev^\nu f^\mu_{\ \nu} \ . \tag{15.244}$$

This field can be expressed as

$$f^{\mu\nu} = F_{\text{in}}^{\mu\nu} + \frac{1}{2} F_{\text{rad}}^{\mu\nu} \ , \tag{15.245}$$

where $F_{\text{in}}^{\mu\nu}$ is the incident field, i.e., any fields whose sources are other than the electron itself. Of course, in the special case we are considering, namely a uniform acceleration, $F_{\text{rad}}^{\mu\nu} = 0$ and we retrieve the standard Lorentz force law with no radiation reaction.

One argument in favour of $F_{\text{rad}}^{\mu\nu}$ as the definition of the radiation field is that it turns out to be finite on the charge worldline, one of the main things proven by Dirac in [14] and a first step toward classical renormalisation. Indeed, one can see that the subtraction in (15.243) is liable to remove the Coulomb parts of the advanced and retarded fields. Another key thing about $F_{\text{rad}}^{\mu\nu}$ is that it does generally give just the usual radiation field from $F_{\text{ret}}^{\mu\nu}$ in regions where one is likely to measure the radiation field.

How does the new definition of the radiated field compare with the usual one for the radiation produced by an accelerating charge? We consider a charge initially moving with constant velocity, then undergoing an acceleration and finally going into a state of constant velocity once more. In the usual theory, the radiation emitted by the charge while it is accelerating will be given by the value of $F_{\text{ret}}^{\mu\nu}$ at great distances from the charge and at correspondingly great times after the time of the acceleration. But $F_{\text{adv}}^{\mu\nu}$ will be zero in this region of spacetime, because the further we move from the charge in space, the longer before the acceleration we must look in order to find the advanced field. So the expression (15.243) will give precisely the same expression, namely $F_{\text{ret}}^{\mu\nu}$, in that region of spacetime where the usual radiated field is defined.

The advantage of this new definition (15.243) is thus that it provides a definite value for the field of radiation throughout spacetime, in particular, giving a meaning to the radiation field close to the charge. The usual theory gives a field that is inextricably mixed up with the Coulomb field. The disadvantage is that this result also attributes a meaning to the radiation field before its time of emission, when it can have no physical significance. Dirac merely says that this is unavoidable if we are to have a well defined radiation field near the charge.

The disadvantage of the definition (15.243) in the present case of a hyperbolically moving charge is undoubtedly that it gives zero for the radiation field everywhere, whereas one can see from results like (15.240) above that there is in fact a radiation field. However, it is indeed $F_{\mathrm{rad}}^{\mu\nu}$ that occurs in the renormalised equation of motion (15.244), which thus reduces to the usual Lorentz force law without radiation reaction.

15.11.4 Comparing Radiated and Coulomb Fields in Region I

We come to what Boulware considers to be the crux of the matter, where we look seriously at the supposed radiated field and try to decide whether we could detect it experimentally for this uniformly accelerated source charge. It is claimed that:

- no limit can be taken such that the radiation field is large compared with the Coulomb field,
- the field expressed in terms of the distance to the worldline of the charge does not drop off as $1/l$ but rather as $1/l^2$.

We have

$$|\mathbf{E}_{\mathrm{rad}}| = \frac{e}{4\pi}\frac{g}{r}\sin\theta\,, \qquad |\mathbf{E}_{\mathrm{Coulomb}}| = \frac{e}{4\pi r^2}\,, \qquad (15.246)$$

whereupon

$$\frac{|\mathbf{E}_{\mathrm{rad}}|}{|\mathbf{E}_{\mathrm{Coulomb}}|} = rg\sin\theta\,. \qquad (15.247)$$

We might expect this, because $g\sin\theta$ is the component of the acceleration perpendicular to the line of sight to the relevant retarded point. Furthermore, on the face of it, it looks as if the ratio goes as r, so that $|\mathbf{E}_{\mathrm{rad}}|$ gets rapidly bigger than the Coulomb field as we move away from the source point.

However, there is something about region I that spoils this. In fact, we have the constraint $0 < z - t$ on the field point. Note that this holds no matter how long the vector ρ may be. We are not just restricted to the spacelike sector of the origin in the Minkowski spacetime. No matter what values of x and y we may have for the field point, z has to be greater than t. We have $t = r$ and $z = g^{-1} + r\cos\theta$, hence,

$$0 < z - t = g^{-1} + r(\cos\theta - 1)\,. \qquad (15.248)$$

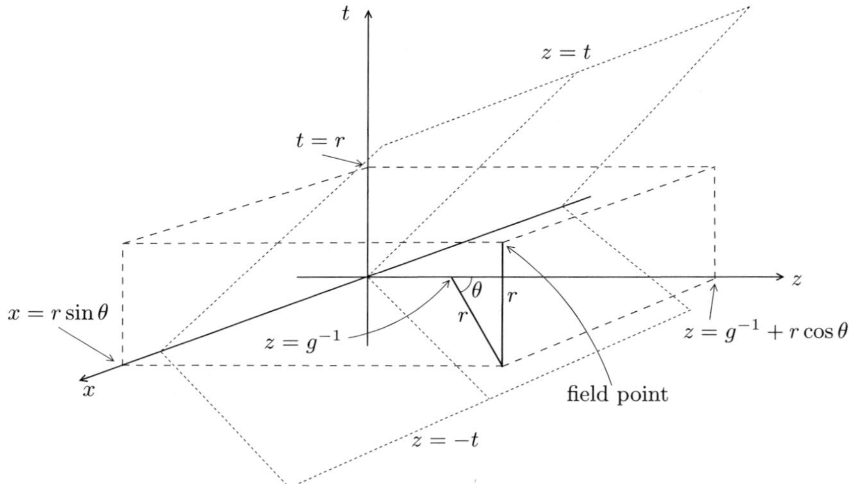

Fig. 15.20 Understanding why the radiation field never looks big compared with the Coulomb field in region I. For fixed viewing angle θ, the variable r can only be extended until the field point hits the plane $z = t$

This puts an upper limit on gr for given $\theta \neq 0$, viz.,

$$gr < \frac{1}{1 - \cos\theta}, \qquad (15.249)$$

and hence an upper limit on the ratio of the radiated and Coulomb fields, viz.,

$$\frac{|\mathbf{E}_{\mathrm{rad}}|}{|\mathbf{E}_{\mathrm{Coulomb}}|} < \frac{\sin\theta}{1 - \cos\theta}. \qquad (15.250)$$

Suddenly, the ratio no longer goes as r because, provided that θ has been fixed, it will always remain below $\sin\theta/(1 - \cos\theta)$, even though this may be very large for some θ close to zero. (In fact, this ratio goes to infinity as $2/\theta$ as $\theta \to 0$.)

Let us see what has happened here. It is important to visualise region I on a diagram (see Fig. 15.20). In an x, z, t diagram, suppressing only one space dimension, it lies between the two perpendicular planes $z = t$ and $z = -t$, with $z > t$. We mark the point $z = g^{-1}$ on the z axis and a line in the x, z plane with length r, making an angle θ with the z axis. The field point lies distance r up the t axis from the end of this line. The z value of the field point is then $g^{-1} + r\cos\theta$ and the x value is $r\sin\theta$, whilst the t value is r. Requiring the t value to be r amounts to requiring the field point to be on the forward light cone of the point where the charge was stationary, i.e., the point $(t, x, y, z) = (0, 0, 0, g^{-1})$.

Now imagine extending r, always for the same θ. Then t grows as r, but z only grows as $r\cos\theta$, until there comes a point, which we may call $r_{\mathrm{lim}}(\theta)$, where t equals z, and thereafter, for all bigger r, t is greater than z. In fact,

$$r_{\lim}(\theta) = \frac{1}{g(1 - \cos\theta)} \, . \tag{15.251}$$

When θ is small, this is big, so we can extend r out a long way. But we nevertheless have a limit, unless $\theta = 0$. In the other extreme, when $\theta = \pi/2$, we have $r_{\lim}(\pi/2) = g^{-1}$. So when the line of sight is set to optimize the radiation, i.e., perpendicular to the axis of acceleration, this is precisely when we are held closest to the source and least able to appreciate how the radiated field will outgrow the Coulomb field.

It is important to see that what we have here is not a bound on the radiated field, but a bound on the distance at which we may measure the radiated field whilst remaining within region I, i.e., a bound that is particularly relevant only to an observer comoving with the charge. We have to go into region II, above the plane $z = t$, to appreciate how the radiated field outgrows the Coulomb field. So there is radiation in region I, but it will be impossible to detect if one is looking for a field that rapidly outgrows the Coulomb field.

There is more to support this claim in Boulware's paper. We now consider the distance of the field point from the charge worldline. What we have done so far is this. Given a field point (x, y, z, t), we have found the retarded point and changed to a frame in which this point was at rest in order to determine what Dirac called the invariant retarded distance r. We expressed the fields at this field point in the rest frame of the charge at the retarded point. But now we are going to change to another frame to express the so-called invariant distance l from the field point to the worldline. So we shall express the field components in one frame in terms of the distance measured in another! It certainly seems an unusual approach, but let us bear with it.

What is this other frame? Given the field point (x, y, z, t), there is a unique point on the worldline such that this field point is on the hyperplane of simultaneity (HOS) of the instantaneously comoving inertial observer (ICIO) at that point on the worldline. So we seek the Minkowski time $s(z, t)$ such that the HOS of the ICIO at this time runs through (z, t). This is precisely the construction used to find semi-Euclidean (SE) coordinates. The invariant distance is then just the distance from the point on the worldline to the field point, now simultaneous for this ICIO, as measured by the ICIO. In other words, it is just the space coordinate of this observer, or the space coordinate of the SE observer.

We can start with a formula like

$$y^1 = \left[\left(x + \frac{c^2}{g}\right)^2 - c^2 t^2\right]^{1/2} - \frac{c^2}{g} \, ,$$

as found in (2.13) on p. 9. In the present case, we replace $c \equiv 1$ and $x + c^2/g \to z$, since in that analysis we had the SE observer passing through the space origin of the Minkowski frame. Not forgetting that we have some x and y in our field point, this means that the invariant distance is given by

$$l^2 = \rho^2 + \left[(z^2 - t^2)^{1/2} - g^{-1}\right]^2 = \rho^2 + (Z - g^{-1})^2 \,. \tag{15.252}$$

This is indeed what Boulware intends. Now the rest of Boulware's argument is straightforward. We have

$$Z = \left(g^{-2} + 2g^{-1}r\cos\theta - r^2\sin^2\theta\right)^{1/2} ,$$

so that

$$\begin{aligned}
(Z - g^{-1})^2 &= Z^2 - 2g^{-1}Z + g^{-2} \\
&= 2g^{-2} + 2g^{-1}r\cos\theta - r^2\sin^2\theta \\
&\quad - 2g^{-1}\left(g^{-2} + 2g^{-1}r\cos\theta - r^2\sin^2\theta\right)^{1/2} ,
\end{aligned}$$

and since $\rho^2 = r^2\sin^2\theta$,

$$l^2 = 2g^{-1}\left(g^{-1} + r\cos\theta\right) - 2g^{-1}\left(g^{-2} + 2g^{-1}r\cos\theta - r^2\sin^2\theta\right)^{1/2} . \tag{15.253}$$

Hence,

$$\frac{1}{2}(gl)^2 = 1 + gr\cos\theta - \left[(1 + gr\cos\theta)^2 - (gr)^2\right]^{1/2} . \tag{15.254}$$

Consider the quantity

$$\alpha := 1 + gr(\cos\theta - 1) \,.$$

Of course, $\cos\theta - 1 < 0$, so $\alpha < 1$. But in region I, $z > t$ and therefore

$$0 < z - t = g^{-1} + r\cos\theta - r \,,$$

as observed previously, and finally, $\alpha > 0$. Furthermore,

$$\frac{1}{2}(gl)^2 = gr + \alpha - (2gr\alpha + \alpha^2)^{1/2} , \tag{15.255}$$

where we bear in mind that $0 < \alpha < 1$.

Now we noted earlier that, for fixed θ,

$$gr < \frac{1}{1 - \cos\theta} \,.$$

We can get large gr by choosing θ small, i.e., a lot more z than x or y in the field point, so that we are examining the fields close to the axis of acceleration. But when gr is large, our formula for l says that

$$\frac{1}{2}(gl)^2 \sim gr \,, \tag{15.256}$$

precisely because $0 < \alpha < 1$. Conversely, when we have the latter, we must have $\theta \sim 0$, i.e., we must have a field point close to the axis of acceleration.

Now we said before that, if we are in region I and if θ is well away from zero, the radiation field is comparable in magnitude with the Coulomb field. The only place where it might get significantly bigger, whilst remaining in region I, is near the axis of acceleration. But now we see that, in this region,

$$|\mathbf{E}_{\text{rad}}| \sim \frac{e}{4\pi}\frac{g\sin\theta}{r} \sim \frac{e}{4\pi}\frac{2\sin\theta}{l^2} \,. \tag{15.257}$$

Boulware concludes that this displays the characteristic $1/l^2$ behaviour of a Coulomb field. As mentioned above, however, it is strange to express the field components in one frame as functions of a distance in another. Is that how we would measure things?

Boulware's statement of intent is as follows:

> Although the field in region I is the Coulomb plus outgoing radiation field, the experimental situation is complicated. First, detailed measurements of the field must be made to determine that fact; second, the measurements will be consistent with the field's being the Coulomb plus incoming radiation field; third, except very near the direction of the acceleration, no limit can be taken such that the radiation field is large compared with the Coulomb field; and fourth, the field, as measured by the distance to the world line of the charge, does not drop off as $1/l$ but rather as $1/l^2$ in region I.

One can see where this kind of argument is leading. One is trying to convince oneself that, even though there is a radiation field, it might be difficult for an observer comoving with the charge to find out about it. In a way, one is wishing that the radiation field were not there. This brings us back to the discussion in the introduction to this chapter (see p. 157 ff).

The present interpretation of what Boulware is trying to do can be stated as follows. He believes Bondi and Gold's claim that, in the case of a static, homogeneous (non-tidal) gravitational field, a charge that is stationary relative to coordinates in which the metric is static cannot radiate. By application of a loosely stated version of the equivalence principle, the situation for an observer comoving with a uniformly accelerating charge and using the semi-Euclidean coordinate system (for which the Minkowski metric is static) must be indistinguishable. The word 'indistinguishable' is taken to allow some leeway over the presence or otherwise of a radiation field. Rather than meaning that there cannot be a radiation field in some objective sense, it is taken to mean that there might be a radiation field, but that the comoving observer must not be able to detect it.

As noted already, there are two difficulties with this stance:

- There is no particular reason to concord with Bondi and Gold on their opinion that, in the case of a static, homogeneous (non-tidal) gravitational field, a charge that is stationary relative to coordinates in which the metric is static cannot radiate.

- The word 'indistinguishable' cannot be allowed to give this leeway. Strict application of WEP and SEP tells us that there could not be radiation, in an objective, observer-independent sense, for a uniformly accelerating charge in a flat spacetime without gravitational effects (not even non-tidal ones), if one accepts Bondi and Gold's opinion stated in the first point.

This therefore serves also to illustrate the dangers of using a version of the equivalence principle that says whether things can be distinguished or not.

Finally, there is another serious argument against Boulware's conclusions. An observer moving with the charge could presumably still carry out careful measurements and discover the form (15.240)–(15.242) of the fields on p. 233 in some neighbourhood. She would then deduce that there was radiation, even despite the results of Sect. 15.11.4. Once the fields are known in the neighbourhood of the source, presumably there is a unique analytic continuation of these fields, i.e., one could detect the fact that they were going to look like radiating fields further away. This is no problem for the view presented in the present book because we are not constrained by the rather arbitrary opinion voiced by Bondi and Gold that a charge that is stationary relative to coordinates in which the metric is static cannot radiate.

15.11.5 Situation in Region II

This section is a discussion of one paragraph on p. 183 of Boulware's paper. First we are assured that the field is the retarded field of the charge. One could retort that this was also true in region I, although the retarded field happened to equal the advanced field there too. The point is of course that the field is unambiguously the retarded field since the advanced field is zero in region II. But was there really any ambiguity? We merely had a choice in interpreting our solution to Maxwell's equations. It is not explained why this choice might have created an ambiguity. Surely there would be more ambiguity if the advanced and retarded solutions were different, since one might then hesitate over choosing one of them. (But not many would hesitate long in choosing the retarded solution, unless experiment were able to demonstrate that one should not. Notions of causality are deeply ingrained.)

The next point is that the radius of the light cone can be arbitrarily large for any nonzero angle θ (see Fig. 15.20 and note that r can be made arbitrarily large). The radiation field can thus be made arbitrarily large compared to the Coulomb field. What we are saying is that we would not dispute there being a radiation field here. It is also true that there is no HOS of an ICIO for the semi-Euclidean observer which gets to enter region II. In fact, we know that all planes of simultaneity of the accelerating observer in her SE system pass through the origin of the Minkowski coordinate system. So we cannot find an invariant distance from a field point in region II to the worldline. At least, we cannot find the equivalent of the distance l discussed in the last section, although we can, of course, find an invariant retarded distance, and this enters into our formulas for the fields in region II. As a result, we are told that we cannot misidentify the fields as $1/l^2$ fields. This is indeed a weak argument. Even if we misidentified the fields, this in itself would not mean that the fields were not radiated fields. But worse, it leaves open the possibility that an astute observer comoving with the charge might not misidentify the fields! In other words, such an observer might well establish that there was radiation in her region of spacetime by careful measurement, even if it proved to be tricky for the reasons given by Boulware in the last section.

One should also abhor the loose way of making the last claim about there being no invariant distance. Boulware says that there is no Lorentz frame which passes through a field point in region II and in which the charge is instantaneously at rest. But a Lorentz frame does not pass through an event. What he means is that there is no Lorentz frame in which the charge is instantaneously at rest at some event on its worldline and the field point is simultaneous with it in the reckoning of that frame.

We now come to the crucial point where an equivalence principle is mentioned. The idea is that there is no coordinate frame in which the charge is at rest, which is static (i.e., the metric components assume a static form), and which covers region II. What we are talking about here is an accelerating frame, like the semi-Euclidean frames, adapted to the accelerating observer. Now we have seen that the semi-Euclidean frame covers only region I. At this point we need to recall the discussion following the proposed coordinate transformations (15.11) on p. 164. Boulware claims that the SE frame can be extended analytically to region II, provided that we accept the non-static metric of (15.12), viz.,

$$ds^2 = \varepsilon(-g^2Z^2 d\tau^2 + dZ^2) + dx^2 + dy^2 , \qquad (15.258)$$

where

$$\varepsilon = \begin{cases} 1 & \text{for } x \text{ in I and III} , \\ -1 & \text{for } x \text{ in II and IV} . \end{cases}$$

Since Boulware refers to 'the analytic continuation of the accelerated frame to region II', we deduce that it must be unique. As pointed out earlier, however, it is not at all clear what is meant here. What is the analytic continuation of a frame? Does it mean the analytic continuation of the coordinates? In fact, what we appear to have from (15.11) are four charts in an atlas that happens to cover each of the regions I, II, III and IV of Minkowski time separately. This observation is a long way from being a proof that there is no coordinate frame in which the charge is at rest, which is static (in the sense that the metric components have static form), and which covers region II.

What we actually have is a coordinate transformation of the Minkowski coordinates on region II to the same coordinate set in $\mathbb{R}^4$ that was used to coordinatise region I. Under this coordinate transformation, the Minkowski metric transforms to a metric that no longer has the static form, the reason being that the time coordinate is then Z in this region, and the metric components depend on Z. However, this really has nothing of an analytic continuation of anything and it certainly does not constitute a proof that there is no coordinate frame in which the charge is at rest, which is static and which covers region II, even though this may well be true.

We still need to see why such a result would be relevant for Boulware. We are told that there can be no equivalence principle argument against radiation in region II. We have to go back to the initial discussion to understand this claim. The argument goes as follows (Boulware's own formulation [6, p. 170]): By the equivalence principle, a uniformly accelerated frame must be indistinguishable from a gravitational field. However, a charged particle at rest in a static gravitational field cannot radiate, hence a uniformly accelerated particle cannot radiate. Leaving aside the propriety of this argument, the idea is that, since the metric is not static in region II, we cannot rule out a stationary particle producing radiation. We are pleased about this because it looks like there really is radiation in region II:

> [...] there is no coordinate frame in which the charge is at rest, which is static and which covers region II; the analytic continuation of the accelerated frame from region I to region II yields the coordinates (15.11) in which the metric is time (Z) dependent. In summary, there is no equivalence principle argument against radiation in region II nor is there any difficulty in identifying the radiation. It is there and is precisely the radiation of the accelerated charge.

So the first problem with this is that we do not appear to have an analytic continuation of the accelerated frame, whatever that might be. If we take the coordinate transformations from the Minkowski coordinates on regions I and II, as given by the inverses of the first two relations of (15.11), we find that each point of the (τ, Z, x, y) space maps to one point in each region and that no points map to the horizon $t = z$ separating the two regions. This can hardly be described as an analytic continuation of the semi-Euclidean coordinates in region I.

Perhaps there is such a continuation, perhaps there is not. It is difficult to see how one could continue beyond the horizon $t = z$, since it is given by $Z = 0$ in these coordinates, and the metric components relative to these coordinates are then singular. But even supposing that one could contrive some coordinates that coincide with the SE coordinates in region I and extend also over region II, would the fact that the charge happens to be at rest in the spatial hypersurfaces of these coordinates (in region I) really allow us to make any deductions about what is happening in region II? Would those coordinates in region II really describe the view of some observer, viz., the observer moving with the charge? For that matter, do the SE coordinates in region I really describe the view of any observer? Coordinates have no a priori physical significance. We shall have to face this question again shortly, because it will be shown that, when the definition of the Poynting vector for the fields is extended to the SE coordinates, this quantity is identically zero, and we shall want to interpret this in terms of what the hyperbolically moving observer will observe.

At least it is now clear what Boulware is trying to do here: he aims to show that there is radiation from the uniformly accelerating charge, that it cannot be detected in region I by an observer moving with the charge (in order to save Bondi and Gold's opinion that a charge that is stationary relative to coordinates in which the metric is static cannot radiate), but that it is really there anyway because it can be detected in region II by an observer moving with the charge (where there is no constraint due to Bondi and Gold's opinion because the metric turns out not to be static for this same observer in region II). This statement shows how odd the whole argument really is: the observer is not actually in region II but these are her coordinates. There is even the idea that she has no choice over her coordinates, because the analytic continuation of the SE coordinates to region II given in (15.11) (if indeed it could be interpreted in this way) is 'the' analytic continuation, and by implication unique. But who said the observer had to use the SE (Rindler) coordinates in region I? Surely she could use any coordinates she liked. One cannot draw conclusions because of coordinates.

Therefore the main criticism in this section is that there is no reason to think that the coordinates τ, Z, x, y proposed for region II are in any sense coordinates in which the hyperbolically moving charge is stationary, so the fact that the Minkowski metric in region II turns out to have non-static components relative to these coordinates is really irrelevant, even if one wanted to save the competing claims that the uniformly accelerating charge is actually radiating and a particle at rest in a static gravitational field cannot radiate, or cannot be seen to radiate by an observer using coordinates in which the particle is at rest and the metric components have static form.

Boulware makes a final point in his comments concerning region II, namely the observation that the fields in region II are invariant under a certain transformation of the data, viz., replacing the charge e by a charge $-e$ and having it follow the mirror-symmetrical worldline in region III. This can be seen by a suitable examination of the fields in Sect. 15.9.2. The conclusion from this is that no measurements restricted to region II can distinguish between the Coulomb plus outgoing radiation field of a uniformly accelerated charge e in region I and the Coulomb plus outgoing radiation field of a uniformly accelerated charge $-e$ in region III.

So observers in region II will proclaim the presence of the radiation but be totally unable to determine whence it came. Indeed, we have to look in other regions of spacetime to find the source. Boulware gives us three things that could signal the location of the charge in region I:

- the presence of the field in region I,
- the absence of the field in region III,
- the delta function field along the surface $z + t = 0$, rather than along the surface $z - t = 0$.

He appears to have overlooked another possibility: we could merely look at the charge.

It is not at all clear what one is supposed to deduce from this. However, it does once again raise the question of what is meant by an observer 'in' region II, or even 'an observer'. Who are these people? Are they just sets of coordinates and components of tensors relative to these coordinates? Are they people who refuse to communicate or cannot communicate with other observers in other regions? And if they will not or cannot communicate, does that really matter for a theory that purports to model spacetime?

15.12 Stress–Energy Tensor

We take the fields (15.178)–(15.182) on p. 216 and calculate the stress–energy
tensor from

$$T^{\mu\nu} = F^{\mu\lambda}F_{\lambda}{}^{\nu} - \frac{1}{4}g^{\mu\nu}F^{\lambda\sigma}F_{\lambda\sigma}\,. \tag{15.259}$$

This is, of course, a straightforward but tedious calculation. The results according
to Boulware are:

$$T^{tt} = \frac{e^2}{16\pi^2}\left\{\frac{g^2}{8R^6}\left[(\rho^2+g^{-2}-Z^2)^2+4\rho^2(t^2+z^2)\right]\theta(z+t)\right.$$

$$\left.+\frac{16g^2z\rho^2}{(1+g^2\rho^2)^4}\delta(z+t)+\frac{4g^4\rho^2}{(1+g^2\rho^2)^2}\delta(0)\delta(z+t)\right\}\,, \tag{15.260}$$

$$T^{tz} = \frac{e^2}{16\pi^2}\left[\frac{g^2tz\rho^2}{R^6}\theta(t+z)-\frac{16g^6z\rho^2}{(1+g^2\rho^2)^4}\delta(z+t)-\frac{4g^4\rho^2}{(1+g^2\rho^2)^2}\delta(0)\delta(z+t)\right]\,, \tag{15.261}$$

$$T^{t\rho} = \frac{e^2\rho}{16\pi^2}\left[\frac{g^2(\rho^2+g^{-2}-Z^2)t}{2R^6}\theta(z+t)-\frac{4g^4}{(1+g^2\rho^2)^3}\delta(z+t)\right]\,, \tag{15.262}$$

$$T^{zz} = \frac{e^2}{16\pi^2}\left\{\frac{g^2}{8R^6}\left[4\rho^2(t^2+z^2)-(\rho^2+g^{-2}-Z^2)^2\right]\theta(z+t)\right.$$

$$\left.-\frac{16g^6t\rho^2}{(1+g^2\rho^2)^4}\delta(z+t)+\frac{4g^4\rho^2}{(1+g^2\rho^2)^2}\delta(0)\delta(z+t)\right\}\,, \tag{15.263}$$

$$T^{z\rho} = \frac{e^2\rho}{16\pi^2}\left[\frac{g^2(\rho^2+g^{-2}-Z^2)z}{2R^6}\theta(z+t)+\frac{4g^4}{(1+g^2\rho^2)^3}\delta(z+t)\right]\,, \tag{15.264}$$

$$T^{\rho\rho} = \frac{e^2\rho}{16\pi^2}\left(-\frac{g^2Z^2\rho\rho}{R^6}+\frac{I}{2R^4}\right)\theta(z+t)\,. \tag{15.265}$$

The appearance of the square of $\delta(z+t)$, written $\delta(0)\delta(z+t)$, is, of course, a serious
mathematical problem, although some physical sense can still be squeezed out of it.
Like the delta function itself, it arises here because of the extreme situation we are

considering, which is at best an approximation to anything that could really happen. We note that the formula

$$\delta(z+t)\theta(z+t) = \frac{1}{2}\delta(z+t)$$

from distribution theory has been used to arrive at the above results. The stress–energy tensor is conserved, except along the worldline of the charge. It must even be (formally) conserved along the null surface $z+t = 0$, because Maxwell's equations are (formally) satisfied there and there is no source of EM effects actually on $z+t = 0$. Apparently, the term arising from the square of the delta function is formally conserved by itself, but we shall skate over this question.

15.12.1 Stress–Energy Tensor in Accelerating Frame

We now examine the stress–energy tensor expressed relative to the coordinate frame of the uniformly accelerated observer, which is restricted to region I and therefore does not involve any distributions. The following result is reached by standard coordinate transformation of the relevant parts:

$$T^{\tau\tau} = \frac{e^2}{32\pi^2}\frac{1}{Z^2 g^2 R^4}\,, \qquad T^{\tau Z} = 0 = T^{\tau\hat{\rho}}\,, \qquad T^{ZZ} = -\frac{e^2}{32\pi^2}\frac{1}{R^4}\,, \qquad (15.266)$$

$$T^{Z\hat{\rho}} = \frac{e^2}{16\pi^2}\frac{\rho}{R^6}g^2 Z(\rho^2 + g^{-2} - Z^2)\,, \qquad (15.267)$$

$$T^{\hat{\rho}\hat{\rho}} = \frac{e^2}{16\pi^2}\left(\frac{I}{2R^4} - \frac{g^2 Z^2 \rho\rho}{R^6}\right)\,. \qquad (15.268)$$

What is striking about these expressions is their relative simplicity.

We note immediately that the Poynting vector in this frame, which formally has components $T^{\tau Z}$ and $T^{\tau\hat{\rho}}$, is zero. This happens because the magnetic field is zero in this frame, as can be confirmed by transforming the field components of (15.220)–(15.222) on p. 229 to the semi-Euclidean system or calculating directly from the Rindler components of the EM vector potential, as was done on p. 231. Now Boulware says that there is no energy flux and no radiation. This is, of course, an interpretation. We know that there is energy flux and radiation when we think like a Minkowski observer, and we do not know a priori how to interpret these things in some other kind of frame. This is one of the key problems of general relativity: how to interpret coordinates, or the components of vectors and tensors, in a more general context than Minkowski spacetime. Our link with reality is the version of the equivalence principle which tells us to refer things to a local inertial frame, which is roughly Minkowskian. In this case, this principle would just refer us back to the original Minkowski frame and we would start talking about radiation again. There is more about this point later.

15.12.2 Energy Flux

Indeed, we now refer back to the Minkowski frame, examining the fields on the forward light cone of the spacetime point $z = 1/g$, $t = 0$, when the charge is at rest relative to those coordinates. We use the parameters r and θ as on p. 232 (see Fig. 15.18), where r is the radius of the light cone at the field point and θ is the angle between the radial vector from $z = 1/g$ to the field point and the z axis (the axis of acceleration). Recall that we then have

$$r = t , \qquad z = g^{-1} + r\cos\theta , \qquad \rho = \hat{\rho}\, r\sin\theta . \qquad (15.269)$$

We calculate the value of the Rindler coordinate Z for the field point from

$$Z^2 = z^2 - t^2 = g^{-2} + 2g^{-1}r\cos\theta - r^2\sin^2\theta . \qquad (15.270)$$

We also have the value of the variable R defined earlier, viz.,

$$R = \frac{g}{2}\left[(g^{-2} + \rho^2 + Z^2)^2 - 4g^{-2}Z^2\right]^{1/2} = r , \qquad (15.271)$$

after a little algebra. Note that R is just a function of the field point, which happens to equal r. This gives us an interpretation of R as the radius at the field point of the light cone centered on the retarded point, when the frame is such that the charge is at rest.

Substituting these parameters into the terms of T^{tz} and $T^{t\rho}$ that are relevant for region I, i.e., the terms with no delta functions, and dropping the factor $\theta(t+z)$, which is always unity in region I, we easily find that

$$T^{tz} = \frac{e^2}{16\pi^2}\left(\frac{g\sin^2\theta}{r^3} + \frac{g^2\sin^2\theta\cos\theta}{r^2}\right) , \qquad (15.272)$$

$$T^{t\rho} = \frac{e^2\hat{\rho}}{16\pi^2}\left(-\frac{g\sin\theta\cos\theta}{r^3} + \frac{g^2\sin^3\theta}{r^2}\right) . \qquad (15.273)$$

We can now calculate the flux of energy out of the sphere of radius r as

$$\int d\mathbf{S}\cdot\mathbf{T}^t = \frac{e^2}{4\pi}\frac{2g^2}{3} , \qquad (15.274)$$

where $\mathbf{T}^t$ is the Poynting vector (T^{tx}, T^{ty}, T^{tz}) in the Minkowski frame. Note that this is an integral over a spherical surface in a spacelike hypersurface, i.e., the hypersurface $t = r/c$.

Having reached this point, Boulware declares that this determination of the energy flow emanating from the charge is non-local in the sense that one must carefully construct a sphere on the light cone centered on the worldline of the particle. There is of course the problem already mentioned several times in the present book that one can make the sphere as small as one likes, i.e., r can be chosen arbitrarily small, provided we adjust t accordingly, that is, provided we integrate over the sphere in the appropriate spacelike hypersurface. One always obtains the same result, viz., (15.274) above, simply because it turns out to be independent of r. This is indeed what suggests that it was the actions of the source at the event $(t,z) = (0, 1/g)$ which caused the radiation. There does not seem to be anything non-local about that.

Energy Flowing into the Charge

A related calculation of great interest is this. We consider the sphere of radius r on the backward light cone of the spacetime event with coordinates $z = 1/g$, $t = 0$ and integrate the flux of the Poynting vector through this sphere. Note that we are still in the Minkowski frame in which the charge is at rest at the event with coordinates $z = 1/g$, $t = 0$. Introducing the parameter r and the angle θ as above, we now have

$$r = -t\,, \qquad z = g^{-1} + r\cos\theta\,, \qquad \rho = \hat{\rho}r\sin\theta\,. \tag{15.275}$$

The only difference is that t has changed sign relative to r, because we go back in time from the event where the charge is at rest to the sphere that might affect it precisely at the moment when it comes to rest $(t = 0)$. The functions Z and R of r and θ turn out to be unchanged. When we work out the Poynting vector from the relevant terms of (15.261) and (15.262), we obtain precisely the reverse of the Poynting vector given in (15.261) and (15.262) for the forward light cone. This means that energy is flowing in through the sphere on the backward light cone and focussing in on the charge. The amount of energy focussing in is exactly equal to the amount that radiates out.

So energy flows radially in towards the charge at exactly the same rate as it flows radially out from the charge in what is usually referred to as its electromagnetic radiation. (Note that the Poynting vector is not radial but our flux integration works out the radial flow.) We shall shortly address the question of where this inward flow of energy comes from. In passing, however, note that one should be wary of conclusions drawn from the Poynting vector, which sometimes leads to strange energy flows, as shown in [15, Sect. 27.5].

Energy Flowing Through a Fixed z Plane

The idea here is to take a look at energy flows through other surfaces in the Minkowski frame. We take a surface of positive z. There are two cases: $z < g^{-1}$, when the charge does not intersect the plane, and $z > g^{-1}$, when the charge does intersect the plane (see Fig. 15.21). When we integrate $d\mathbf{S} \cdot \mathbf{T}^t$ over all ρ for some fixed value of t, we obtain the rate at which energy passes through the surface. We can then integrate over all $t \in (-z, z)$ to find the total energy which has passed through the part of the surface accessible to the accelerated observer.

The basic idea here is that

$$\mathbf{dS} \cdot \mathbf{T}^t = \mathrm{d}S T^{tz} \,,$$

and since we are not concerned with the delta function terms in the proposed region of integration, we are basically integrating

$$\mathrm{d}S \frac{e^2}{16\pi^2} \frac{g^2 t z \rho^2}{R^6} \,.$$

But the region of integration is symmetrical about $t = 0$ so the integral is zero, whether the charge runs through it or not. If the charge does run through this region of spacetime, it comes in and out of it with the same kinetic energy, by the symmetry of the motion about $t = 0$. The integral will include the field self-energy of the charge as it enters and the same as it leaves, so the two should cancel, meaning that we are left only with radiation energy. The idea is that the total energy passing through such surfaces, insofar as they are accessible to the accelerating observer, is zero. Once again, we seem to be trying to show that there is no radiation in the Minkowski picture.

There is another point. The symmetry in the range of the t integration is critical. If we integrated from time $t = 0$, we would not find zero. We could well have some energy crossing one way before this time, then the same energy crossing back after this time. Contrast this careful choice of surface with Boulware's comment that, in our first integration to find the energy flowing out of the charge, we must carefully construct a sphere on the light cone centered on the worldline of the particle. One would not expect to have to require the surface to be either a sphere or centered, since the stress–energy tensor is conserved here.

Energy Flowing Through Segments of the Null Surfaces $z = \pm t$

In Fig. 15.21, we have two surfaces of fixed z, one to the left of $z = g^{-1}$ and one to the right, labelled $z = z_2$ and $z = z_1$, respectively. These surfaces are only considered where they intersect region I. We then join the ends of the lines $z = z_2$ and $z = z_1$ along each of the null surfaces $z = -t$ and $z = t$. We do not take t right up or down to the values $\pm z_1$ and $\pm z_2$, but consider surfaces just in the interior of region I. What about the integral of $\mathrm{d}S^\mu T^t_{\ \mu}$ over these two pieces of null surface?

For the portion of the null surface $z = t$ between z_2 and z_1, we obtain

$$I^{z=t} := \int \mathrm{d}^2\rho \int_{z_2}^{z_1} \mathrm{d}z (T^{tt} - T^{tz})(\rho,z,z) \,, \tag{15.276}$$

where the symbol (ρ,z,z) at the end indicates that we evaluate T^{tt} and T^{tz} at points with $t = z$. The integrand is the difference between T^{tt} and T^{tz} because $\mathrm{d}S^\mu = (1,-1,0,0)\mathrm{d}S$, using the obvious notation. On the lower null surface $z = -t$, we have $\mathrm{d}S^\mu = (1,1,0,0)\mathrm{d}S$ and hence,

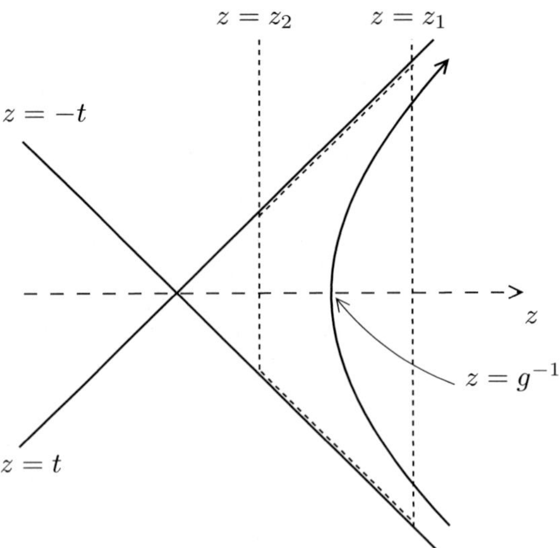

Fig. 15.21 Finding the energy flow out of a region bounded by two surfaces of fixed z and two portions of the null surfaces $z = t$ and $z = -t$

$$I^{z=-t} := \int d^2\rho \int_{z_2}^{z_1} dz (T^{tt} + T^{tz})(\rho, z, -z) \,. \tag{15.277}$$

The vectors giving the directions of the surface elements are formally normal to the surfaces, as though we had a Cartesian metric on spacetime. This kind of situation is examined in some detail in [13, Sect. 3.8].

Once again, we carefully avoid the step functions and delta functions in (15.260) and (15.261) and we soon find that

$$\begin{aligned}
I^{z=t} &= (z_1 - z_2)\frac{e^2 g^2}{2\pi} \int_0^\infty \frac{d(\rho^2)g^2}{(1+g^2\rho^2)^4} \\
&= (z_1 - z_2)\frac{e^2}{4\pi}\frac{2}{3}g^2 \,. \tag{15.278}
\end{aligned}$$

It turns out that we also have

$$I^{z=-t} = (z_1 - z_2)\frac{e^2}{4\pi}\frac{2}{3}g^2 \,, \tag{15.279}$$

since exactly the same terms cancel and we end up doing the same integration.

We need not be concerned with the technical details here. The problem is rather how to interpret these results. We are told that the integrals $I^{z=\pm t}$ represent the energy passing through the portions of the null surfaces $z = \pm t$ between $z = z_2$ and $z = z_1$. If this were a correct interpretation, we would deduce that the same amount of energy enters through the relevant portion of the surface $t = -z + 0$ as leaves through the relevant portion of the surface $t = z - 0$. We note that the two integrals did not have to be equal because the stress–energy tensor is not conserved throughout region I, even though we found equal amounts of energy coming in through the plane $z = z_2$ as going out through $z = z_1$. It is on the particle worldline that the stress–energy is not conserved, and this expresses the fact that the fields are interacting with the charge, so the fields may lose energy and the charge gain, or vice versa. However, the particle enters and leaves our region of spacetime with the same kinetic energy, so we might indeed expect the two integrals to be equal.

But do the above integrals $I^{z=\pm t}$ really represent energy passing through null surfaces? This case is considered on p. 107 of Parrott's book [13]. Parrott says that there is no really satisfying physical interpretation of the integral through a piece of null surface. We shall bear this in mind as we proceed with Boulware's account.

According to Boulware, the fact that $I^{z=t}$ and $I^{z=-t}$ have the same absolute value is a consequence of the radiation reaction being zero. As much energy is being absorbed by the charge as is being emitted. This is when he mentions the flux calculation through the sphere of radius r centered on the rest position of the charge in this frame at the time $t = -r$. The latter seems to be a much better way of understanding why the radiation reaction might be zero: the charge absorbs and reemits equal amounts of energy, in this picture. It is not clear that the integrals $I^{z=\pm t}$ throw much light on that particular interpretation, precisely because it is difficult to interpret the integrals over the null boundaries of our region.

Now Boulware asks where the inward flow of energy comes from. This is where it may be worth remembering the strange answers sometimes given by the Poynting vector [15, Sect. 27.5]. However, let us consider the arguments put forward. We may perhaps consider that the energy flows into the region $t > -z$ from the $t = -z$ surface, because there is no stress–energy in the region $t < -z$, so it could not come from there. The stress–energy confined to the $t = -z$ surface consists of two terms:

- There is an undefined term involving the square of the delta function. According to Boulware, it does not vary along the surface. In fact, it does vary over the surface, although not with t (or z), because it varies with x and y. Still, we choose to forget it, saying that the energy associated with this term simply passes through the region $z_1 > z > z_2$, which is indeed plausible.
- There is a term involving an ordinary delta function. We can see from the calculation of the stress–energy tensor that this term arises from the interference between the continuous field in the $t > -z$ region and the delta function field. It varies as t (or z), decreasing as t increases and z decreases. The energy released propagates from the null surface $z + t = 0$ to the worldline of the charge to be absorbed and reemitted.

This is certainly an interesting picture, but once again, we have to face the problem of interpreting energy flows in electromagnetic fields and the difficulty of physical interpretation when the surface is null.

15.12.3 Boulware's Conclusion about Energy Flow

In any radiation process, Boulware proposes two equivalent pictures:

- One may say that the charge is accelerated and that this acceleration produces the radiation.
- Equivalently, one may say that the charge before its acceleration is supporting an electromagnetic field. (This was his explanation of the delta function terms on the null surface $z = -t$.) After the acceleration, its state of motion is different and it will support a different field. However, the field does not change instantaneously into the field which the charge in its new state of motion will support. The difference of the two fields is propagated away as radiation.

Once again, one wonders whether this sophisticated second picture is useful. Boulware's picture goes like this:

> The release of energy along the delta function is precisely the phenomenon of the original Coulomb field being converted to radiation. The energy flow is just sufficient, for $z > 0$, to produce the advanced field of the charge including the radiation focussed on the worldline of the charge. For $z < 0$, the energy flow from the $z = -t$ surface is the same as that which would be associated with the field produced by the charge moving in region III which, as was discussed earlier, is exactly the same as the field produced by the charge in region I.

Now how do we understand this? The delta function releases energy here, a vague statement of what might be supposed to be happening physically. On the other hand, it does seem that, mathematically, there could be a causal relation between the delta function part of the field on $z = -t$ and the flow of energy in through the sphere on the past light cone of the charge. But how do we envisage this as a physical process? Concerning the energy flow from the $z = -t$ surface into region II (this is presumably the region Boulware is referring to when he speaks of the region $z < 0$), we do not appear to have analysed that. One might well find that the energy flow in region II is from the charge. And once again, what is the relevance of the fact that an imaginary charge $-e$ in region III could generate the same fields in region II?

To conclude, these energy flow considerations are certainly of interest, but it is not at all clear what can be deduced from them. However, it does look as though the fields accumulated on the null surface $z = -t$ from the infinitely long early stages of the charge motion may act as a kind of cause for an incoming radiation field focussed back on the charge at later stages and exactly cancelling the usual outward radiation. This description is intended to suggest a mechanism for the fact that there is no radiation reaction in the case of an eternally uniformly accelerating charge.

15.13 General Conclusions

It looks as though Boulware has transformed the original problem raised by Bondi and Gold. Instead of seeing a difficulty for any version of the equivalence principle, he seems to have taken Bondi and Gold's idea that a charge that is stationary relative to coordinates in which the spacetime metric has static form cannot radiate and, noting that this would appear to be false in the case of an SHGF, he then proceeds to try to save it by showing that an observer moving with the charge would never be able to detect the radiation. Not only do the arguments used along the way fail, but the very principle of his attempt is untenable. Let us deal with the latter point first.

The problem is that it is really irrelevant what a comoving observer would be able to detect. The equivalence principles WEP and SEP that are essential to the whole theory, and which he abundantly uses, imply that since the uniformly accelerating charge does radiate, the stationary charge relative to the semi-Euclidean coordinates for a spacetime containing an SHGF must also radiate. So Bondi and Gold's idea is actually wrong, unless one rejects SEP or declares it inapplicable, as they rightly pointed out in their original paper [2]. Indeed, Boulware accepts that there is radiation and hence that Bondi and Gold's idea is wrong. He does not need to show that a comoving observer will not detect radiation, and indeed, it would make no difference whether she could detect it or not, since we agree that it is there.

Of course, it may still be an interesting question in its own right. However, Boulware's arguments are inadequate to show this. The first stage of the argument, that the comoving observer should only have access to data in region I, seems valid enough, since no information can ever reach this observer from the other region affected by EM fields sourced by the charge, viz., region II. It does make the whole attempt look a little odd, however, because one knows very well that radiation can be detected in region II and it seems rather strange to think that such a fundamental theory as this should be dependent in any way on the fact that one particular observer is partly isolated from most of spacetime by an event horizon.

In any case none of the three arguments put forward to show that the comoving observer cannot detect radiation within region I is actually valid:

- The first argument points out that the ratio $|\mathbf{E}_{\mathrm{rad}}|/|\mathbf{E}_{\mathrm{Coulomb}}|$ of the radiated field to the Coulomb field does not go as r in the region accessible to the comoving observer. This was discussed in Sect. 15.11.4, in particular (15.250) on p. 237. The problem is that the observer need in no sense depend on such a simplistic criterion as this to decide whether there is radiation or not. In particular, the observer does not have to examine the fields a long way from the charge worldline in order to determine how the fields are going to evolve.
- The second argument points out that

$$|\mathbf{E}_{\mathrm{rad}}| \sim \frac{e}{4\pi} \frac{2\sin\theta}{l^2} \, ,$$

where l is called the invariant distance from the field point to the worldline. This is also discussed in Sect. 15.11.4, in particular (15.257) on p. 240. It is interpreted to mean that the radiated part of the field seems to vary like a Coulomb field with distance from the source. This argument comes to grief in the same way as the last. The operative word is 'seems', for there is no particular reason why the observer should be duped, especially if she can make accurate measurements nearby. Indeed, it seems highly probable that the observer could accurately predict the fields even into region II on the basis of sufficiently accurate measurement of the fields in the close neighbourhood of the charge worldline. This too reveals something of the oddity of the idea that the limitations on what one particular observer can measure might be relevant.

- The third argument, often cited as definitive proof that the comoving observer could not detect radiation, or even that there is no radiation, is the fact that the generalisation of the Poynting vector to the semi-Euclidean coordinate system is identically zero. This was shown in Sect. 15.12.1 on p. 246. However, no attempt is made to establish the physical meaning of this generalisation of the Poynting vector to coordinates other than inertial coordinates. In fact, it does not give an energy flow when integrated over a spacelike hypersurface.

The last item is discussed in greater detail in the next chapter, in the light of Parrott's criticism [7].

To end this rather bleak conclusion, we should therefore ask what motivated Boulware to try to show that a comoving observer would not detect the radiation from the charge. The argument proposed in this book goes as follows. He believes Bondi and Gold's claim that, in the case of a static, homogeneous (non-tidal) gravitational field, a charge that is stationary relative to coordinates in which the metric is static cannot radiate. He then applies a subjective form of the equivalence principle, whereby the situation for an observer comoving with a uniformly accelerating charge, in a flat spacetime without even non-tidal gravitational effects, and using the semi-Euclidean coordinate system, for which the Minkowski metric has static form, must be indistinguishable. But the subjective nature of the word 'indistinguishable' is taken to mean, not that there cannot be a radiation field in some objective sense, but rather that there might be a radiation field, but that the comoving observer must not be able to detect it.

As noted already, the difficulties with this are firstly that there is no particular reason to accept Bondi and Gold's opinion that, in the case of a static, homogeneous (non-tidal) gravitational field, a charge that is stationary relative to coordinates in which the metric is static cannot radiate, and secondly that the word 'indistinguishable' cannot be understood in the above subjective manner, i.e., strict application of WEP and SEP tells us that there could not be radiation, in an objective, observer-independent sense, for the uniformly accelerating charge in a flat spacetime without gravitational effects (not even non-tidal ones), if one accepts Bondi and Gold's opinion stated in the first point. This conclusion is taken in the present thesis as a warning against versions of the equivalence principle that speak about whether things can be distinguished or not.

Chapter 16
Interpretation of Physical Quantities in General Relativity

This chapter discusses the paper [7] by Parrott. The present book considers that the first statement in the abstract of [7] is correct with a slight alteration, while fully supporting the second:

> We argue that purely local experiments can distinguish a stationary charged particle in a static gravitational field from an accelerated particle in (gravity-free) Minkowski space. Some common arguments to the contrary are analysed and found to rest on a misidentification of 'energy'.

The qualification required in the first sentence is to replace 'can' by 'could' and to add the phrase 'if there really were any distinction to be made' at the end. Note that Parrott inserts the qualifier 'gravity-free' because the static (homogeneous) gravitational field (SHGF) is also modelled in GR by Minkowski spacetime. In fact the model for the two is precisely the same in every detail, as explained in Chap. 2, up to the question of whether one accepts to apply the strong equivalence principle (SEP) to the SHGF. The present view is that one should apply SEP, because if one did not, there would be no simple way of doing electromagnetism in the SHGF case.

Once again, what blocks Parrott from applying SEP is the idea put forward by Bondi and Gold that a charge that is stationary relative to coordinates in a spacetime for which the metric has static components cannot radiate. This will be an opportunity to look more closely at this idea, the root of all the debate. In any case, this explains why Parrott declares in his introduction:

> The equivalence principle does not apply to charged particles.

There is still a certain ambiguity about this statement. He may not intend to say that the principle does not always work, but rather that the very statement of it includes conditions for it to work and that these conditions are not fulfilled here. A condition might be: the relevant physical effects must be truly local. Here, we are perhaps concerned with remote field values, so we cannot appeal to the principle.

In any case, he clearly does not mind if the equivalence principle he is referring to does not apply to charged particles:

> We do not think there is any 'paradox' remaining, unless one regards the inapplicability of the equivalence principle to charged particles as a 'paradox'. Even if the equivalence principle does not apply to charged particles, no known mathematical result or physical observation is contradicted.

One might well agree with this. The problem is just that one cannot do electromagnetism in general relativity if SEP does not apply. This seems a serious enough problem in itself to want to rescue SEP. But one then has to sacrifice Bondi and Gold's idea that a stationary charge in a static gravitational field, such as a Schwarzschild spacetime, for example, does not radiate energy.

Note that Parrott does not state the version of the equivalence principle he is using, which makes it difficult for someone coming on this paper without prior knowledge of the debate to understand what is at issue. The same lack of a clear statement of the equivalence principle can be deplored in all the papers cited here. We have to work on the hypothesis that we are using the kind of thing discussed in Chap. 2. However, the implication from use of the word 'distinguish' in the above statement of the abstract reminds us of the risk in formulating the principle in this way, viz., the risk of accepting the subjective problems of local observers with event horizons as sufficient reason for declaring two things indistinguishable.

The present view does agree that purely local experiments can detect radiation emission, but it disagrees with the idea that such experiments would distinguish between the cases that Parrott is referring to in the quote at the top of p. 255. This position should be clear from the discussion in previous chapters.

The present view also agrees that energy is misidentified by Boulware and this part of Parrott's paper is discussed in detail. The basic issue is this (summarised at the very end of the last chapter): one can extend the notion of electric and magnetic fields to quite arbitrary coordinates in arbitrary spacetimes, and one can extend the notion of the Poynting vector with it, but one does not understand what these things mean physically. In other words, the extension is purely formal. This is a very general problem in GR, where little time is usually spent interpreting the components of tensor quantities, i.e., linking theory with what could be measured. Boulware's mistake, so clearly exposed by Parrott, serves as a perfect illustration of the dangers.

So there are basically two points to watch out for in this chapter:

- Further discussion of Bondi and Gold's idea, supported by Parrott, that a charge that is stationary relative to coordinates in which the spacetime metric is static cannot radiate.
- An explanation of why the fact that the Poynting vector is zero for an observer comoving with a uniformly accelerating charge does not mean that the charge is not radiating, and nor does it mean that it will not appear to radiate, in some sense, to the comoving observer.

The next chapter deals with the last part of Parrott's paper [7]:

- A detailed discussion of the motion of a charged rocket in a gravity-free Minkowski spacetime, with conclusions regarding the reliability of the Lorentz–Dirac equation.

The latter is connected to the problem of whether or not one can accept the Lorentz–Dirac equation, rather than the issue over the equivalence principle, although Parrott motivates this analysis as follows (introduction to [7]):

> To put the matter in an easily visualised physical framework, imagine that the acceleration of a charged particle in Minkowski space is produced by a tiny rocket engine attached to the particle. Since the particle is radiating energy which can be detected and used, conservation of energy suggests that the radiated energy must be furnished by the rocket – we must burn more fuel to produce a given accelerating worldline than we would to produce the same worldline for a neutral particle of the same mass. Now consider a stationary charge in Schwarzschild spacetime, and suppose a rocket holds it stationary relative to the coordinate frame (accelerating with respect to local inertial frames). In this case, since no radiation is produced, the rocket should use the same amount of fuel as would be required to hold stationary a similar neutral particle. This gives an experimental test by which we determine locally whether we are accelerating in Minkowski space or stationary in a gravitational field – simply observe the rocket's fuel consumption.

The present view is to agree with everything here except the claim that one can say a priori without further discussion, as everyone seems to, that the charged particle held stationary relative to Schwarzschild coordinates in the Schwarzschild spacetime will not radiate. Why is there never any proof of this? It is discussed further below.

In fact, it is a shame that, in the last quote from the introduction to [7], Parrott mixes up two different spacetimes to make his point. What about the rocket holding a charge in position relative to the semi-Euclidean coordinates in an SHGF? By an application of SEP *par excellence* (because the local inertial frames happen to be global), this rocket will also have to do more work than one that was just holding a neutral particle. That is what the theory says anyway, because the mathematics is strictly identical to the case of the gravity-free Minkowski spacetime with the charged or neutral particles being uniformly accelerated. Then one does agree with Parrott that the idea of the little rocket doing the necessary acceleration in each case is a perfect local test of whether there is radiation. The disagreement is with the claim that one would distinguish the two cases. The theory (with SEP) definitely says that one would not.

The Schwarzschild spacetime is then another situation and a priori one has no reason to suppose that the charge held stationary relative to the Schwarzschild coordinates will not be found to radiate in the sense detectable by the little rocket. After all, they are only coordinates. This is discussed in Sect. 16.9, while the final section of this chapter refutes the standpoint adopted by Rohrlich in his book [22].

16.1 Definition of Energy

The next few sections discuss some simple metrics and associated Killing vector fields. A Killing (co)vector field K is one that satisfies the equation

$$K_{a;b} + K_{b;a} = 0 \,. \tag{16.1}$$

This ensures that the Lie derivative of the metric along the flow of K is zero, so that the flow of K is in some sense a symmetry of the metric. The reason for discussing such things is that, if one has a zero-divergence symmetric tensor T^{ij}, i.e., satisfying

$$T^{ij} = T^{ji} , \qquad T^{ij}{}_{;j} = 0 , \tag{16.2}$$

then one can construct a vector field

$$v^i := T^{ij} K_j \tag{16.3}$$

with zero covariant divergence, viz.,

$$v^i{}_{;i} = 0 , \tag{16.4}$$

a simple matter to demonstrate. Then by Gauss' theorem, the integral of the normal component of v over the 3D boundary of any 4D region of spacetime is zero.

It is well worth reading Sect. 3.8 of Parrott's book [13], which discusses the interpretation of energy–momentum tensors. Since these are symmetric and divergence-free (when all sources are included), one gets a divergence-free vector field for every Killing vector field of the metric. This does not mean that every such divergence-free vector field can be interpreted as the rate of flow of field energy–momentum. Highly symmetric spacetimes like Minkowski spacetime have several Killing vector fields and only ∂_t gives the energy–momentum of the field via this construction. What would work in the case of Schwarzschild spacetime, if anything, is quite another matter.

16.2 Lorentz Boost Killing Vector Field in Minkowski Spacetime

It is claimed that the one-parameter family $\lambda \mapsto \phi_\lambda$, where

$$\phi_\lambda(t,x,y,z) := (t\cosh\lambda + x\sinh\lambda, t\sinh\lambda + x\cosh\lambda, y, z) , \tag{16.5}$$

is the flow of a Killing vector field for the Minkowski metric. In Minkowski spacetime with inertial coordinates, the Killing equation becomes

$$K_{a,b} + K_{b,a} = 0 . \tag{16.6}$$

Due to the symmetry in a and b, there are ten equations here, rather than sixteen. Since

$$\frac{\partial K_0}{\partial t} = 0 , \quad \frac{\partial K_1}{\partial x} = 0 , \quad \frac{\partial K_2}{\partial y} = 0 , \quad \frac{\partial K_3}{\partial z} = 0 , \tag{16.7}$$

we can write the functional dependences

$$K_0(x,y,z)\,,\quad K_1(t,y,z)\,,\quad K_2(t,x,z)\,,\quad K_3(t,x,y)\,. \tag{16.8}$$

This leaves the six equations

$$\frac{\partial K_0}{\partial x}=-\frac{\partial K_1}{\partial t}\,,\quad \frac{\partial K_0}{\partial y}=-\frac{\partial K_2}{\partial t}\,,\quad \frac{\partial K_0}{\partial z}=-\frac{\partial K_3}{\partial t}\,,$$
$$\frac{\partial K_1}{\partial y}=-\frac{\partial K_2}{\partial x}\,,\quad \frac{\partial K_1}{\partial z}=-\frac{\partial K_3}{\partial x}\,,\quad \frac{\partial K_2}{\partial z}=-\frac{\partial K_3}{\partial y}\,. \tag{16.9}$$

Now to show that (16.5) is indeed a flow, according to the definition of these diffeomorphisms, we have to show among other things that

$$\phi_{\lambda+\sigma}=\phi_\lambda\circ\phi_\sigma\,. \tag{16.10}$$

In other words, we have a group homomorphism from $\mathbb{R}$ under addition into the group of diffeomorphisms of the manifold under composition, since it is immediately obvious that $\phi_0=$ identity. To show (16.10), we examine

$$\phi_{\lambda+\sigma}(t,x,y,z):= \tag{16.11}$$
$$\bigl(t\cosh(\lambda+\sigma)+x\sinh(\lambda+\sigma),t\sinh(\lambda+\sigma)+x\cosh(\lambda+\sigma),y,z\bigr)\,,$$

and use the addition laws for the hyperbolic functions, viz.,

$$\cosh(\lambda+\sigma)=\cosh\lambda\,\cosh\sigma+\sinh\lambda\,\sinh\sigma\,, \tag{16.12}$$

$$\sinh(\lambda+\sigma)=\sinh\lambda\,\cosh\sigma+\cosh\lambda\,\sinh\sigma\,. \tag{16.13}$$

The rest is simple algebra.

For fixed $x\in\mathbb{M}$, $\lambda\mapsto\phi_\lambda(x)$ is a curve (where $\mathbb{M}$ denotes the Minkowski spacetime manifold). The vector field in question is the tangent to this, i.e., $K(x)$ is tangent to the curve $\lambda\mapsto\phi_\lambda(x)$ at $\lambda=0$. To find $K(y)$ for another $y\in\mathbb{M}$, we take the tangent to $\lambda\mapsto\phi_\lambda(y)$ at $\lambda=0$. Looking at (16.5), we immediately obtain

$$K(x)=\begin{pmatrix}x\\t\\0\\0\end{pmatrix}=x\frac{\partial}{\partial t}+t\frac{\partial}{\partial x}\,. \tag{16.14}$$

It is a simple matter to check that the Killing equations mentioned above are satisfied.

In the inertial coordinates (t,x,y,z), we can also check that the flow lines are described by the formula $(x^1)^2-(x^0)^2=$ constant, since

$$(t\sinh\lambda+x\cosh\lambda)^2-(t\cosh\lambda+x\sinh\lambda)^2=x^2-t^2\,.$$

The notation leaves something to be desired. However, if we choose any x,t and consider the curve passing through (t,x) when λ is zero, it is just the hyperbola

$$(x^1)^2 - (x^0)^2 = x^2 - t^2 .$$

We can look, for example, at the curves passing through points $(0,X)$ when $\lambda = 0$. These are the hyperbolas

$$(x^1)^2 - (x^0)^2 = X^2 .$$

It is also claimed that each curve in the flow is the worldline of a uniformly accelerated particle. There is something intuitively obvious about this (in a dangerous way), because ϕ_λ boosts its argument point in spacetime by a Lorentz boost with rapidity λ.

To analyse this crucial claim, consider the curve through $(0,X)$. Its tangent with respect to the parameter λ is

$$\frac{\partial}{\partial \lambda} = \begin{pmatrix} X\cosh\lambda \\ X\sinh\lambda \\ 0 \\ 0 \end{pmatrix} = \begin{pmatrix} x \\ t \\ 0 \\ 0 \end{pmatrix} . \tag{16.15}$$

This vector is not a unit vector, so it is not the 4-velocity until normalised. Now the pseudolength of ∂_λ calculated using the Minkowski metric is just X. So the 4-velocity is

$$u := \partial_\tau = \frac{1}{X}\partial_\lambda , \tag{16.16}$$

whence the proper time along the worldline is

$$\tau = \lambda X . \tag{16.17}$$

Since $t = X\sinh\lambda$ from the equation for the flow, we have

$$\tau = X\sinh^{-1}\frac{t}{X} , \tag{16.18}$$

where t is the coordinate time elapsed since the particle crossed the x axis, which was when $\lambda = 0$. Also from (16.15) and (16.16), we deduce the relation

$$u(\tau) = \left(\cosh\frac{\tau}{X}, \sinh\frac{\tau}{X}, 0, 0 \right) . \tag{16.19}$$

There are many other ways to arrive at this formula. We now have the four-acceleration as

$$a(\tau) = \frac{1}{X}\left(\sinh\frac{\tau}{X}, \cosh\frac{\tau}{X}, 0, 0 \right) . \tag{16.20}$$

Naturally, there is nothing constant about this. What is constant is its magnitude (or pseudolength), which is

$$a^2 = a^i a_i = -1/X^2 \,. \tag{16.21}$$

We can deduce what is called the scalar proper acceleration for the specific case of 1D motion, as Parrott does on many occasions, by defining a unit spacelike vector w that is always orthogonal to the motion by

$$w(\tau) = \left(\sinh\frac{\tau}{X}, \cosh\frac{\tau}{X}, 0, 0 \right) \,. \tag{16.22}$$

We know that a has to be orthogonal to u, hence parallel to w, in this 2D world. We define the scalar proper acceleration A by $a = Aw$. We immediately have $A = 1/X$, and this, like the magnitude of a^i, is constant.

What we have here is precisely the kind of motion, viz., uniformly accelerated motion, that has been the subject of most of this book.

16.3 Killing Vector Field for Static Spacetime

First of all, a static spacetime is one whose metric can be written in the form

$$ds^2 = g_{00}(x^1,x^2,x^3)(dx^0)^2 + \sum_{I,J=1}^{3} g_{IJ}(x^1,x^2,x^3)dx^I dx^J \,, \tag{16.23}$$

for some coordinate system. We shall now show that ∂_{x^0} is a Killing vector field for this spacetime. The Killing equations are

$$K_{a,b} + K_{b,a} = 2\Gamma_{ab}^c K_c \,. \tag{16.24}$$

The K we are testing has components

$$K^0 = 1\,, \quad K^1 = 0\,, \quad K^2 = 0\,, \quad K^3 = 0\,. \tag{16.25}$$

The associated 1-form is

$$K_0 = g_{00}(x^1,x^2,x^3)\,, \quad K_1 = 0\,, \quad K_2 = 0\,, \quad K_3 = 0\,. \tag{16.26}$$

Four of the Killing equations are

$$K_{0,b} = 2\Gamma_{0b}^c K_c = 2\Gamma_{0b}^0 g_{00}(x^1,x^2,x^3)\,, \tag{16.27}$$

which will only be true if we can show that

$$\Gamma_{00}^0 = 0\,, \quad \partial_b g_{00} = 2\Gamma_{0b}^0 g_{00}(x^1,x^2,x^3) \quad \text{for } b = 1,2,3\,. \tag{16.28}$$

The diagonal Killing equations (i.e., those with $a = b$) read

$$\Gamma^0_{00}K_0 = 0 \,, \quad \Gamma^0_{11}K_0 = 0 \,, \quad \Gamma^0_{22}K_0 = 0 \,, \quad \Gamma^0_{33}K_0 = 0 \,, \tag{16.29}$$

since $K_{a,a} = 0$ (no summation) for all a. Since one of these is the same as one of (16.27), this accounts for seven equations, leaving the three in which a and b are both in $\{1,2,3\}$, viz.,

$$\Gamma^0_{ab} = 0 \,, \quad a,b \in \{1,2,3\} \,. \tag{16.30}$$

It is clear that the problem is resolved by calculating these connection coefficients. We use the standard Levi-Civita formula

$$\Gamma^i_{jk} = \frac{1}{2}g^{il}\left(\partial_j g_{lk} + \partial_k g_{jl} - \partial_l g_{jk}\right) \,. \tag{16.31}$$

Considering (16.30) to begin with, we examine

$$\Gamma^0_{jk} = \frac{1}{2}g^{0l}\left(\partial_j g_{lk} + \partial_k g_{jl} - \partial_l g_{jk}\right) \,. \tag{16.32}$$

In the sum over l, the only term contributing is $l = 0$. The term in ∂_l will thus give nothing. The other terms are derivatives of g_{0k} and g_{j0}. But these are zero unless $k = 0$ and $j = 0$, respectively. In (16.30), neither k nor j is going to be zero so this equation is indeed satisfied.

We also have

$$\Gamma^0_{0b} = \frac{1}{2}g^{00}\partial_b g_{00} \,. \tag{16.33}$$

Since $g^{00} = 1/g_{00}$ for this kind of metric, we have shown (16.27). The three coefficients $\Gamma^0_{11}, \Gamma^0_{22}, \Gamma^0_{33}$ are all zero by the same considerations, and this shows that the three last equations of (16.29) are all satisfied. Note that this analysis only requires us to consider the connection coefficients of the form Γ^0_{jk}. This is precisely because the putative Killing vector field only had a zero component. We are through.

16.4 Killing Vector Fields for Schwarzschild Spacetime

The metric is

$$g_{ij} = \begin{pmatrix} 1 - \dfrac{A}{r} & & & \\ & -\left(1 - \dfrac{A}{r}\right)^{-1} & & \\ & & -r^2 & \\ & & & -r^2 \sin^2\theta \end{pmatrix} \,,$$

where $A = 2GM$. It is convenient to display the connection coefficients in the form of four 4×4 matrices:

$$\Gamma^0_{ij} = \begin{pmatrix} 0 & \dfrac{A}{2r^2}B(r)^{-1} & & \\ \dfrac{A}{2r^2}B(r)^{-1} & 0 & & \\ & & 0 & 0 \\ & & 0 & 0 \end{pmatrix},$$

$$\Gamma^1_{ij} = \begin{pmatrix} \dfrac{A}{2r^2}B(r) & 0 & 0 & 0 \\ 0 & -\dfrac{A}{2r^2}B(r)^{-1} & \dfrac{1}{r} & 0 \\ 0 & \dfrac{1}{r} & -rB(r) & 0 \\ 0 & 0 & 0 & -rB(r)\sin^2\theta \end{pmatrix},$$

$$\Gamma^2_{ij} = \begin{pmatrix} 0 & 0 & & \\ 0 & 0 & & \\ & & 0 & 0 \\ & & 0 & -\sin\theta\cos\theta \end{pmatrix}, \qquad \Gamma^3_{ij} = \begin{pmatrix} 0 & 0 & 0 & 0 \\ 0 & 0 & 0 & \dfrac{1}{r} \\ 0 & 0 & 0 & \cot\theta \\ 0 & \dfrac{1}{r} & \cot\theta & 0 \end{pmatrix},$$

where $B(r) = 1 - A/r$ is just a device to get the second matrix onto one line.

We have already seen that ∂_t is a Killing vector field, since this metric is static in the sense outlined above. It can be shown that the only other Killing vector fields have the form

$$K_\theta(r,\theta,\phi)\partial_\theta + K_\phi(r,\theta,\phi)\partial_\phi , \tag{16.34}$$

for some functions K_θ and K_ϕ, which are not functions of t. These functions satisfy some other constraints, which are the remnants of the Killing equations. The aim here is to find these constraints and one simple solution.

The Killing equations are

$$K_{a,b} + K_{b,a} = 2\Gamma^c_{ab}K_c . \tag{16.35}$$

We shall not attempt to show here that (16.34) gives the only other solutions, but merely plug in this form of solution and find what the Killing equations require. We have 10 independent equations for K_c. We think of these as the top right-hand half of a matrix of 16 equations indexed by a,b. Let us establish these quite generally for the Schwarzschild spacetime and then apply to the above proposal for K_c.

Along the top row we obtain

$$\frac{\partial K_0}{\partial r} + \frac{\partial K_1}{\partial t} = 2\Gamma_{01}^c K_c = \frac{2M}{r^2}\left(1 - \frac{2M}{r}\right)^{-1} K_0 \,, \tag{16.36}$$

since the connection matrix Γ^0 is the only one with a nonzero entry in the 01 position, then

$$\frac{\partial K_0}{\partial \theta} + \frac{\partial K_2}{\partial t} = 2\Gamma_{02}^c K_c = 0 \,, \tag{16.37}$$

and

$$\frac{\partial K_0}{\partial \phi} + \frac{\partial K_3}{\partial t} = 2\Gamma_{03}^c K_c = 0 \,, \tag{16.38}$$

since none of the connection matrices have entries at 02 and 03.

The diagonal Killing equations, with $a = b$, are

$$\frac{\partial K_0}{\partial t} = \frac{1}{r^2}\left(1 - \frac{2M}{r}\right) K_1 \,, \quad \frac{\partial K_1}{\partial r} = -\frac{1}{r^2}\left(1 - \frac{2M}{r}\right)^{-1} K_1 \,, \tag{16.39}$$

$$\frac{\partial K_2}{\partial \theta} = -r\left(1 - \frac{2M}{r}\right) K_1 \,, \quad \frac{\partial K_3}{\partial \phi} = -r\left(1 - \frac{2M}{r}\right)\sin^2\theta \, K_1 - \sin\theta\cos\theta \, K_2 \,. \tag{16.40}$$

The three last Killing equations, for a, b between the diagonal and the top row of the array, are

$$\frac{\partial K_1}{\partial \theta} + \frac{\partial K_2}{\partial r} = 2\Gamma_{12}^c K_c = \frac{2}{r} K_1 \,, \tag{16.41}$$

$$\frac{\partial K_1}{\partial \phi} + \frac{\partial K_3}{\partial r} = 2\Gamma_{13}^c K_c = \frac{2}{r} K_3 \,, \tag{16.42}$$

$$\frac{\partial K_2}{\partial \phi} + \frac{\partial K_3}{\partial \theta} = 2\Gamma_{23}^c K_c = 2\cot\theta \, K_3 \,. \tag{16.43}$$

We now turn to the proposed solution (16.34). This is the contravariant vector. The associated one-form has components

$$K_0 = 0 \,, \quad K_1 = 0 \,, \quad K_2 = -r^2 K_\theta(r,\theta,\phi) \,, \quad K_3 = -r^2 \sin^2\theta K_\phi(r,\theta,\phi) \,. \tag{16.44}$$

Equation (16.36) is immediate because K_0 and K_1 are both zero. Equations (16.37) and (16.38) are also immediate because K_0 is zero and K_2 and K_3 do not depend on t. Equations (16.39) are satisfied because they only involve K_0 and K_1.

We now begin to obtain further constraints. The first equation of (16.40) tells us that K_2 is independent of θ, which means that K_θ is independent of θ. The second equation here is

$$\frac{\partial K_3}{\partial \phi} = -\sin\theta\cos\theta\, K_2 \ . \tag{16.45}$$

This can be written

$$-r^2\sin^2\theta\,\frac{\partial K_\phi}{\partial \phi} = r^2\sin\theta\cos\theta\, K_\theta \ ,$$

whence

$$\frac{\partial K_\phi}{\partial \phi} = -\cot\theta\, K_\theta \ . \tag{16.46}$$

Equation (16.41) tells us that K_2 is independent of r, and hence K_θ is proportional to $1/r^2$. It thus has the form

$$K_\theta(r,\theta,\phi) = \frac{K'_\theta(\phi)}{r^2} \ , \tag{16.47}$$

for some function K'_θ of ϕ alone. (We saw above that K_θ had to be independent of θ.) Equation (16.42) implies that

$$\frac{\partial K_3}{\partial r} = \frac{2}{r}K_3 \ , \tag{16.48}$$

which can be written

$$-\sin^2\theta\,\frac{\partial}{\partial r}(r^2 K_\phi) = -\frac{2}{r}r^2\sin^2\theta\, K_\phi \ ,$$

whence

$$\frac{\partial}{\partial r}(r^2 K_\phi) = 2rK_\phi \ . \tag{16.49}$$

Finally, this becomes

$$\frac{\partial K_\phi}{\partial r} = 0 \ . \tag{16.50}$$

The last Killing equation, viz., (16.43), tells us that

$$-r^2 \frac{\partial K_\theta}{\partial \phi} - r^2 \frac{\partial}{\partial \theta} \left(\sin^2 \theta \, K_\phi \right) = -2r^2 \cos \theta \sin \theta \, K_\phi \; .$$

This gives

$$\frac{\partial K_\theta}{\partial \phi} = -\sin^2 \theta \, \frac{\partial K_\phi}{\partial \theta} \; . \tag{16.51}$$

Let us summarise these relations, putting $K_\theta = K'_\theta(\phi)/r^2$ and $K_\phi = K_\phi(\theta, \phi)$, with no r dependence due to (16.50). We then have

$$K'_\theta = -r^2 \tan \theta \, \frac{\partial K_\phi}{\partial \phi} \; , \qquad \frac{\mathrm{d} K'_\theta}{\mathrm{d} \phi} = -r^2 \sin^2 \theta \, \frac{\partial K_\phi}{\partial \theta} \; . \tag{16.52}$$

There is an obvious solution, viz.,

$$K'_\theta = 0 \; , \qquad K_\phi = \text{constant} \; . \tag{16.53}$$

We have thus shown that ∂_ϕ is a Killing vector field in the Schwarzschild spacetime. Another two linearly independent Killing vector fields are

$$\cos \phi \cot \theta \, \frac{\partial}{\partial \phi} + \sin \phi \, \frac{\partial}{\partial \theta} \quad \text{and} \quad -\sin \phi \cot \theta \, \frac{\partial}{\partial \phi} + \cos \phi \, \frac{\partial}{\partial \theta} \; . \tag{16.54}$$

It is a straightforward matter to check that

$$K_\phi = \cos \phi \cot \theta \; , \quad K'_\theta = r^2 \sin \phi \; , \tag{16.55}$$

and

$$K_\phi = -\sin \phi \cot \theta \; , \quad K'_\theta = r^2 \cos \phi \; , \tag{16.56}$$

do both satisfy the two relations in (16.52). Referring to a standard textbook [20], we see that these three Killing vector fields are precisely the ones we find for the 2-sphere with its standard metric when embedded in the Euclidean 3-space.

Commutation of Killing Fields

Parrott mentions that the Killing fields

$$\partial_t \; , \qquad A := K_\theta \partial_\theta + K_\phi \partial_\phi$$

commute, and it is important to remember what this means. In fact, this refers to the fact that the Lie bracket of the two vector fields is zero, i.e.,

$$[\partial_t, A] = 0 \; .$$

This question is discussed in many standard textbooks [20]. For two vector fields v and w with components v^i and w^j, the Lie bracket is defined to be

$$[v,w] := \left(v^i \frac{\partial}{\partial x^i} w^j - w^i \frac{\partial}{\partial x^i} v^j \right) \frac{\partial}{\partial x^j} . \tag{16.57}$$

In the present case, we have $v = \partial_t$, whence $v = (1,0,0,0)$ in component form. In this case, for any w,

$$[v,w] = \frac{\partial w^j}{\partial t} \frac{\partial}{\partial x^j} . \tag{16.58}$$

But when we insert $w = A$, none of the components of w depend on t so the right-hand side of the last relation is indeed zero.

16.5 Another Metric

We consider a simple static metric of the form

$$ds^2 = c(x)^2 dt^2 - dx^2 - dy^2 - dz^2 . \tag{16.59}$$

We can consider $c(x)$ to be the x-dependent speed of light as observed from the coordinate frame. The metric is supposed to describe a gravitational field in the x direction. Indeed, a stationary particle relative to these coordinates has an acceleration in the x direction. On the other hand, the Riemann curvature tensor is zero for some functions c. An example is the functions that arise when we obtain the semi-Euclidean coordinates for an accelerating observer, since this construction is always made in a Minkowski spacetime. Let us examine these points.

It is a simple matter to show that the only nonzero connection coefficients are

$$\Gamma^0_{01} = \Gamma^0_{10} = \frac{c'}{c} , \qquad \Gamma^1_{00} = c'c . \tag{16.60}$$

This already gives us the acceleration of a stationary particle. It has four-velocity $u = (1/c,0,0,0)$ and

$$a^i = (D_u u)^i = u^j \frac{\partial u^i}{\partial x^j} + u^j \Gamma^i_{jk} u^k = \frac{1}{c^2} \Gamma^i_{00} , \tag{16.61}$$

whence

$$a = D_u u = \frac{c'}{c} \partial_x . \tag{16.62}$$

Let us ascertain which functions $c(x)$ lead to flat spacetimes.

The Riemann tensor is given by

$$R^a_{bcd} = \frac{\partial \Gamma^a_{db}}{\partial x^c} - \frac{\partial \Gamma^a_{cb}}{\partial x^d} + \Gamma^a_{cf}\Gamma^f_{db} - \Gamma^a_{df}\Gamma^f_{cb} \; . \tag{16.63}$$

There are twenty independent components. We first note that if any one index a,b,c,d is equal to 2 or 3, the corresponding component is zero. Even without considering symmetries, this reduces the number of nonzero components to less than $2^4 = 16$.

By the antisymmetry $R^a_{bcd} = -R^a_{bdc}$, we have $R^a_{b00} = 0 = R^a_{b11}$, for all values of a,b. Now

$$R^a_{b01} = -R^a_{b10} = -\frac{\partial \Gamma^a_{0b}}{\partial x^1} + \Gamma^a_{0f}\Gamma^f_{1b} - \Gamma^a_{1f}\Gamma^f_{0b} \; .$$

These are the last eight components. We have

$$\begin{aligned}
R^1_{001} &= -\frac{\partial \Gamma^1_{00}}{\partial x^1} + \Gamma^1_{0f}\Gamma^f_{10} - \Gamma^1_{1f}\Gamma^f_{00} \\
&= -\frac{\mathrm{d}}{\mathrm{d}x}(c'c) + c'^2 = c''c \; .
\end{aligned}$$

This gives the two components

$$R^1_{001} = -R^1_{010} = c''c \; . \tag{16.64}$$

Likewise, the formula above gives

$$R^0_{101} = -R^0_{110} = -\frac{c''}{c} \; . \tag{16.65}$$

Finally,

$$R^0_{001} = -R^0_{010} = 0 = R^1_{101} = -R^1_{110} \; . \tag{16.66}$$

So in the end, only four of the components are nonzero.

It is easy to see now which functions $c(x)$ give flat spacetimes. The only solution is $c'' = 0$, and this implies

$$c(x) = Ax + B \; , \quad \text{for some constants } A, B \; . \tag{16.67}$$

These metrics then take the form

$$\mathrm{d}s^2 = (Ax + B)^2 \mathrm{d}t^2 - \mathrm{d}x^2 - \mathrm{d}y^2 - \mathrm{d}z^2 \; . \tag{16.68}$$

An example would be a metric of the form

$$\mathrm{d}s^2 = \left(1 + \frac{gy^1}{c^2}\right)^2 (\mathrm{d}y^0)^2 - (\mathrm{d}y^1)^2 - (\mathrm{d}y^2)^2 - (\mathrm{d}y^3)^2 \; , \tag{16.69}$$

the metric of (2.16) on p. 9.

16.6 And Another Metric

This time we consider a simple static metric of the form

$$ds^2 = X^2 d\lambda^2 - dX^2 - dy^2 - dz^2 . \tag{16.70}$$

This is one of the metrics we have just shown to be flat. This metric is just the Minkowski metric in disguise, because the coordinates are not inertial coordinates. It therefore has the hidden symmetries associated with Lorentz boosts analysed above. It is very close to the semi-Euclidean form of the metric in (16.69), except that it has an apparent singularity at $X = 0$, whereas (16.69) has its apparent singularity at $y^1 = -c^2/g$.

We note in passing that the connection coefficients are

$$\Gamma^0_{01} = \frac{1}{X} = \Gamma^0_{10} , \qquad \Gamma^1_{00} = X , \tag{16.71}$$

and the curvature components are $R^a_{bcd} = 0$. We can show directly that the transformation

$$t = X \sinh \lambda , \quad x = X \cosh \lambda , \tag{16.72}$$

returns inertial coordinates t, x, y, z by using the ploy of inserting

$$\begin{cases} dt = dX \sinh \lambda + X \cosh \lambda \, d\lambda , \\ dx = dX \cosh \lambda + X \sinh \lambda \, d\lambda , \end{cases} \tag{16.73}$$

into the invariant ds^2 constructed from the Minkowski metric, viz.,

$$ds^2 = dt^2 - dx^2 - dy^2 - dz^2 . \tag{16.74}$$

We have

$$dt^2 - dx^2 = X^2 d\lambda^2 - dX^2 . \tag{16.75}$$

Killing Vector Fields

We know that ∂_λ is a Killing field for the above metric (16.70), because we proved this quite generally for static metrics of this kind. Let us show that it is precisely the Lorentz boost Killing field of (16.14), viz.,

$$K(x) = x\frac{\partial}{\partial t} + t\frac{\partial}{\partial x} . \tag{16.76}$$

We shall transform the vector v^μ with components $(1,0,0,0)$ with respect to coordinates λ, X, y, z, to get $v'^\mu = (x, t, 0, 0)$ with respect to coordinates t, x, y, z. Indeed,

$$v'^{\mu} = \frac{\partial x'^{\mu}}{\partial x^{\nu}} v^{\nu} = \frac{\partial x'^{\mu}}{\partial \lambda} = \begin{pmatrix} \partial t/\partial \lambda \\ \partial x/\partial \lambda \\ 0 \\ 0 \end{pmatrix} = \begin{pmatrix} X\cosh\lambda \\ X\sinh\lambda \\ 0 \\ 0 \end{pmatrix} = \begin{pmatrix} x \\ t \\ 0 \\ 0 \end{pmatrix}.$$

We may also ask what the Killing vector field ∂_t becomes when converted to coordinates λ, X, y, z. The inverse coordinate transformation is

$$\lambda = \tanh^{-1}\frac{t}{x}, \qquad X = (x^2 - t^2)^{1/2}. \tag{16.77}$$

We start with $w = \partial_t = (1,0,0,0)$ in inertial coordinates. The transformation is

$$w'^{\mu} = \frac{\partial x'^{\mu}}{\partial x^{\nu}} w^{\nu} = \frac{\partial x'^{\mu}}{\partial t} = \begin{pmatrix} \partial \lambda/\partial t \\ \partial X/\partial t \\ 0 \\ 0 \end{pmatrix}.$$

Now

$$\frac{\partial X}{\partial t} = -\frac{t}{X} = -\sinh\lambda\,,$$

and $\tanh\lambda = t/x$ so

$$(1 - \tanh^2\lambda)\frac{\partial \lambda}{\partial t} = \frac{1}{x}\,,$$

whence

$$\frac{\partial \lambda}{\partial t} = \frac{\cosh\lambda}{X}\,.$$

So the components of the Killing vector field ∂_t in the λ, X, y, z coordinates are

$$\partial_t \longleftrightarrow \left(\frac{\cosh\lambda}{X}, -\sinh\lambda, 0, 0\right) \quad \text{in } \lambda, X, y, z. \tag{16.78}$$

16.7 Rindler or Elevator Coordinates

The coordinates λ, X, y, z giving the metric

$$ds^2 = X^2 d\lambda^2 - dX^2 - dy^2 - dz^2 \tag{16.79}$$

are sometimes called Rindler coordinates or elevator coordinates. The idea is that X, y, z may be viewed as space coordinates as seen by occupants of a rigidly accelerated elevator. But why should we take these as space coordinates? Of course, there are some interesting features, well known from our earlier study of semi-Euclidean coordinates. But surely, the question is whether we would set up such coordinates if we were accelerating. In other words, are these physically the same as accelerating coordinates, in some sense, for some operational way of setting up coordinates? Indeed, is there any privileged set of coordinates for an accelerating observer, as there is for an inertial observer?

This is an important question here, because a lot of this debate ought to have been about how one should interpret this general theory of relativity in specific cases, where one happens to have a coordinate system on a manifold, and a metric in component form. The last few sections show how easy it is to run out mathematical formulas. But this really is the easy part of the theory, even if the mathematics soon gets tough. Once one has raised the question of how to interpret coordinates and metric components, one finds that there is little reference to any kind of answer in textbooks and papers, and that authors do fall into traps as a result.

To return to the relatively simple case of an accelerating observer in a flat space-time, where we may or may not consider that there is a non-tidal gravitational field, we are asking whether Rindler's coordinates, which are obviously closely related to our semi-Euclidean coordinates, are in fact a privileged set in some way. Well in fact it really looks as though they are not. When we set them up as in Chap. 2, we can see why. It is because we lay down a set of conditions for the coordinates to be suitably adapted to the accelerating worldline that we see just how artificial they actually are. Indeed, why should there be privileged coordinates for accelerating observers? There are such coordinates for inertially moving observers because it turns out that inertial motion is physically irrelevant. But accelerated motion makes a difference, because acceleration is not relative but absolute, for reasons we still do not understand.

We can see from the metric that the proper time along any worldline with constant X, y, z is $\tau = \lambda X$. We cannot therefore consider the worldline of the origin of this coordinate system, where there is a singularity. The proper time along the worldline of $X = 1$ is the coordinate λ. This corresponds to the fact that, when we construct the semi-Euclidean metric (16.69), we arrange for the origin of the coordinate system to have proper time equal to the coordinate time. This is one of the features that might make the semi-Euclidean coordinate system a candidate for the privileged accelerating frame, if indeed there were such a thing. So the coordinates we have here are the semi-Euclidean coordinates for the observer at $X = 1$.

We also observe that any worldline with X, y, z constant is the orbit of the point $t = 0$, $x = X$ under the flow (16.5), i.e., the worldline of a uniformly accelerated particle with scalar proper acceleration $1/X$. The idea behind the Rindler frame then is to regard all curves with X, y, z constant as the worldlines of a collection of uniformly accelerated observers, all of whom are at rest in the Minkowski frame at time $t = 0$.

Now Parrott says on p. 8 of [7] that an observer following such a worldline sees at any given moment other such worldlines at a constant distance in her rest frame at that moment. We must deplore the use of the word 'sees'. What the observer sees is light emitted from those worldlines and that is a quite different matter. We have to replace 'sees' by something like 'conceives', since the observer with these coordinates is going to lay her interpretation on the story told by the coordinates. She has agreed that all points with the same value of λ are 'simultaneous', although she ought to be cynical about the value of this concept since she is only borrowing the notion of simultaneity from an instantaneously comoving *inertial* observer. But this does explain what is meant by 'at any given moment' in the above statement.

It is then claimed that we may take a collection of such worldlines and imagine connecting them with rigid rods, and that the rods here can be rigid because the proper distances are constant, so that we obtain a rigid accelerating structure which we might call an elevator. The point is that we can propose a quite convincing equation for the length of an accelerating rigid rod, and it turns out that it corresponds precisely to the kind of rod defined as extending between two fixed spatial coordinates along the axis of acceleration in the semi-Euclidean system [19]. However, there are problems with this approach to rigidity.

Let us analyse this notion briefly to see what is involved. We consider a rod with points

$$(\lambda, X_1, 0, 0) \,, \ \ldots \,, \ (\lambda, X, 0, 0) \,, \ \ldots \,, \ (\lambda, X_2, 0, 0) \,, \quad X \in [X_1, X_2] \,.$$

We have simplified without loss of generality by choosing y and z to be zero. At any Rindler time λ, the rod is straight in the sense that it lies along the X axis for the Rindler observer, with proper length $X_2 - X_1$ relative to the Rindler simultaneity. But what does this rod look like in Minkowski spacetime?

The point at Rindler X follows

$$(t, x) = X(\sinh \lambda, \cosh \lambda)$$

as λ varies. Hence, $x^2 - t^2 = X^2$ or

$$x(t) = (X^2 + t^2)^{1/2} \,.$$

As we know, each point on the rod has a different proper acceleration. It may seem difficult to understand in what sense the rod is rigid in this description, and one might expect to get a safer opinion in Minkowski spacetime. The Minkowski distance between points at Rindler X_1 and X_2 (the ends of the rod) at a given value of t is

$$\mathscr{L}_t = x_2(t) - x_1(t) = (X_2^2 + t^2)^{1/2} - (X_1^2 + t^2)^{1/2} \,,$$

which is not constant in time. Of course the Rindler observer attributes a constant proper length, but proper length is observer dependent, because it depends on the choice of spacelike hypersurfaces that are considered to be planes of simultaneity. We can also say that, at Minkowski time t, the rod occupies the points

$$\mathscr{R}_t = \left\{ (X^2 + t^2)^{1/2} : X \in [X_1, X_2] \right\} \,.$$

Note that, since it lies along the x axis, it does remain 'straight' in the usual understanding of the term. We are talking about a rod moving in the direction of its own length.

Does the rod contract or expand in the Minkowski description as Minkowski time t goes by? We have

$$\frac{\mathrm{d}\mathscr{L}_t}{\mathrm{d}t} = \frac{t}{(X_2^2 + t^2)^{1/2}} - \frac{t}{(X_1^2 + t^2)^{1/2}} \,. \tag{16.80}$$

Since $X_2 > X_1$, we have

$$\frac{\mathrm{d}\mathscr{L}_t}{\mathrm{d}t} < 0 \,, \tag{16.81}$$

for all values of t. This rod contracts in the Minkowski view. However, it is not a Fitzgerald contraction. Of course, the points on $\mathscr{R}_t$ have different accelerations. Indeed, the mere fact that they are accelerating means that this situation differs from the usual context for talking about Fitzgerald contraction. The latter is a phenomenon which affects rods with constant velocity relative to some Minkowski coordinate frame.

In fact, the rigidity here is just this: at any given instant, the proper distance from one particle in the rod to its nearest neighbour as measured in the inertial frame instantaneously comoving with either particle is exactly the Fitzgerald contracted distance, as compared with when the whole system is at rest in an inertial frame, for the instantaneous value of the speed of the relevant particle. However, it is not obvious that one could ever get a rod to behave like this. We have to accelerate each point of it in the right way, and it must distort along its length, so there is a real sense in which it is not rigid. At least, we may well define 'rigid' to suit our purposes, but could anything ever behave rigidly?

Could a rod behave like this in our GR model of a static homogeneous gravitational field, i.e., a flat spacetime with coordinates (λ, X, y, z) such that the metric takes the form (16.79), if it were supported at one end, in the sense that one end was held at a fixed value of X, with the rod axis along the X axis? One would be applying a force to this end of the rod to keep it off its geodesic, and internal forces in the rod would have to keep the other particles in the rod off geodesics in an appropriate way, so that each had the uniform acceleration appropriate to its value of the X coordinate. Without making any assumptions about those interparticle forces, this looks a tall order. But note that, if we wished to investigate the effect of a uniform gravitational field, we could attempt to carry out the kind of analysis that Bell has done [18] in the case of special relativity, looking at the effect of an acceleration on the electric fields produced by atomic nuclei, and so on, and this does seem a feasible project.

To put the problem another way, one could ask how a rod would behave in a special relativistic model of a static homogeneous gravitational field, but in free fall this time, of course, so that each particle is subjected to a constant gravitational acceleration, the same as all its neighbours. Once again, the problem here is to assess the effects of interparticle forces. In the extreme case where the particles of the rod are merely juxtaposed, but otherwise free, i.e., not subject to any interparticle forces, one finds that this 'rod' gets longer for the instantaneously comoving inertial observer, which is clearly not what is usually intended by the notion of a rigid rod!

In any case, when we observe an inertially moving rod from a Minkowski frame, we can define it to be rigid if the only distortion is the Fitzgerald contraction. If the rod is accelerated from one inertial motion to another, it is not obvious what its length will be during the transition. Presumably, even rigid rods according to this definition could behave differently depending on their microscopic composition and structure. For example there might be some transient oscillation in the length, superposed on the overriding Fitzgerald effect.

There is one other intriguing question about a rod moving in the way described in Parrott's paper: if the rod were oriented suitably relative to its direction of motion, it would bend, as viewed in the Minkowski frame. We have considered a rod moving along its own axis, which merely distorts in length. If the rod were parallel to the y axis, with constant X values at each point of its length, then each point of it would have the same proper acceleration (which depends only on X), so it would not even change length. But suppose it lay askew in the X,y plane. Now different points of it would have different proper accelerations and it would bend, as viewed in the Minkowski frame!

The point of this discussion is just to remind ourselves that things are vastly more complex in relativity theory, even special relativity theory, than they are in classical physics. All the more reason to treat electromagnetic field components and Poynting vectors with some respect when the coordinate frame is not inertial.

16.8 The Problem with the Poynting Vector

Enough has already been said to warn anyone about rash interpretation of the electric and magnetic field components, or the corresponding Poynting vector, when the coordinate frame is not inertial. Because the EM field tensor is a two-form (antisymmetric, covariant, and rank two), its component matrix can always be written in the form

$$(F_{\mu\nu}) = \begin{pmatrix} 0 & E_1 & E_2 & E_3 \\ -E_1 & 0 & -B_3 & B_2 \\ -E_2 & B_3 & 0 & -B_1 \\ -E_3 & -B_2 & B_1 & 0 \end{pmatrix},$$

for some functions E_i and B_j on the spacetime, whatever coordinate system one chooses. Does this mean that $\mathbf{E}$ and $\mathbf{B}$ are electric and magnetic fields as we know them?

One could of course make this the definition, but it should be borne in mind that they only correspond to our usual understanding in a locally inertial frame, and then only locally, i.e., with decreasing reliability as one moves away from the spacetime event at which one has made the connection components equal to zero. This is the utility of the strong principle of equivalence: physics looks roughly as we know it in locally inertial frames. Without this principle, one could not even begin to interpret the constructions one has made on these manifolds. The weak and strong principles of equivalence taken together allow one to interpret motion relative to arbitrary coordinates, and they also allow one to interpret other fields defined relative to arbitrary coordinates. It is worth remembering that one could say nothing about the link between the spacetime manifold and the spacetime observed without having principles like these. Without SEP, one would require some other principle that made the link.

We now examine Parrott's argument appearing in Sect. 4 of [7]. This discussion overlaps what was said in Chaps. 6, 8, and 11 of the present book. A charged particle is unaccelerated in the distant past, put into accelerated motion for a while, and reenters unaccelerated motion in the distant future. It is surrounded by an imaginary worldtube across which one can calculate the flow of energy–momentum. The ends of the tube can have any convenient geometrical shape in the rest frame of the unaccelerated particle in the distant past (or future), say a sphere of given radius ε. The field energy–momentum inside the sphere is the infinite energy of the Coulomb field, which is discarded in a mass renormalisation. Although this energy is infinite, it is well understood and can be unambiguously calculated in this special case. If we refer to Parrott's book [13, Sect. 4.4], it is explained that this is the reason for insisting that the particle be unaccelerated in the distant past and future.

Here is a brief version of the analysis. We can picture the particle as enclosed in a spherical elevator of radius ε that moves along with it. Both are initially at rest, then nudged into uniform acceleration. The state of rigidly accelerated motion is maintained for an arbitrarily long time, after which the acceleration is gently removed and the elevator thereafter enters a state of uniform motion. The integral giving the energy–momentum radiated through the walls of the elevator between proper times τ_1 and τ_2 is

$$\int_{S(\tau_1,\tau_2)} T^{i\alpha}\,\mathrm{d}S_\alpha = -\frac{2}{3}q^2 \int_{\tau_1}^{\tau_2} a(\tau)^2 u^i(\tau)\,\mathrm{d}\tau + \frac{q^2}{2\varepsilon}\left[u^i(\tau_2) - u^i(\tau_1)\right], \qquad (16.82)$$

where u and a are the 4-velocity and 4-acceleration, respectively. The coefficient of the 4-velocity in the second term on the right-hand side is absorbed into the inertial mass of the charge in the process known as renormalisation. The energy component of the first term is always positive. This is what tells us that energy is radiated out through the walls of the elevator.

According to Parrott's idea (see the quote on p. 257), if we believe in conservation of the usual Minkowski energy, the pilot of the rocket driving the charge will observe an additional fuel consumption when his payload is a charged particle, relative to the corresponding identical motion of an uncharged particle, the additional fuel consumption exactly compensating for the radiated energy. The details of how (and, in particular, when) this borrowed energy must be repaid are discussed in Chap. 17.

We might also imagine dividing the elevator walls into a large number of small coordinate patches with an observer stationed on each patch. The observers would measure the fields, calculate the corresponding energy–momentum tensor, and subsequently approximate the energy component of the above integral to arbitrary accuracy. The key point now is that this procedure is not the same as the following one: each of the observers calculates her own local energy outflow

$$\mathbf{n} \cdot \frac{\mathbf{E} \times \mathbf{B}}{4\pi} \Delta S \Delta \tau \, ,$$

where $\mathbf{n}$ is the outward unit normal vector to the wall in the observer's rest frame, ΔS is the area of her patch, $\Delta \tau$ the increment in her proper time, and $\mathbf{E}$ and $\mathbf{B}$ her electric and magnetic fields, respectively. Generally speaking, we could not attach any covariant meaning to the sum of these quantities over the different observers. This would amount to adding up quantities referring to different rest frames, which would not give energy as we know it. Furthermore, if we changed elevators to calculate the outflow of this quantity, e.g., having a bigger elevator, containing the first, then we would not come up with the same total: there is no reason to suppose that the observers on the larger elevator would obtain the same number for the energy radiation using this definition as those on the smaller one. Worse still, neither number would be expected to be related in any simple way to the additional energy required by the rocket for a charged versus uncharged payload.

In the procedure just described, the observers are not measuring what is usually called energy. They are measuring something else, a kind of pseudo-energy. It may be tempting to call it something like energy as measured in the elevator frame, but it is conceptually and experimentally distinct from the usual Minkowski energy. For arbitrary motion, it is not a conserved quantity and does not therefore even deserve the name 'energy'. For the special case of an elevator occupying constant Rindler coordinates, it does happen to be independent of the elevator shape, because it is in fact zero for all such elevators, but this does not in itself imply that it is energy in the usual sense of the term.

We shall show in what follows that it is in fact the conserved quantity constructed from the Killing vector field ∂_λ corresponding to the flow of the one-parameter family $\lambda \mapsto \phi_\lambda$ of Lorentz boosts, where

$$\phi_\lambda \left(t, x, y, z \right) := \left(t \cosh \lambda + x \sinh \lambda , t \sinh \lambda + x \cosh \lambda , y, z \right) . \tag{16.83}$$

In the Rindler coordinates (λ, X, y, z) defined from the Minkowski coordinates by

$$t = X \sinh \lambda \, , \qquad x = X \cosh \lambda \, , \tag{16.84}$$

the Minkowski components of this Killing vector field are

$$\partial_\lambda = (X \cosh \lambda, X \sinh \lambda, 0, 0) \, , \tag{16.85}$$

and the Rindler components are of course $\partial_\lambda = (1, 0, 0, 0)$. Finally, the conserved quantity in question is $T^{i\alpha} K_\alpha$, where $K = \partial_\lambda$.

Now Boulware has proven that the Poynting vector is zero in the Rindler frame, and Parrott suggests that Boulware might try to use this fact to show that there is no radiation for an observer accelerating with the charge. Let us see how this works. The pseudo-energy defined physically by the above measuring procedure for a spherical elevator S of radius R in Rindler coordinates is given mathematically by the following integral in spherical Rindler coordinates R, θ, ϕ, which bear the same relation to rectangular Rindler coordinates X, y, z as ordinary spherical coordinates bear to Euclidean coordinates x, y, z, with the angle θ being measured from the X axis. In the integral $u = u(\tau, R, \theta, \phi)$ denotes the 4-velocity of the point of the elevator located at Rindler spherical coordinates R, θ, ϕ at its proper time τ, i.e., at Rindler time coordinate $\lambda = \tau/X$, and $n = n(R, \theta, \phi)$ is the spatial unit normal vector to the sphere at the indicated point, i.e., n is orthogonal to u and also normal to the sphere in the spacelike hypersurface, so that in Rindler coordinates $n = (0, \mathbf{n})$. Then Parrott says that

$$\text{pseudo-energy radiation} = \int_{\tau_1}^{\tau_2} d\tau \int_0^\pi d\theta \int_0^{2\pi} d\phi \; R^2 \sin\theta \, u_\alpha T^{\alpha\beta}(-n_\beta) , \quad (16.86)$$

with a minus sign because the inner product induced on spacelike hypersurfaces is negative definite. The measure of the surface element is just $R^2 \sin\theta \, d\theta d\phi$ because the line element has the form

$$ds^2 = X^2 d\lambda^2 - dX^2 - dy^2 - dz^2 \qquad (16.87)$$

in Rindler coordinates, inducing the ordinary negative definite Euclidean metric on spacelike hypersurfaces $\lambda = $ constant. The contraction $u_\alpha T^{\alpha\beta}$ picks out what would be the zeroth row of the energy–momentum tensor in the frame of the local observer, and this row is the energy–momentum flow rate for that observer. The further contraction of this vector with $-n_\beta$ gives the rate at which electromagnetic energy is flowing across a unit area normal to the vector $\mathbf{n}$ [13, pp. 117–8].

We now wish to use the fact that the 4-velocity u^α of any point (which one might say loosely to be) moving with the Rindler frame, such as the points on the elevator surface, is given by

$$u^\alpha = \partial_\tau = \frac{1}{X}\partial_\lambda . \qquad (16.88)$$

In Rindler coordinates, $\partial_\lambda =: \partial_0$, we have $u^0 = 1/X$ and $u_0 = X$. Hence,

$$u_\alpha T^{\alpha\beta} n_\beta = X T^{0\beta} n_\beta = X \sum_{J=1}^{3} T^{0J} n_J .$$

We also have $\tau = \lambda X$ and $K_0 = X^2$. Equation (16.86) would then read

$$\text{pseudo-energy radiation} = \int_{\lambda_1}^{\lambda_2} d\lambda \int_0^\pi d\theta \int_0^{2\pi} d\phi \, R^2 \sin\theta \sum_{J=1}^{3} \left(-K_\alpha T^{\alpha J} n_J \right) ,$$

$$(16.89)$$

where we have defined $\lambda_i := \tau_i/X$ for $i = 1, 2$. Equation (16.89) would then tell us that (16.86) is actually calculating the radiation of the conserved quantity corresponding to the Killing vector $K = \partial_\lambda$.

There seem to be two slight problems with Parrott's derivation:

- λ_1 and λ_2 depend on our position on the sphere, but we have already integrated over position.
- The charge is located at $X = 1$, not the forbidden $X = 0$, so we require spherical coordinates centered on $X = 1$.

These can be dealt with as follows.

The above formula (16.86) is the result of measurements made locally by a host of observers, each located at a point on a Rindler sphere centered on the charge. The first difficulty shows up because we have defined the limits of the λ integral to be $\lambda_i := \tau_i/X$, and these depend on X, which is a variable of integration as we move around the sphere. This draws attention to the fact that we have not specified how our host of observers will synchronise their ranges of proper times $[\tau_1, \tau_2]$. This omission is what saves Parrott's argument. However, we do need to put the proper time integrals before the angle integrals, like this:

$$\text{pseudo-energy radiation} = \int_0^\pi d\theta \int_0^{2\pi} d\phi \int_{\tau_1}^{\tau_2} d\tau \, R^2 \sin\theta \, u_\alpha T^{\alpha\beta} (-n_\beta) , \quad (16.90)$$

where the range $[\tau_1, \tau_2]$ may now be allowed to vary with θ and ϕ, i.e., with the location of the observer on the Rindler sphere, in a way that we shall soon specify. There is a certain improvement in the logic here, because we always said that we would sum over the host of observers, and in order to do so, we have to wait for them to finish their measurements.

What we would like is to end up with each range $[\lambda_1, \lambda_2]$ being the same. We can then bring the λ integration outside the integral over the Rindler sphere. This too is logical enough: it is the observer with the charge, at the sphere centre, who is coordinating these measurements, and she could instruct each observer on the sphere as to how long (in their individual proper times) they should measure the pseudo-energy flux across their particular surface element.

At this point, we make a slight adjustment to Parrott's notion of the spherical Rindler coordinates to deal with the second problem above, because we would like the coordinates to be centered on the charge at $X = 1$, whereas according to what he says they are centered on the forbidden $X = 0$. Now the local observer with coordinate θ has Rindler space coordinate $X = 1 + R\cos\theta$ (see Fig. 16.1), and since $\lambda = \tau/X$, the central observer will instruct her to measure over her proper time range (see Fig. 16.2)

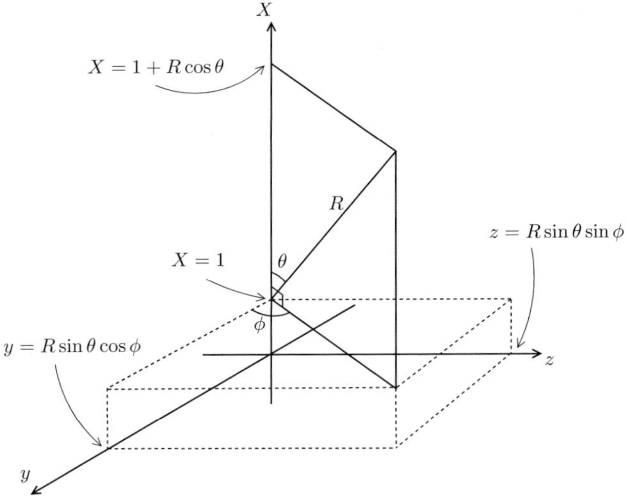

Fig. 16.1 Shifting the origin of the spherical Rindler coordinates

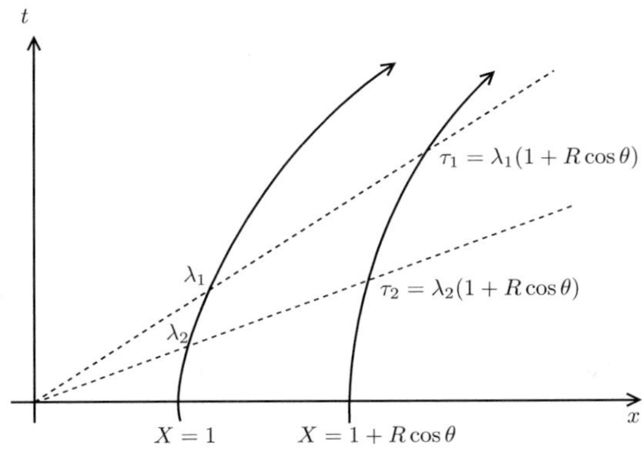

Fig. 16.2 Correcting the Parrott argument. The observers with coordinate θ (and arbitrary value of ϕ) all have X coordinate equal to $1 + R\cos\theta$ so they all follow a curve like the right-hand curve in this Minkowski diagram (but at different values of coordinates y and z). The observer at $X = 1$, who follows the left-hand curve, coordinates the measurement effort over her proper time interval $[\lambda_1, \lambda_2]$. This corresponds to the proper time interval $[\tau_1, \tau_2]$ on the right-hand curve. *Dotted lines* through the Minkowski origin are hyperplanes of simultaneity for the coordinating observer at $X = 1$, but also for the other observers (see Chap. 15). The observer at $X = 1$ merely borrows the spatial metric of an ICIO on any given HOS. But so do all observers with coordinate θ, for each given HOS. This explains why the area measure $R^2 \sin\theta\, d\theta\, d\phi$ imposed by the observer at $X = 1$, which is the proper area for this observer relative to her HOS, happens to be precisely the proper area measure for each observer at θ

$$\tau \in \left[\lambda_1(1 + R\cos\theta), \lambda_2(1 + R\cos\theta)\right],$$

revealing the position dependence of τ_1 and τ_2 in (16.90).

It now looks as though Parrott's argument goes through and we arrive at (16.89). One might wonder about two outstanding questions concerning Parrott's analysis:

- Why should the host of local observers all use a range of proper times to suit the central observer? Once again, the answer must be that the observations have to have a minimum of coordination, and this is probably the most obvious way to achieve it. If each observer measured for a random duration, what could we possibly do with the results?
- Why should the host of local observers all use the measure of their surface element proposed by the central observer? This was the justification of the measure in (16.86) due to the fact that the SE interval is (16.87), back on p. 277. The answer is that, for a given value $\lambda \in [\lambda_1, \lambda_2]$ of the proper time of the observer at $X = 1$, this measure of their surface element is its proper area as reckoned by that observer in her corresponding HOS. But we showed in Chap. 15 that this was also the HOS for any observer at coordinate θ, corresponding to the proper time $\tau = \lambda (1 + R \cos \theta)$ of that observer, if both observers set up SE coordinates. Indeed, in that case, both would borrow the Euclidean metric of the appropriate ICIO, and they have the same ICIO at these respective proper times (see Fig. 16.2). This means that the proper area for one at her proper time λ would be the proper area for the other at her proper time τ. The central observer merely has to coordinate the observations so that no two observers' patches actually overlap.

The whole argument was perhaps contrived to produce (16.89) and one may suspect that Parrott started with the latter and then worked backwards. However, it would appear to be perfectly relevant in the present context, if only to illustrate the pitfalls of not being careful about the meaning of quantities expressed relative to non-inertial coordinates.

After all, the question remains: what should one do with the zero Poynting vector observed by a semi-Euclidean observer? What would this observer calculate in order to claim that there was no radiation in her frame? The expression in (16.89) is indeed the obvious thing to work out from a formal, i.e., naive point of view. The reverse argument, arriving at (16.86), then looks the more logical. Insofar as we consider (16.89) to be an odd thing to evaluate, this could be taken as an argument against the naive approach. Indeed, we still have the point made by Parrott that it may be possible to construct several conserved vectors from a conserved stress–energy tensor.

And the reader should note in passing how useful it is to view what is happening in a Minkowski frame, like the one used in Fig. 16.2. Indeed, one of the themes of this book is that it is not merely useful, but essential. The thesis here is that one cannot just make a naive interpretation of quantities expressed relative to coordinates like the SE coordinates, considering that an SE observer who happened to be using such coordinates would somehow take them for a Minkowski system. Primarily, one should say that, since they are not Minkowski coordinates, they should not be treated as such. But worse for the other position, which would appear to be advocated by Rohrlich (see Sect. 16.10), there is no obvious reason why an accelerating observer should adopt SE coordinates, despite certain convenient features. Even though this particular problem can be considered entirely within special relativity theory, the lesson from general relativity is surely that coordinates are just coordinates.

But to return to Parrott's argument, it was said a little earlier that, for arbitrary motion, the quantity calculated by (16.86) is not a conserved quantity and does not therefore even deserve the name 'energy', but that, for the special case of an elevator occupying constant Rindler coordinates, it does happen to be independent of the elevator shape. Concerning the first of these two points, we note that our charge sits at a fixed spatial point in the Rindler coordinates, because its motion can be described as a uniform acceleration. This in turn means that the elevator chosen to surround it can consist of points with like motion, all in uniform acceleration, so that they sit at fixed spatial points in the Rindler coordinates. We can therefore replace u by ∂_λ/X and τ by λX. Without these replacements, we could not generally get (16.86) into a form where it becomes an integral of a conserved quantity.

Anyway, once we have these results, a sufficient condition for (16.89), and hence (16.86), to be zero is just $T^{0J} = 0$ for all spatial indices J in Rindler coordinates. This is established in Boulware's paper for the fields due to a uniformly accelerated charge, sitting therefore at a fixed spatial coordinate relative to the Rindler system.

Now one can also view the charge here as a stationary charge relative to coordinates for which the metric is static. It was shown in Chap. 13 that there are retarded field solutions for such a charge with zero magnetic field, assuming only that the electric field is static. If the retarded field solution is unique, this means that the retarded field solution for such a charge always has zero magnetic field, provided that one can assume that the retarded electric field is static. (But note that the electric field is not generally Coulomb, as exemplified on p. 231 ff.) Then one can deduce that (16.86) is zero. There is probably a general result lurking behind this kind of reasoning, with the proviso that we do not conceal any of the assumptions going into it, which says that a stationary charged particle in a static spacetime does not radiate the pseudo-energy defined as the conserved quantity corresponding to translation by the formal time coordinate in this spacetime. For we observe that the time coordinate provides a Killing vector field for the static metric, and this Killing vector field provides a conserved quantity in conjunction with the electromagnetic energy–momentum tensor.

Parrott agrees with Boulware that there is no radiation of the conserved quantity corresponding to the Killing vector ∂_λ, but says that this is irrelevant to questions concerning physically observed radiation. Parrott also says [7, p. 13] that it is an experimental question whether it is Minkowski energy radiation or pseudo-energy radiation which corresponds to energy that must be furnished by the driving forces. He would settle the question by uniformly accelerating a large charge in a rocket and observing whether more fuel were required than for a neutral payload of the same mass. He would expect more fuel to be used, implying that Minkowski energy is the physically relevant energy.

However, there is no attempt anywhere to prove that the Boulware pseudo-energy has any connection with driving forces on the charge. It may have been surreptitiously assumed somewhere, but an actual demonstration is nevertheless absent. On the other hand, there are theoretical reasons for expecting an extra driving force to provide for Minkowski energy radiation. It is only this that would be tested by an experiment. That is to say, we never discuss how a driving force might be connected with pseudo-energy radiation, so there is no theoretical claim there to test by the proposed experiment. The experiment described would help us to understand the Lorentz–Dirac theory and the radiation formulas found in Minkowski coordinates.

Another question to raise at the end of all this is the following: is there any direct connection between the fact that the radiation reaction terms in the Lorentz–Dirac (LD) equation are zero and the fact that the Poynting vector is zero in the semi-Euclidean (SE) frame? The word 'direct' is probably operative here, because we know that there is a connection: both facts occur only when the acceleration is uniform. However, we know that there is radiation in the Minkowski frame. So even though the LD radiation reaction is zero, we do not need to show that the SE observer will have zero Poynting vector. The two facts in the above question look more or less like two accidents that happen to occur when the acceleration is uniform.

16.9 Schwarzschild Spacetime Revisited

We have seen in Sect. 16.4 that the only Killing vector fields in Schwarzschild spacetime are linear combinations of ∂_t and an angular momentum Killing field with the form

$$A = K_\theta(r, \theta, \phi)\partial_\theta + K_\phi(r, \theta, \phi)\partial_\phi \, ,$$

where K_θ and K_ϕ satisfy some additional conditions. Furthermore, ∂_t and A commute and so do their flows. The only Killing symmetries in Schwarzschild spacetime are thus the expected ones arising from rotational and time invariance of the metric.

Here is what Parrott says about this:

In this situation, the only natural mathematical candidate for an 'energy' is the conserved quantity corresponding to the Killing field ∂_t; for one thing, it is the only rotationally invariant choice. It is also physically reasonable in our context of analyzing the motion and fields of charged particles. If we surround a stationary charged particle by a stationary sphere which generates a three-dimensional 'tube' as it progresses through time, the integral of the normal component of T^{0i} over the tube between times t_1 and t_2 physically represents the outflow of the conserved quantity corresponding to ∂_t between these times. When the calculation is carried out, it is seen to be the same as integrating the normal component of the Poynting vector $\mathbf{E} \times \mathbf{B}/4\pi$ over the sphere and multiplying by a factor proportional to $t_2 - t_1$. It is usually assumed that the field produced by a stationary charged particle may be taken to be a pure electric field, and [Chap. 13] proves this under certain auxiliary hypotheses. In other words, $\mathbf{B} = 0$, so the integral vanishes, and there is no 'radiation' of our conserved quantity. We *expect* no energy radiation; otherwise we would be able to garner an unlimited amount of 'free' energy, since it takes no energy to hold a particle stationary in a [static] gravitational field.

He explains that a stationary particle is one whose worldline is

$$x^0 \longmapsto (x^0, c^1, c^2, c^3)$$

relative to the usual coordinate system in which the metric has the static form, where the c^i are constants independent of x^0. The word 'static' has been added to the last sentence since it is clearly intended.

This brings us back to that idea by Bondi and Gold which triggered all this discussion in the first place. One has the impression that Parrott has struggled somewhat to convince himself of the content of the last quote and arrives at the final conclusion with some relief. However, it does indeed seem right to expect that a charged particle sitting still relative to a static gravitational field will not require an energy supply to do so, and nor will it do anything. This is certainly what one would say in Newtonian gravity theory, or special relativity theory with the obvious naive theory of gravity as a force field. A stationary particle is inertial in SR and so Maxwell's theory says there will be no radiation.

But in order to understand the physical implications of a scenario in general relativity theory, one is supposed (presumably) to look at what is happening in the locally inertial frame guaranteed by the weak equivalence principle (WEP) and deduce things about electromagnetic fields by applying the strong equivalence principle (SEP). If one did not apply SEP, one would have to find some other principle for relating the mathematics of the theory to physical observations. But when one looks at the static charge mentioned above as it would be described in a locally inertial frame, one finds that it is accelerating. There is then nothing obvious at all about the above conclusions. Indeed, the minimal extension of Maxwell's equations (MEME) to this context, which is an application of SEP, says that this charge will radiate. If energy radiates out, and we have seen that this could indeed be detected locally, within the given locally inertial frame, then it will have to be supplied by whatever is holding the charge up against the gravitational effects.

One then realises that just the fact that the conserved quantity corresponding to the Killing vector field ∂_t happens to be the only natural mathematical candidate for an energy is really not sufficient to make it what one would normally call energy physically. In contrast with a commonly held attitude in GR, mathematical convenience is not sufficient to be sure that one is doing physics, i.e., that one is getting the right relation with what is out there. It is surprising that Parrott accepts this, despite the general scepticism with which he greets the interpretation of Boulware's Poynting vector in the uniformly accelerating frame for Minkowski spacetime.

Of course, there is a major difference between the two scenarios: in the Minkowski spacetime there are several Killing vector fields that might be taken to give the energy, even if one of them undoubtedly should take precedence. But what could we do in Schwarzschild spacetime if we did not use ∂_t to build an energy quantity from energy–momentum tensors? There is nothing else, at least nothing else obvious. On the other hand, even in Schwarzschild spacetime, there are locally Minkowski frames and our physical interpretation of the theory has to pass by them.

It must also be stressed that GR is very different from SR as regards gravitational effects. GR builds in an interaction between gravity and other fields, like electromagnetic fields, which just does not exist in special relativistic versions of these other fields. For example, light is affected by gravitational effects in GR, whereas it really is not in Newtonian gravity theory unless one attributes a mass to photons, but that is not then an interaction between EM fields and gravity. If a charge is stationary relative to the coordinates one happens to have chosen, one must remember that these are only coordinates. It is not stationary relative to the kind of coordinates one is supposed to use to understand the theory physically. And not only is the particle interacting with the gravitational effects via its own passive gravitational mass, but its EM fields are also interacting with the gravitational effects in this theory.

It seems that the case of the uniformly accelerating frame in Minkowski space-time amply illustrates this. If one can interpret this scenario as holding a charge stationary relative to an SHGF, one sees that the stationarity of the particle relative to SE coordinates, in which the appropriate (Minkowski) metric has the static form, means nothing with regard to the question of EM radiation. One has to view the situation from the locally (and in this case, globally) inertial frame, whereupon one finds that the charge is in fact accelerating, and that the minimal extension of Maxwell's theory clearly implies EM radiation. There may still be a question about whether this radiation is a subjective thing, in some sense, that can be observed by the locally inertial observer but not by one accelerating with the charge, but Parrott seems to have answered that with the conclusion that EM radiation is an objective phenomenon, i.e., if the locally inertial observer detects it, then so will the comoving observer, provided the latter looks at the right quantities locally and measures them carefully enough.

Some would say that the scenarios of a non-tidal gravitational field or an eternally uniformly accelerating charge are not physical and cannot be used to illustrate anything, but the present view in this book is quite the opposite: they provide an application of the equivalence principles WEP and SEP *par excellence* which allows one to be very clear about what is different in GR, as compared with SR. It is this clear case that warns us not to just go by the mathematics in a less intuitively obvious case like Schwarzschild spacetime.

So why should the conserved quantity constructed from ∂_t and the energy–momentum tensor of the EM fields be what we expect to call the EM energy, when t is just a convenient coordinate? Should the energy of the EM fields be conserved when these fields are interacting with another field? In other words, should we be able to find a conserved thing to correspond to the energy of the EM fields in Schwarzschild spacetime?

16.10 Antithesis of the Present View

This is a convenient moment to mention the stance adopted in the widely accepted textbook on classical charged particles by Rohrlich [22]. Indeed it is widely enough accepted to have gone into a third edition and is in fact the only textbook known to the present author that devotes any space to the specific question of the validity of the equivalence principle for charged particles. However, the views it advocates are really the antithesis of the ideas put forward in the present book, particularly with regard to the subject of the present chapter, viz., the question of how one should interpret quantities expressed relative to non-inertial coordinate systems. Having said this, it will only be taken to illustrate the kind of points we wish to make here, while this classic textbook is highly recommended to the reader.

As usual, we begin with a few short quotes to set the scene, taken from Sect. 8.3 of [22]:

> [...] we restrict our attention to a constant, static, homogeneous gravitational field (SHGF). This is a field whose lines of force are equidistant parallels, such as the gravitational field in the laboratory. It is known that this type of gravitational field can be simulated by uniform acceleration of a neutral particle in Newtonian mechanics and in special relativity. Is this also true for the motion of a charged particle?

Leaving aside the vagueness in the idea of simulating a gravitational field, Rohrlich proposes the following argument which he claims to be fallacious:

> At first the answer seems to be 'no', because it would require that a neutral and a charged particle fall equally fast in an SHGF. But this cannot be the case, because the charged particle will radiate (being accelerated), will thereby lose energy, and will consequently fall more slowly than the neutral particle.

Having declared this to be fallacious, his statement of intent is as follows:

> [...] we shall prove that a charged and a neutral particle in an SHGF fall equally fast, despite the fact that the charged one loses energy by radiation.

So what makes the other bit of the quote fallacious is not its claim that the charged particle will radiate, or even the claim that it is being accelerated. It looks exactly as though we are discussing special relativity: a freely falling object is being accelerated in SR, in fact, uniformly for an SHGF (see Chap. 3), and the Maxwell equations predict that it will radiate. However, he then goes on to explain why a freely falling observer is inertial, so he must actually be talking about general relativity!

Now comes the proof that a charged and a neutral particle fall equally fast, i.e., they would have the same worldline, in an SHGF. He says that Maxwell's equations can be applied in the frame of a falling observer (the principle called SEP in this book) and that the problem is now trivial:

> [...] the falling observer I is inertial and the falling charge will remain at rest with respect to I, since no forces are present. Thus, a neutral and a charged particle will fall equally fast relative to I.

This is the argument from GR+SEP. The problem that remains, about which Rohrlich keeps very quiet for the moment, is that we have not explained why there should be radiation! The last quote is tantamount to a declaration that there is no radiation. At least, I is not going to see any. And this is not because the charge has a uniform acceleration and there is no radiation reaction, or anything like that. It is because the charge is just sitting still in an inertial frame.

This is why we turn to S, a supported observer. In order to 'transform' from I to S, we have to know what coordinates S is using. We are told that the most general gravitational field which satisfies the requirements of an SHGF, in particular, the requirement of zero curvature, among other things, is given by the timelike line element

$$c^2 d\tau^2 = u^2(z)c^2 dt^2 - \left[c^2 u'(z)/g\right]^2 dz^2 - dy^2 - dx^2 , \qquad (16.91)$$

where $u(z)$ has to approach

$$u(z) = 1 + \frac{gz}{c^2} , \qquad (16.92)$$

in some sense that is not specified. When we have (16.92), the interval in (16.91) reduces to the semi-Euclidean interval for a uniformly accelerating observer. Let us not worry about the possible extra generality in (16.91), for the purposes of illustrating how Rohrlich's view deviates from the one presented here. But the reader should bear in mind that we are just talking about Minkowski spacetime, or at least a piece of Minkowski spacetime, with different possible choices of coordinates.

The first observation Rohrlich makes about (16.91) is that the supported observer who uses the coordinates $\{x^\mu\}_\mu = (ct, x, y, z)$, and who is presumably at rest relative to the spatial coordinates, 'sees' an SHGF whose strength is characterised by the acceleration of gravity g. This is what we discussed in Chap. 2. It is then pointed out that the motion of a freely falling neutral particle is given by the usual geodesic equation [see (2.95) on p. 36] for the relevant connection coefficients derived from the above interval relative to the given coordinates. It is emphasised that this is *not* uniformly accelerated (hyperbolic) motion, but that it can be, provided we make some specific choice of the function $u(z)$. The reader will immediately appreciate how very different this approach is, compared with the one being advocated in the present book.

Of course, geodesic motion is totally unaccelerated motion in general relativity. The geodesic equation says precisely that the 4-acceleration of the particle is zero. No amount of fiddling around with the coordinates, by making alternative choices for the function $u(z)$, will make free fall in this flat spacetime, or indeed in any other spacetime, look like uniform acceleration. Of course, that is not what Rohrlich means. He is saying that, if S should be duped by her coordinate system into thinking that it has any real significance, the freely falling particle may appear to have this or that acceleration. But for Rohrlich, one senses that this *is* supposed to have some real kind of significance, i.e., coordinates can be taken by their resident observer to have some real physical significance. And this is the antithesis of what is being advocated in the present book.

It is significant, however, that Rohrlich should seek coordinates for S relative to which free fall becomes uniform acceleration (making some suitable choice of the function u does amount precisely to making some suitable adjustment to the coordinates for this Minkowski spacetime), because free fall in an SHGF *is* uniform acceleration in the naive special relativistic model of gravity. The reader should turn back to Chap. 3 and reconsider in particular (3.15) on p. 49. There would appear to be some confusion here about the differences between special and general relativity.

Let us now see what is proposed for the transformation of the Coulomb field of the freely falling charge in its inertial system. Rohrlich states the coordinate transformation between the Minkowski coordinates of the inertial system falling freely with the charge and the coordinate system that S has apparently adopted with interval (16.91). We are then told that:

> We are now in a position to solve electrodynamic problems in the non-inertial coordinate system S. All we need to do is to transform the Maxwell–Lorentz equations from I to S by means of [the coordinate transformations]. Furthermore, if we know the solutions of these equations in I, we also know them in S from [the coordinate transformations].

Of course, the relevant physical quantities are tensors and the Maxwell–Lorentz equations can be expressed in tensorial form in SR, so we get them in tensorial form in this general relativistic context by replacing coordinate derivatives relative to the Minkowski coordinates by covariant derivatives involving the Levi-Civita connection components for the non-inertial coordinates used by S. In brief, we apply SEP. Alternatively to rewriting the equations, we can solve these equations in the Minkowski coordinate system, then transform the relevant tensor (the electromagnetic field tensor F_{ij}) in the usual way to obtain its components relative to the non-inertial system.

We opt for the latter, since we already know that the solution for the fields in the freely falling frame is just the Coulomb field. Rohrlich now declares that:

> Transformation via [the coordinate transformations] shows that the observer sees this charge radiating. Nevertheless, the charged particle remains at rest in I, just as a falling neutral particle would. It is, of course, clear that the radiation rate $\mathscr{R}$ is Lorentz invariant, but it is not invariant under [the transformation to the non-inertial system].

The calculation of the radiation rate $\mathscr{R}$ seen by S is left as an exercise to the reader [see, for example, (5.7) on p. 63 and the following discussion]. But are we really going to find that the charge radiates for S? The problem is this: we have a formula for $\mathscr{R}$ in ordinary Minkowski coordinates in special relativity, whereby it is found from the components of the EM field tensor, but when we transform to arbitrary coordinates for this spacetime, do we really expect the same algebraic combination of the new components of the EM field tensor to deliver a rate of radiation of energy? This is the disagreement with Rohrlich's account. Note in passing that Rohrlich cites the paper by Mould [5], discussed in Chap. 14, confirming Mould's view that a radiation detector which moves in uniform acceleration past a Coulomb field will record radiation.

Rohrlich now asks how we know that a charged particle at rest relative to S does not radiate. The view in the present book is that it does, and that S will spot this. After all, in GR we understand what is happening by looking first in the inertial frame, relative to which the charge is now accelerating. SEP tells us that there will be radiation in this frame. Although S happens to be accelerating relative to the inertial frame, and in fact, moving with the charge, the view in the present book is that S should be able to detect that radiation. Rohrlich does indeed follow the first part of this procedure. He finds that the charge has hyperbolic motion, i.e., uniform acceleration, relative to the Minkowski coordinates for the freely falling frame I.

Then since *I* is an inertial system, he deduces that the supported charge as seen by the falling observer radiates at the constant rate (5.14) on p. 65 (see Chap. 5 for the discussion of Rohrlich's paper with Fulton).

We must now transform this to the non-inertial coordinate system used by *S*. Of course, Rohrlich finds no radiation. His deduction is based on the discovery that $\mathbf{B} = 0$ in *S*. In the present view, there is no reason to treat $\mathbf{B}$ as the kind of magnetic field we know and love from our school days, precisely because it is not expressed relative to inertial coordinates. Rohrlich does note that the presence of the SHGF makes a difference to the electric field, which is not a pure Coulomb field. This is hardly surprising, given the complexity of both the EM fields in the inertial system, and the transformation to the non-inertial system. If the reader has any doubt here, it is worth asking what would happen for *S* if this observer used some other coordinate system adapted to the supported point.

Chapter 17
Charged Rocket

Parrott begins his paper [7] with the idea of accelerating a charged particle in Minkowski spacetime by means of a tiny rocket engine attached to the particle. If the particle radiates energy which can be detected and used, then energy conservation suggests that the radiated energy must be provided by the rocket, i.e., it must burn more fuel to produce a given accelerating worldline than a rocket driving a neutral particle of the same mass. This scenario is examined quantitatively in his Appendix 2 and the present chapter closely follows Parrott's calculation.

17.1 Preamble

Recall that Boulware showed (see Sect. 15.12) [6] that EM energy flows from the hypersurface $z = -t$ onto the charge in just such a way as to equal the energy that is supposed to flow out of it when it radiates. This energy ultimately comes from when the rocket was accelerating (decelerating) the charge over the infinite range of very early times. One might therefore expect some strange answers in such a highly idealistic scenario.

Parrott also mentions, without detailed analysis, the case of a stationary charge in Schwarzschild spacetime, where 'stationary' means 'not moving relative to the spacelike coordinates'. Here, too, we can imagine a rocket holding it thus, so that it is accelerating relative to local inertial frames. If no radiation is produced, as he seems to imply, the rocket should use the same amount of fuel as would be required to hold an otherwise identical neutral particle to this worldline.

In principle this would provide a local test for the existence of radiation, in either Minkowski or Schwarzschild spacetime. He does not analyse the Schwarzschild case, but in the flat spacetime, we have the following problem (see Parrott's Appendix 1). In Minkowski spacetime, when a charge is uniformly accelerated (in the sense that its acceleration has constant pseudolength), the radiation reaction term in the Lorentz–Dirac equation is zero. Let us first show this. The Lorentz–Dirac equation for a particle of mass m and charge q in an external field F is

$$m\frac{\mathrm{d}u}{\mathrm{d}\tau} = qF(u) + \frac{2}{3}q^2\left(\frac{\mathrm{d}a}{\mathrm{d}\tau} + a^2 u\right),\tag{17.1}$$

where τ is proper time, $u = u^i$ is the 4-velocity, $a = \mathrm{d}u/\mathrm{d}\tau$ is the 4-acceleration, with $a^2 := a_i a^i$, and $F(u)^i = F^i{}_j u^j$. The left-hand side is the rate of change of mechanical energy–momentum, $qF(u)$ is the Lorentz force due to the external field, and the remaining term is the so-called radiation reaction term, supposed to describe the effect of the particle radiation on the particle motion.

The fact that this radiation reaction term is zero for uniform acceleration is shown as follows. We write $u = (\gamma, v\gamma)$, dropping the irrelevant components. The quantity v is the naive 3-velocity. Then $w = (v\gamma, \gamma)$ is an orthogonal unit vector. The proper acceleration a must be orthogonal to u, and hence parallel to w, so that there is some scalar quantity A such that $a = Aw$. Then $A^2 = -a^2$, since $w^2 = -1$, and we are saying that A is constant as proper time goes by. This is the hypothesis of uniform acceleration. It implies that

$$\frac{\mathrm{d}a}{\mathrm{d}\tau} = \frac{\mathrm{d}A}{\mathrm{d}\tau}w + A\frac{\mathrm{d}w}{\mathrm{d}\tau} = A\frac{\mathrm{d}w}{\mathrm{d}\tau}.\tag{17.2}$$

Now $\mathrm{d}w/\mathrm{d}\tau$ is orthogonal to w, because w is unit length, and this means that it is parallel to u. Hence $\mathrm{d}a/\mathrm{d}\tau$ is also parallel to u. The constant of proportionality (and it is constant!) is found from

$$u^i\frac{\mathrm{d}a_i}{\mathrm{d}\tau} = \frac{\mathrm{d}(u^i a_i)}{\mathrm{d}\tau} - a_i\frac{\mathrm{d}u^i}{\mathrm{d}\tau} = -a^2.$$

We conclude that

$$\frac{\mathrm{d}a}{\mathrm{d}\tau} + a^2 u = 0,$$

as claimed. The converse is immediate. The last relation implies that $\mathrm{d}a/\mathrm{d}\tau$ is parallel to u, and then (17.2) implies that $\mathrm{d}A/\mathrm{d}\tau$ is zero.

As observed in earlier chapters, this is a striking result. The radiation reaction in the Lorentz–Dirac equation is zero in precisely those cases where the charge is accelerated as it would be by a static, homogeneous gravitational field in a special relativistic version of a gravitational theory (see Chap. 3). This adds to the surprising result that the retarded field generated by a charge in such motion has zero magnetic field and time-constant electric field in the semi-Euclidean coordinates of an observer moving with the charge, as well as the fact that the semi-Euclidean metric for precisely this observer is static (see Sects. 15.11.1 and 2.2, respectively).

But to return to the above proof, does this mean, as it appears, that there is no radiation reaction for a uniformly accelerated charged particle, i.e., that a rocket-driven uniformly accelerated charge requires no more energy from the rocket than an otherwise identical neutral particle, despite the fact that one gathers radiation energy in the charged case? We return to the idea of energy conservation. If the uniformly accelerated charged particle does indeed radiate energy into Minkowski spacetime, which can subsequently be collected and used, then this radiated energy must be provided by some decrease in energy of other parts of the system. In the case where the uniform acceleration is eternal, Fulton and Rohrlich say that it may come from the field energy, but they do not explain how. In this same case, Boulware discovered a flow of energy that focuses in on the charge from the hypersurface $z = -t$, which might answer Fulton and Rohrlich's question. But one can also consider this zero radiation reaction in a situation where the uniform acceleration occurs over a finite period of time. In any case, it is understood here that the uniformly accelerated charge will radiate as described, so that there would then be a problem for energy conservation.

Now the derivation of the Lorentz–Dirac equation is motivated by energy–momentum conservation: the change of energy–momentum of the particle over a given proper time interval should equal the energy–momentum provided by the external forces driving the particle minus the radiated energy–momentum. (We have to assume that infinite terms of a certain structure can be absorbed by mass renormalisation.) Despite this, and as Parrott discusses in [13, Sect. 4.3], the Lorentz–Dirac equation does not generally guarantee conservation of energy–momentum! One special case where it does is the situation when the particle is asymptotically free, i.e., its proper acceleration is zero in the infinite past and future. Conversely, it is not clear that the equation should apply to a particle which is not asymptotically free, such as a particle that is uniformly accelerated for all time. At any rate, the equation does not guarantee energy–momentum conservation in this case, by any of the reasoning that went into its derivation. In this sense, the fact that the radiation reaction term is zero implies nothing about the additional force which the rocket must provide for perpetually uniformly accelerated motion. But what about the asymptotically free case?

Parrott asks what happens in the case of a particle which is unaccelerated in the distant past, nudged into uniform acceleration, uniformly accelerated for a long time, then finally nudged back into an unaccelerated state. In this case, the Lorentz–Dirac equation does imply conservation of energy–momentum. However, the radiation reaction term is zero for the period of uniform acceleration, no matter how long it is! So if the radiation reaction term in the Lorentz–Dirac equation were the correct radiation reaction, one could say that all the energy that goes into radiation must therefore be provided at the beginning and end of the trip, while the particle is nudged into and out of its uniformly accelerated state. This is what Parrott analyses in his Appendix 2, and which is discussed in full below.

Note in passing that, apart from the nudging into and out of the accelerated state, the other, uniform acceleration does not incur a radiation reaction, whence one might think that it is the changing of the acceleration which leads to radiation. It would then not be a second time derivative but a third time derivative that counts here. However, the third derivative is nonzero even during the uniform acceleration, as can be seen from (17.2).

We are saying that, if we believe in the Lorentz–Dirac equation, we need to add a bit of energy to start the uniform acceleration, slightly more than if the rocket were uncharged, and thereafter the radiation is free, even if it persists for an arbitrarily long time and adds up to an arbitrarily large amount! Of course, one still needs energy to keep the charged rocket in uniform acceleration, but no more than if it were neutral, according to the radiation reaction term provided by the Lorentz–Dirac equation. In effect, we can borrow an arbitrarily large amount of radiated energy, which can in principle be collected and used by other observers in Minkowski spacetime, although we shall see below that we do have to pay it back at the end of the trip, whenever the uniform acceleration is switched off. As Parrott says, this is hard to accept physically, and it seems to be a good reason for abandoning the Lorentz–Dirac equation.

It is worth recalling the theoretical reasons for thinking that there is radiation from a uniformly accelerated charge. One has to go back to the derivation of the Lorentz–Dirac equation as described in Chap. 11 and [13, Sect. 4.3]. In the case of an asymptotically free particle, it is established that

$$m_{\mathrm{ren}} u_{\mathrm{out}} - m_{\mathrm{ren}} u_{\mathrm{in}} = - \int_{\mathrm{side}} {}_{\perp} T \ , \tag{17.3}$$

where m_{ren} is the mass renormalised by an infinite contribution from the tube caps, but not yet completely renormalised. The right-hand side is an integral over the tube walls. Equation (17.3) is the condition for energy–momentum conservation. We seek an equation of motion that guarantees this.

We put (17.3) in differential form, viz.,

$$m_{\mathrm{ren}} \frac{\mathrm{d}u}{\mathrm{d}\tau} = - \lim_{\Delta\tau \to 0} \frac{1}{\Delta\tau} \lim_{r_0 \to 0} \int_{S_{r_0}(\tau, \tau + \Delta\tau)} {}_{\perp} T \ , \tag{17.4}$$

whose integral with respect to τ gives the limit as $r_0 \to 0$ of (17.3). We have to take the limit here, because the left-hand side of (17.4) cannot depend on r_0. We then calculate the integral over the piece of tube wall. This leads to the proposed equation of motion

$$\bar{m} \frac{\mathrm{d}u}{\mathrm{d}\tau} = q F_{\mathrm{ext}}(u) + \frac{2}{3} q^2 \langle a, a \rangle u \ , \tag{17.5}$$

where the mass has been renormalised yet again to give $\bar{m}$, due to more infinite terms (of the right structure) arising in the integral over the tube walls. This equation is impossible to satisfy unless $\langle a, a \rangle = 0$, because the first two terms are orthogonal to u and the last is parallel to it.

This is where we note that (17.4) implies (17.3), but not conversely. In other words, there may still be an equation of motion similar to (17.5) which implies energy–momentum conservation as expressed by (17.3). We therefore aim to see what would lead to the integrated form

$$\bar{m} u_{\mathrm{out}} - \bar{m} u_{\mathrm{in}} = q \int_{-\infty}^{\infty} F_{\mathrm{ext}}\big(u(\tau)\big) \, \mathrm{d}\tau + \frac{2}{3} q^2 \int_{-\infty}^{\infty} \langle a(\tau), a(\tau) \rangle u(\tau) \, \mathrm{d}\tau \ . \tag{17.6}$$

The last integral can be rewritten as an integral of something else, i.e.,

$$\int_{-\infty}^{\infty} \langle a, a \rangle u \, d\tau = \int_{-\infty}^{\infty} \left(\frac{da}{d\tau} - \left\langle \frac{da}{d\tau}, u \right\rangle u \right) d\tau , \tag{17.7}$$

provided that we have $a(-\infty) = a(\infty)$, something we already assumed (in fact that both were zero) in order to deal with the integrals over the caps (the assumption of an asymptotically free particle). This is how we can say that the proposed equation of motion

$$\bar{m} \frac{du}{d\tau} = q F_{\text{ext}}(u) + \frac{2}{3} q^2 \left(\frac{da}{d\tau} - \left\langle \frac{da}{d\tau}, u \right\rangle u \right) \tag{17.8}$$

implies the equation (17.3) of energy–momentum conservation. Equation (17.8) is just another writing of the Lorentz–Dirac equation.

Now it is the integral of the energy–momentum tensor (or something derived from it) over the tube walls which leads to the idea of radiated energy–momentum. Parrott exposes the calculation in great detail in [13, Sect. 4.4], where he shows that (see also p. 67 in Chap. 6 and the more detailed discussion in Chap. 11)

$$\int_{S_r(\tau_1,\tau_2)} {}_{\perp} T = \frac{q^2}{2r} \left[u(\tau_2) - u(\tau_1) \right] \quad \text{(mass renormalisation)}$$

$$- \frac{2}{3} q^2 \int_{\tau_1}^{\tau_2} u(\tau) \langle a(\tau), a(\tau) \rangle \, d\tau \quad \text{(radiation)}$$

$$- q \int_{\tau_1}^{\tau_2} \hat{F}_{\text{ext}}(u) \, d\tau \quad \text{(Lorentz force)} . \tag{17.9}$$

So we interpret the term

$$- \frac{2}{3} q^2 \int_{\tau_1}^{\tau_2} u(\tau) \langle a(\tau), a(\tau) \rangle \, d\tau \tag{17.10}$$

as energy–momentum radiation. It is because this term arises that we expect accelerating charged particles to radiate energy and momentum.

So what happens to this term in the case of a uniformly accelerating charge? Why do we say that this is nonzero, whilst the term it leads to (indirectly) in the Lorentz–Dirac equation is zero? Now the point is that the integral in (17.10) is between any two proper times, and we can check from Parrott's derivation [13, pp. 140–1] that

$$\int_{\tau_1}^{\tau_2} u(\tau) \langle a(\tau), a(\tau) \rangle \, d\tau = -a(\tau_2) + a(\tau_1) + \int_{\tau_1}^{\tau_2} \left(\frac{da}{d\tau} - \left\langle \frac{da}{d\tau}, u \right\rangle u \right) d\tau . \tag{17.11}$$

For uniform acceleration between these two proper times, this gives

$$\int_{\tau_1}^{\tau_2} u(\tau) \langle a(\tau), a(\tau) \rangle \, d\tau = -a(\tau_2) + a(\tau_1) \neq 0 . \tag{17.12}$$

This brings out quite clearly why we still have grounds for claiming that there is radiation, whilst the proposed radiation reaction term is identically zero for this kind of motion.

The point is then that the acceleration changes during a period of uniform acceleration, because the word 'uniform' only refers to the constancy of the pseudolength $\langle a, a \rangle$. This is very clear from (16.20) on p. 260 or (17.2) on p. 290. Indeed, the condition of uniformity on the acceleration only requires $da/d\tau$ to be orthogonal to a, not necessarily zero, and we have the explicit formula in (16.20) to confirm this.

Therefore, in this version, the uniformly accelerated charge is radiating energy–momentum at the rate implied by (17.12), and this despite the fact that the terms in the Lorentz–Dirac equation that are usually interpreted as describing the radiation reaction are actually zero. The Lorentz–Dirac equation only implies the more fundamental energy–momentum conservation relation over a proper time interval $[\tau_1, \tau_2]$ when the acceleration is the same at the beginning and end of the interval. One could even say that the Lorentz–Dirac equation is wrong if this is not the case.

Indeed, we can see from the very derivation of the Lorentz–Dirac equation why it makes the strange prediction mentioned earlier in the context of a particle that is unaccelerated in the distant past, nudged into uniform acceleration, uniformly accelerated for a long time, then finally nudged back into an unaccelerated state. In this case, the Lorentz–Dirac equation does imply conservation of energy–momentum. However, the radiation reaction term is zero for the period of uniform acceleration, no matter how long it is. All the radiation energy must therefore be provided at the beginning and end of the trip, while the particle is nudged into and out of its uniformly accelerated state.

As noted above, the acceleration itself, because uniform, does not incur a radiation reaction, at least according to the terms in the Lorentz–Dirac equation. It is the non-uniformity of the acceleration which does, while the rocket is nudging itself into a uniform acceleration in the present example. But once again, let us stress that one should not consider that it is not a second derivative but a third derivative that counts here, because the third derivative $da/d\tau$ is nonzero in the uniform acceleration case, a point that the appellation 'uniform' tends to conceal. On the other hand, the whole question of higher derivatives does suggest that it might be worth considering a model in which the charge is non-pointlike: in that case, although it is a difficult calculation, we find a self-force on the charge that not only avoids the need for renormalisation, but potentially explains inertial mass, explains how the radiation reaction occurs, and includes higher order derivatives than just the acceleration.

Anyway we said above that, if we believe in the Lorentz–Dirac equation, we need to add a bit of energy to start the uniform acceleration, slightly more than if the rocket were uncharged, and thereafter the radiation will be free, even if it persists for an arbitrarily long time and adds up to an arbitrarily large amount. No more energy is needed to keep the charged rocket in uniform acceleration than if it were neutral, according to the radiation reaction term provided by the Lorentz–Dirac equation, which suggests that one could borrow an arbitrarily large amount of radiated energy, which could in principle be collected and used by other observers in Minkowski spacetime, with the proviso that one must pay it back at the end of the trip, whenever

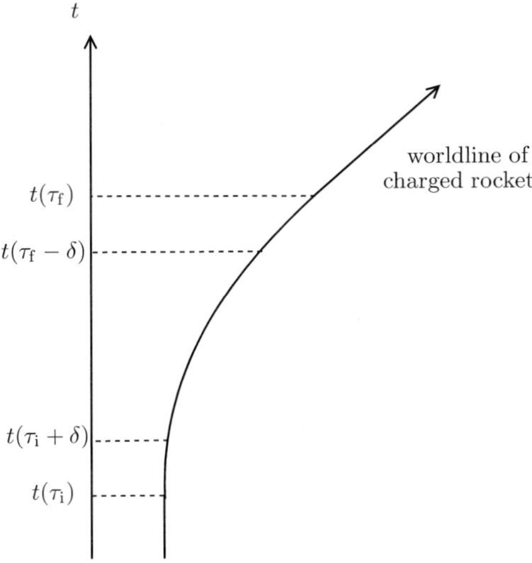

Fig. 17.1 A rocket begins at rest in some Lorentz frame up to an initial time τ_i, before being nudged into uniform acceleration over a small proper time interval $[\tau_i, \tau_i + \delta]$, uniformly accelerated up to proper time $\tau = \tau_f - \delta$, nudged back into an unaccelerated state over the interval $[\tau_f - \delta, \tau_f]$, and continuing unaccelerated for $\tau > \tau_f$. The function $t(\tau)$ expresses the Minkowski time for any proper time of the rocket

the uniform acceleration is switched off. Here is yet another reason for doubting the Lorentz–Dirac equation. Indeed, the equation of energy–momentum conservation arising directly from (17.9) seems to be a superior equation.

As mentioned above, we shall establish the equation of motion of a charged rocket on the basis of the Lorentz–Dirac equation in the following case (see Fig. 17.1): the rocket begins at rest in some Lorentz frame up to an initial time τ_i, before being nudged into uniform acceleration over a small proper time interval $[\tau_i, \tau_i + \delta]$, uniformly accelerated up to proper time $\tau = \tau_f - \delta$, nudged back into an unaccelerated state over the interval $[\tau_f - \delta, \tau_f]$, and continuing unaccelerated for $\tau > \tau_f$. At least, this is the first part of the motion we shall consider.

It turns out that the excess energy required to accomplish the final deceleration of a charged rocket as compared with a neutral rocket (strictly speaking, it is not a deceleration because we allow the rocket to maintain its velocity at whatever value it has reached; what we have is a decrease in acceleration to zero), as measured in the final rest frame at $\tau = \tau_f$, is independent of the period of uniform acceleration. This is in fact obvious, because one refers to the final rest frame. It would thus seem that, from the standpoint of the rocket pilot, only a modest excess amount of fuel must be burned to start the acceleration at the beginning of the trip and stop it at the end, with no excess fuel relative to an uncharged rocket required during the arbitrarily long period of uniform acceleration, during which a correspondingly arbitrary amount of radiation energy can be gathered.

The line of attack used by some proponents of this example is as follows. They suggest that one could allow the uniform acceleration to continue indefinitely without using any excess fuel as compared with the neutral rocket. By extension, they even suggest that perpetual uniform acceleration would require no extra fuel at all as compared with the situation for a neutral particle. If there were radiation energy from the charged rocket but not from the uncharged rocket, and if, as suggested for the latter scenario, no extra energy had to be put in to get this energy, one would obviously have a violation of energy conservation somewhere. One way out is to say that there might be radiation, but that it is not physically accessible, as proposed by Boulware, but rejected here.

Parrott counters this 'free energy' argument by re-examining the more realistic (sic) case of asymptotically free rockets, i.e., with constant velocity outside some finite period of uniform velocity, and showing that, not only does the Lorentz–Dirac equation predict something very strange in the case of the charged rocket, but it does put a limit on the duration of the uniform acceleration if one hopes eventually to remove the acceleration and allow the rocket to continue at constant velocity beyond some point in time. This is what is analysed mathematically below.

This shows how the Lorentz–Dirac equation saves itself from disaster in the case where it is supposed to work, i.e., when the motion is asymptotically free: if we do have to remove the uniform acceleration at some point, then that state of uniform acceleration cannot have been going on beyond some time determined by the initial rest mass of the charged rocket (see below for the quantitative version). Furthermore, even if the uniform acceleration has been going on for some time within the acceptable bound, one should have no doubt that the radiated energy does have to be paid for at the transition from the uniform acceleration to the uniform velocity state. The fact that only a modest amount of extra energy (as compared with the neutral rocket) is required to make this transition, as measured in the final frame of the rocket, does not mean that one is getting most of the radiated energy for nothing: the modest amount of energy used in the final frame, along with its associated momentum, can Lorentz transform into a large amount of energy in the initial rest frame at $\tau = \tau_i$, so energy conservation can indeed hold in the initial frame. After all, we know that the Lorentz–Dirac equation expresses energy conservation in an asymptotically free case like this.

Despite the decisive result that there is a limit on the duration of the uniform acceleration if one hopes eventually to remove the acceleration and allow the rocket to continue at constant velocity beyond some point in time, Parrott nevertheless extends his mathematical analysis in the following way: At time τ_f, we reverse the proper acceleration in a time-symmetric way, so that

$$a(\tau_f + \sigma) = -a(\tau_f - \sigma), \quad \sigma > 0,$$

whence the rocket will eventually come back to rest in the initial frame at $\tau = 2\tau_f$ (see Fig. 17.2). This was in answer to a more sophisticated challenge he mentions in [7], in which it was claimed that the excess fuel used over the whole interval $[\tau_i, 2\tau_f]$

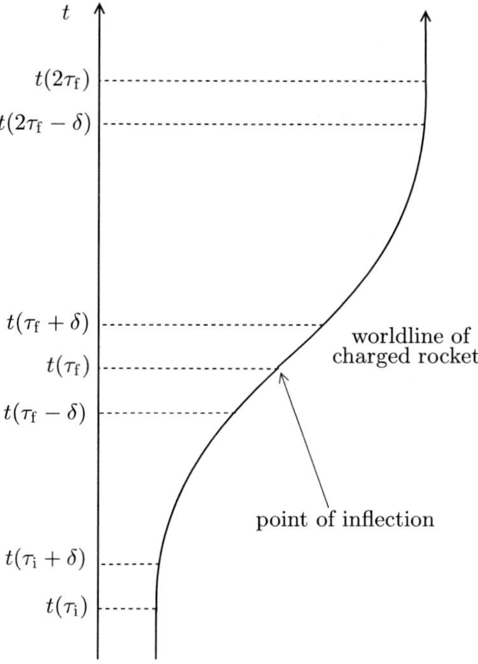

Fig. 17.2 A rocket begins at rest in some Lorentz frame up to an initial time τ_{i}, before being nudged into uniform acceleration over a small proper time interval $[\tau_{\mathrm{i}}, \tau_{\mathrm{i}} + \delta]$, uniformly accelerated up to proper time $\tau = \tau_{\mathrm{f}} - \delta$, nudged back into an unaccelerated state over the interval $[\tau_{\mathrm{f}} - \delta, \tau_{\mathrm{f}}]$, and continuing unaccelerated for $\tau > \tau_{\mathrm{f}}$. At time τ_{f}, we reverse the proper acceleration in a time-symmetric way. The function $t(\tau)$ expresses the Minkowski time for any proper time of the rocket

(as compared with the neutral rocket) is modest and independent of the duration of the uniform acceleration. Parrott says:

> At the beginning and end of the trip, the rocket is at rest in the initial frame, so the energy of the radiation plus the exhaust should equal the rest mass loss of the rocket [i.e., the rest mass of the fuel used]. If the loss of rest mass is finite and independent of the duration of the uniform acceleration, hence independent of the arbitrarily large energy radiation, we have a violation of energy conservation in the initial frame.

Since we have a situation here where we expect the Lorentz–Dirac equation to apply, and since it implies energy conservation when it applies, this would appear to cast a serious shadow over the proceedings. However, we shall now use the Lorentz–Dirac equation to show that, with a fixed amount of initial fuel, one cannot obtain an arbitrarily long period of uniformly accelerated motion, i.e., τ_{f} cannot be arbitrarily large. Put another way, according to the Lorentz–Dirac equation, a charged rocket will run out of fuel if it uniformly accelerates for long enough, so the proposed time-symmetric motion is impossible with fixed initial fuel and arbitrarily large τ_{f}. In contrast, and we shall show this along the way, a uniformly accelerated uncharged rocket can accelerate forever, provided that all its mass can be consumed as fuel. So it looks as though the Lorentz–Dirac equation can save itself from what would here be tantamount to a self-contradiction.

17.2 Calculation

The rocket moves in the positive x direction and the other two constant space coordinates and associated components of vectors are ignored. Let v be the initial frame velocity, corresponding to a rapidity

$$\theta := \tanh^{-1} v \,,$$

so that the initial 4-velocity is

$$u = (\cosh\theta, \sinh\theta) \,. \tag{17.13}$$

The scalar proper acceleration A is defined by

$$a = \frac{du}{d\tau} = Aw \,, \quad \text{where } w := (\sinh\theta, \cosh\theta) \,.$$

But by (17.13),

$$\frac{du}{d\tau} = (\sinh\theta, \cosh\theta)\frac{d\theta}{d\tau} \,,$$

whence

$$A = \frac{d\theta}{d\tau} \,. \tag{17.14}$$

If we assume that the charged rocket is powered in such a way that its acceleration is uniform, then we are imposing $A = g = $ constant. This gives an equation for the rapidity, viz.,

$$\theta(\tau) = g\tau + \theta(0) \,. \tag{17.15}$$

The Lorentz–Dirac equation suggests that the proper time rate of energy–momentum radiation by a charge q is

$$-\frac{2q^2}{3}\left(\frac{da}{d\tau} + a^2 u\right) \,. \tag{17.16}$$

Now

$$\frac{da}{d\tau} = \frac{dA}{d\tau}w + A\frac{dw}{d\tau} = \frac{dA}{d\tau}w + A^2 u \,, \tag{17.17}$$

and $a^2 u = -A^2 u$, so

$$\text{proper time rate of energy–momentum radiation} = -\frac{2q^2}{3}\frac{\mathrm{d}A}{\mathrm{d}\tau}w\,. \tag{17.18}$$

This is zero of course in the case of a uniform acceleration, as discussed earlier. We choose units so that $2q^2/3 = 1$.

We now introduce the time-dependent rest mass of the rocket, denoted by $m(\tau)$. Then $-\mathrm{d}m/\mathrm{d}\tau$ is the rate of ejection of rest mass into the exhaust. We shall see that this is not the same as the rate at which the exhaust acquires rest mass, i.e., rest mass is not conserved in such a system. Two parameters can be used to control the rocket worldline: the exhaust velocity and the rate $-\mathrm{d}m/\mathrm{d}\tau$ at which rest mass is ejected in the exhaust. We shall fix one of these by insisting that the exhaust velocity as seen from the rocket is always constant. Then $\mathrm{d}m/\mathrm{d}\tau$ is varied in just the right way to produce the desired worldline.

The rocket always moves to the right in the initial frame, and the exhaust to the left in the rocket frame. We allow $\mathrm{d}m/\mathrm{d}\tau$ to have either sign. A positive $\mathrm{d}m/\mathrm{d}\tau$ means that the rocket is taking on mass. This could be achieved physically by shooting bullets into it from the right, with the bullets constituting the exhaust.

If $-v$ is the exhaust rapidity in the rocket frame at any instant of time, with $v > 0$, then an exhaust particle has velocity $\tanh(-v)$ in this frame. We shall now find the 4-velocity of the exhaust. We start with the well known velocity addition formula in special relativity, viz.,

$$v_x' = \frac{v + v_x}{1 + v v_x}\,, \tag{17.19}$$

where $v = \tanh\theta$ and $v_x = -\tanh v$. By the addition formulas for the hyperbolic functions, we have

$$
\begin{aligned}
v_x' &= \frac{\tanh\theta - \tanh v}{1 - \tanh\theta\,\tanh v}\\
&= \frac{\sinh\theta\,\cosh v - \sinh v\,\cosh\theta}{\cosh\theta\,\cosh v - \sinh v\,\sinh\theta}\\
&= \tanh(\theta - v)\,.
\end{aligned}
\tag{17.20}
$$

This is also a well known result: rapidities are additive in special relativity. We can now find

$$\gamma(v_x') = (1 - v_x'^2)^{-1/2} = \cosh(\theta - v)\,.$$

This is the zero-component of the 4-velocity of the exhaust in the initial frame, and $\gamma(v_x')v_x'$ is the one-component. Consequently,

$$\text{4-velocity of exhaust in initial frame} = \big(\cosh(\theta - v), \sinh(\theta - v), 0, 0\big)\,. \tag{17.21}$$

Now observe that

$$u\cosh v - w\sinh v = \left(\begin{array}{c} \cosh\theta \\ \sinh\theta \end{array}\right)\cosh v - \left(\begin{array}{c} \sinh\theta \\ \cosh\theta \end{array}\right)\sinh v$$

$$= \left(\begin{array}{c} \cosh(\theta - v) \\ \sinh(\theta - v) \end{array}\right) . \tag{17.22}$$

Hence,

$$\text{4-velocity of exhaust in initial frame} = u\cosh v - w\sinh v . \tag{17.23}$$

We now consider the proper time rate $\rho = \rho(\tau)$ at which the exhaust rest mass increases. Let $Rw = R(\tau)w(\theta(\tau))$ denote the proper time rate at which electromagnetic energy–momentum is being emitted. Then, if the Lorentz–Dirac equation is right, we are saying that

$$R(\tau) = -\frac{\mathrm{d}A}{\mathrm{d}\tau} . \tag{17.24}$$

This is zero whenever the acceleration is uniform. Furthermore, if we consider an uncharged rocket, as we shall, for comparison, then $R \equiv 0$.

We now have an equation of energy–momentum balance, viz.,

$$0 = \frac{\mathrm{d}(mu)}{\mathrm{d}\tau} + (u\cosh v - w\sinh v)\rho + Rw$$

$$= \left(\frac{\mathrm{d}m}{\mathrm{d}\tau} + \rho\cosh v\right)u + (mA - \rho\sinh v + R)w . \tag{17.25}$$

On the right-hand side of the first line, the first term is the proper time rate of change of energy–momentum of the rocket, the second is the proper time rate at which the exhaust is acquiring energy–momentum, and the third is the proper time rate of energy–momentum radiation. This is an adapted version of the Lorentz–Dirac equation, including the radiation reaction term but also what we may consider to be an external force, due to the ejection of some mass. Now since u and w are orthogonal, this yields two equations, viz.,

$$\rho = -\frac{1}{\cosh v}\frac{\mathrm{d}m}{\mathrm{d}\tau} , \quad mA - \rho\sinh v + R = 0 . \tag{17.26}$$

We take the first of these to define ρ, noting that ρ is not just $-\mathrm{d}m/\mathrm{d}\tau$. Indeed,

$$\rho + \frac{\mathrm{d}m}{\mathrm{d}\tau} = \left(1 - \frac{1}{\cosh v}\right)\frac{\mathrm{d}m}{\mathrm{d}\tau} . \tag{17.27}$$

This shows that the total rest mass increases when $\mathrm{d}m/\mathrm{d}\tau$ is positive, and decreases when it is negative, since $\cosh v \geq 1$.

If we define $\lambda := 1/\tanh v$, the second equation of (17.26) gives

$$\frac{\mathrm{d}m}{\mathrm{d}\tau} + \lambda mA + \lambda R = 0 . \tag{17.28}$$

Now $A = \mathrm{d}\theta/\mathrm{d}\tau$, so we have

$$\frac{\mathrm{d}m}{\mathrm{d}\tau} + \lambda m \frac{\mathrm{d}\theta}{\mathrm{d}\tau} + \lambda R(\tau) = 0 . \tag{17.29}$$

Hence,

$$\frac{\mathrm{d}}{\mathrm{d}\tau}\left(\mathrm{e}^{\lambda\theta}m\right) = \mathrm{e}^{\lambda\theta}\frac{\mathrm{d}m}{\mathrm{d}\tau} + \lambda m \mathrm{e}^{\lambda\theta}\frac{\mathrm{d}\theta}{\mathrm{d}\tau} = -\mathrm{e}^{\lambda\theta}\lambda R . \tag{17.30}$$

Finally, the solution with zero initial rapidity, i.e., $\theta(\tau_i) = 0$, is

$$\mathrm{e}^{\lambda\theta(\tau)}m(\tau) = m(\tau_i) - \int_{\tau_i}^{\tau} \lambda R(s)\mathrm{e}^{\lambda\theta(s)}\mathrm{d}s . \tag{17.31}$$

This can be written

$$m(\tau) = \mathrm{e}^{-\lambda\theta(\tau)}m(\tau_i) - \lambda\mathrm{e}^{-\lambda\theta(\tau)}\int_{\tau_i}^{\tau} R(s)\mathrm{e}^{\lambda\theta(s)}\mathrm{d}s . \tag{17.32}$$

In the case of an uncharged rocket, $R \equiv 0$ and we have

$$m(\tau) = \mathrm{e}^{-\lambda\theta(\tau)}m(\tau_i) \qquad \text{uncharged rocket} . \tag{17.33}$$

This says that $m(\tau)$ decreases exponentially with $\theta(\tau)$. We also know that, during the first period of uniform acceleration in the scenario we are considering, we have

$$\theta(\tau) = (\tau - \tau_i - \delta)g + \theta(\tau_i + \delta) , \qquad \tau \in [\tau_i + \delta, \tau_f - \delta] .$$

So for an uncharged rocket, with $R \equiv 0$, equation (17.33) tells us that $m(\tau)$ decreases exponentially with τ during the period $[\tau_i + \delta, \tau_f - \delta]$ of uniform acceleration. This period of uniform acceleration could thus continue forever without $m(\tau)$ ever reaching zero.

What happens to a charged rocket? We now have $R = -\mathrm{d}A/\mathrm{d}\tau$, which is zero in the interval $\tau \in [\tau_i + \delta, \tau_f - \delta]$, but is not zero for $\tau \in [\tau_i, \tau_i + \delta]$, nor for $\tau \in [\tau_f - \delta, \tau_f + \delta]$. It is then zero again for $\tau \in [\tau_f + \delta, 2\tau_f - \delta]$ and nonzero for $\tau \in [2\tau_f - \delta, 2\tau_f]$. In fact, $\tau_i = 0$, so that our time interval $[\tau_i, 2\tau_f]$ is centered on τ_f. Equation (17.32) now reads

$$m(\tau) = \mathrm{e}^{-\lambda\theta(\tau)}m(\tau_i) - \lambda\mathrm{e}^{-\lambda\theta(\tau)}\int_{\tau_i}^{\tau}\left(-\frac{\mathrm{d}A}{\mathrm{d}s}\right)\mathrm{e}^{\lambda\theta(s)}\mathrm{d}s . \tag{17.34}$$

Let us suppose that A has the constant value g over the interval $\tau \in [\tau_i + \delta, \tau_f - \delta]$, then decreases monotonically to zero over the interval $\tau \in [\tau_f - \delta, \tau_f]$. We consider δ to be fixed and show that $m(\tau_f)$ must become negative for large τ_f. This means that, for such an A and for a fixed initial mass $m(\tau_i)$, we cannot find positive mass solutions defined for arbitrarily long proper times τ_f.

We make an estimate using the mean value theorem. For $\tau_f - \delta \le s \le \tau_f$,

$$\frac{\theta(\tau_f) - \theta(s)}{\delta} \le \frac{\theta(\tau_f) - \theta(s)}{\tau_f - s}$$

$$= \frac{d\theta}{d\tau}(\hat{\tau}) \text{ for some } \hat{\tau} \text{ s.t. } s \le \hat{\tau} \le \tau_f$$

$$= A(\hat{\tau})$$

$$\le g.$$

Noting that $-dA/ds \ge 0$ and $\lambda > 0$, this then gives

$$e^{-\lambda\theta(\tau_f)} \int_{\tau_f-\delta}^{\tau_f} \left(-\frac{dA}{ds}\right) e^{\lambda\theta(s)} ds = \int_{\tau_f-\delta}^{\tau_f} \left(-\frac{dA}{ds}\right) e^{-\lambda\left[\theta(\tau_f)-\theta(s)\right]} ds$$

$$\ge e^{-\lambda g\delta} \int_{\tau_f-\delta}^{\tau_f} \left(-\frac{dA}{ds}\right) ds$$

$$= e^{-\lambda g\delta}\left[-A(\tau_f) + A(\tau_f-\delta)\right]$$

$$= e^{-\lambda g\delta} g.$$

We now return to (17.34), from which

$$m(\tau_f) = e^{-\lambda\theta(\tau_f)} m(\tau_i) - \lambda e^{-\lambda\theta(\tau_f)} \int_{\tau_i}^{\tau_i+\delta} \left(-\frac{dA}{ds}\right) e^{\lambda\theta(s)} ds$$

$$- \lambda e^{-\lambda\theta(\tau_f)} \int_{\tau_f-\delta}^{\tau_f} \left(-\frac{dA}{ds}\right) e^{\lambda\theta(s)} ds$$

$$\le e^{-\lambda\theta(\tau_f)} m(\tau_i) - \lambda e^{-\lambda\theta(\tau_f)} \int_{\tau_i}^{\tau_i+\delta} \left(-\frac{dA}{ds}\right) e^{\lambda\theta(s)} ds - \lambda e^{-\lambda g\delta} g.$$

We wish to show that, if τ_f is large, then this must be negative, i.e., that

$$e^{-\lambda\theta(\tau_f)} \left[m(\tau_i) - \lambda \int_{\tau_i}^{\tau_i+\delta} \left(-\frac{dA}{ds}\right) e^{\lambda\theta(s)} ds\right] \tag{17.35}$$

can be made as small as we like. Note that, if we assume that A increases monotonically from 0 to g in the interval $\tau \in [\tau_i, \tau_i + \delta]$, then $-dA/ds$ is negative in the integral of (17.35). Consider the interval $\tau_i \le s \le \tau_i + \delta$, over which θ increases from zero with

$$0 < \frac{d\theta}{ds} < g.$$

Hence,

$$0 < \theta(s) < g\delta \, ,$$

and

$$0 < \int_{\tau_i}^{\tau_i+\delta} \frac{\mathrm{d}A}{\mathrm{d}s} \mathrm{e}^{\lambda\theta(s)} \mathrm{d}s < \mathrm{e}^{\lambda g\delta} \int_{\tau_i}^{\tau_i+\delta} \frac{\mathrm{d}A}{\mathrm{d}s} \mathrm{d}s$$

$$= \mathrm{e}^{\lambda g\delta} \left[A(\tau_i+\delta) - A(\tau_i) \right]$$

$$= \mathrm{e}^{\lambda g\delta} g \, .$$

Hence, the object which we would like to make as small as possible in (17.35) is smaller than

$$\mathrm{e}^{-\lambda\theta(\tau_f)} \left[m(\tau_i) + \lambda \mathrm{e}^{\lambda g\delta} g \right] \, , \tag{17.36}$$

and since the part of this between square brackets is indeed fixed, and since θ always increases right up until proper time τ_f when the uniform acceleration to the right has been brought back down to zero, the result is proven.

Therefore, to obtain a positive mass solution for arbitrarily large τ_f, and hence arbitrarily large radiated energy, we need arbitrarily large rocket mass, i.e., fuel, $m(\tau_i)$ to start with. As Parrott states so clearly:

> In other words, a charged rocket in Minkowski spacetime which starts with a finite amount of fuel cannot uniformly accelerate for an arbitrarily long time, after which the acceleration is removed. Unlike a corresponding uncharged rocket, it must eventually run out of fuel. What is peculiar is that, if it has sufficient fuel to get into the uniformly accelerated state, it will not run out of fuel until after the uniform acceleration has been removed. It can uniformly accelerate for an arbitrarily long period, radiating all the while, but the physical contradiction of running out of fuel followed by the mass going negative will not be revealed until after the uniform acceleration has been removed.

It has to be said that the most doubtful assumption leading to this strange situation would appear to be the expression for the radiation reaction in the Lorentz–Dirac equation. It would be an interesting exercise here to see whether the fuller expression for the self-force on a spatially extended electron (a charge distribution, rather than a point charge) would save us from this situation.

We now investigate what happens in the rest of our scenario, as the rocket is uniformly decelerated over the interval $\tau \in [\tau_f+\delta, 2\tau_f-\delta]$. First of all, the acceleration $A = g$ of the first period of uniform acceleration must be removed over the interval $\tau \in [\tau_f-\delta, \tau_f+\delta]$. In this interval, A decreases from g to $-g$, and this is the only time when $\mathrm{d}A/\mathrm{d}\tau$ is negative.

We consider first the uncharged rocket. Since our formulation assumes that the exhaust velocity cannot be varied, the deceleration after $\tau = \tau_f$ is achieved by taking in mass and momentum, so we have $\mathrm{d}m/\mathrm{d}\tau > 0$ for $\tau_f < \tau < 2\tau_f$, right up until the moment when we once again have $\theta = 0$. In effect, deceleration is accomplished by returning some of the previous exhaust energy–momentum to the rocket. For an uncharged rocket, the symmetry of the situation suggests that this energy–momentum return will be accomplished in a time-symmetric manner, and we can anticipate without calculation that all the exhaust energy–momentum will have been returned to the rocket at the final resting time $\tau = 2\tau_f$. In particular, the final rest mass should equal the initial rest mass. It is easy to see that this is the case from (17.34), when the integral on the right-hand side is zero, i.e.,

$$m(2\tau_f) = e^{-\lambda\,\theta(2\tau_f)}m(\tau_i) = m(\tau_i) \quad \text{(uncharged rocket)}\,. \tag{17.37}$$

The situation is quite different for a charged rocket.

Now $m(2\tau_f)$ differs from $m(\tau_i)$ by the amount of the second term in (17.34):

$$m(2\tau_f) = m(\tau_i) - \lambda \int_{\tau_i}^{2\tau_f} \left(-\frac{\mathrm{d}A}{\mathrm{d}s}\right) e^{\lambda\,\theta(s)}\,\mathrm{d}s\,. \tag{17.38}$$

We write this in the form

$$m(\tau_i) - m(2\tau_f) \tag{17.39}$$

$$= -\lambda \int_{\tau_i}^{\tau_i+\delta} \frac{\mathrm{d}A}{\mathrm{d}s} e^{\lambda\,\theta(s)}\,\mathrm{d}s - \lambda \int_{\tau_f-\delta}^{\tau_f+\delta} \frac{\mathrm{d}A}{\mathrm{d}s} e^{\lambda\,\theta(s)}\,\mathrm{d}s - \lambda \int_{2\tau_f-\delta}^{2\tau_f} \frac{\mathrm{d}A}{\mathrm{d}s} e^{\lambda\,\theta(s)}\,\mathrm{d}s\,.$$

Note that $\mathrm{d}A/\mathrm{d}s$ is positive in the first and third integrals on the right-hand side, and negative in the second integral. Furthermore, $e^{\lambda\theta}$ is small in the first and third integrals, and large in the middle integral. Indeed, θ reaches its maximum at τ_f, and the bigger we make τ_f, i.e., the longer the uniform acceleration, the bigger this maximum will be. So it is the positive second integral in the last relation that dominates and there is indeed a mass deficit when we reach $2\tau_f$, i.e.,

$$m(2\tau_f) < m(\tau_i)\,.$$

The pilot sees a modest excess fuel loss over the interval $[\tau_f - \delta, \tau_f + \delta]$, as compared with the pilot of an uncharged rocket, but because she is going very fast in the initial frame, she understands that it corresponds to a large energy loss in the initial frame. Moreover, as she decelerates back to rest at $\tau = 2\tau_f$, this modest rest mass loss grows exponentially to an excess rest mass loss large enough to pay for the radiated energy in the initial frame reckoning.

To see this, we note from (17.33) that the rest mass of an uncharged rocket will increase exponentially during the interval $[\tau_f + \delta, 2\tau_f - \delta]$ when the rocket is uniformly decelerating. The same is true for the charged rocket during this period. In fact, over this period, the charged rocket behaves in exactly the same way as an uncharged rocket with the same rest mass at time $\tau = \tau_f + \delta$. However, the uncharged rocket which started with initial rest mass m_i at $\tau = 0$ does not have exactly the same rest mass at $\tau = \tau_f + \delta$ as the charged rocket with the same worldline and initial mass. There is a difference due to the radiation in the time interval $[\tau_i, \tau_f + \delta]$. This difference is modest even when the radiation is large. If τ_f is large enough to give large radiation, this modest difference in rest masses at $\tau = \tau_f + \delta$ is amplified by the exponential growth to a correspondingly large difference in rest masses at $\tau = 2\tau_f - \delta$.

This seems to show that there is a physical radiation reaction for a particle which is uniformly accelerated for a finite time even though the Lorentz–Dirac radiation reaction expression is identically zero during the period of uniform acceleration. However, if we believe the Lorentz–Dirac equation, we must accept the strange conclusion that all of this radiation reaction occurs at the beginning ($\tau \approx \tau_i$) and end ($\tau \approx \tau_f$) of the trip, while the particle is being nudged into or out of its uniformly accelerated state.

Whether there is a radiation reaction for a perpetually uniformly accelerated particle depends on the definition of the radiation reaction. The radiation reaction term in the Lorentz–Dirac equation is identically zero, but there is no good physical reason to identify this term with the physically observed radiation reaction. It seems more reasonable to obtain the answer for perpetual uniform acceleration as a limit of whatever answer is eventually accepted for uniform acceleration over a finite time. The Lorentz–Dirac equation probably has to be rejected in the process, if only because it accounts so strangely for the events of the finite-duration uniform acceleration in the above discussion.

17.3 Conclusion

One might say that the most revealing part of this last discussion occurs on p. 293, with regard to (17.11) and (17.12). The radiation through the tube walls between τ_1 and τ_2 is the left-hand side of (17.11), according to the careful calculation in Parrott's book. For uniform acceleration throughout the period $\tau \in [\tau_1, \tau_2]$, we obtain (17.12), which says that the radiation is numerically equal to the nonzero quantity $-a(\tau_2) + a(\tau_1)$. Now in the present context, we have arranged for $a(\tau_1) = a(\tau_2)$, but this does not mean that the radiation given by the right-hand side of (17.11) will be zero, because we do not have uniform acceleration throughout the interval $[\tau_1, \tau_2]$ when we put $\tau_1 = \tau_i$ and $\tau_2 = \tau_f$. As explained above, the putative radiation reaction term in the Lorentz–Dirac equation makes its contribution to radiation (as calculated through the tube walls) only when the acceleration is non-uniform.

A second comment here concerns (17.25) on p. 300. As observed there, this is an adapted version of the Lorentz–Dirac equation, including the radiation reaction term but also what we may consider to be an external force, due to the ejection of some mass. In the usual Lorentz–Dirac equation, the external force, if there is one, comes from other electromagnetic fields. Here we are using mass ejection to drive the charge along some worldline. The problem that arises then is that we find ourselves with a negative mass in some circumstances, and we use this to rule out certain scenarios. Speculatively, one wonders what problem arises (or whether there is a problem) when the charge is driven by electromagnetic fields.

Let us just see explicitly how this negative mass problem arises. Whilst $\tau < \tau_f - \delta$ and the charge is still undergoing uniform acceleration, (17.34) tells us that

$$m(\tau) = e^{-\lambda\theta(\tau)}m(\tau_i) - \lambda e^{-\lambda\theta(\tau)} \int_{\tau_i}^{\tau_i+\delta} \left(-\frac{dA}{ds}\right) e^{\lambda\theta(s)}ds .\tag{17.40}$$

Now the right-hand side here is always positive because $dA/ds > 0$ in the interval $s \in [\tau_i, \tau_i + \delta]$. Even though it tends to zero due to the factor $e^{-\lambda\theta(\tau)}$ in which $\lambda > 0$ and $\theta(\tau)$ is increasing, it never reaches zero or goes negative. However, the analysis shows that it has to go negative if we uniformly accelerate for too long and then bring the acceleration back down to zero over some period $[\tau_f - \delta, \tau_f]$.

It looks as though we are condemned to go on accelerating for ever by mass ejection (and that this is theoretically possible, with ever smaller mass ejection rate as the charge gets lighter), if we wish to gather an unbounded quantity of radiated energy. Physically, this shows that the Lorentz–Dirac equation in the form (17.25) is not correctly accounting for the energy balance. The only really questionable part of the equation is the so-called radiation reaction term. But we may also ask how this would work if the charge motion were subsumed under the genuine Lorentz–Dirac equation, being driven by electromagnetic fields rather than mass loss.

Chapter 18
Summary

Such a lot has been said that it is essential to put the main points down without the historical aspect or the details. This is therefore an abstract and could be read first, or without the rest, but cross-references serve as a guide to the rest.

There is now some consensus, on purely theoretical grounds, that a uniformly accelerating charge in flat spacetime will radiate electromagnetic (EM) energy, despite the fact that the radiation reaction terms in the Lorentz–Dirac equation are zero. This is the view in this book and it is the view of all the authors cited. The Lorentz–Dirac equation is not always justified and one cannot interpret the vanishing of what are called the radiation reaction terms as meaning that there is no radiation [see Chap. 17 and in particular (17.12) on p. 293]. The Lorentz–Dirac equation only implies the more fundamental energy–momentum conservation relation [(17.9) on p. 293] over a proper time interval $[\tau_1, \tau_2]$ when the acceleration is the same at the beginning and end of the interval, and the fact that it implies that relation does not necessarily mean that it gives a correct blow-by-blow account of what the particle is doing in the meantime.

The present book adopts two equivalence principles in the context of general relativity (GR) (see Chap. 2):

- The weak equivalence principle (WEP) just observes that (see p. 15), if one uses the Levi-Civita connection derived from the metric, then for any event (point) P in spacetime, there is always a choice of coordinates in some neighbourhood of that event for which the connection coefficients are zero at P and the metric takes the Minkowski form η at P. By continuity, the connection remains close to zero and the metric very close to η over some small enough neigbourhood of P.
- The strong equivalence principle (SEP) says that (see p. 52), if one uses the locally inertial coordinates provided by WEP, any other kind of physics than gravitational physics will look approximately the same as it does in special relativity (SR). This is formulated by replacing partial coordinate derivatives in the SR version by components of covariant derivatives, where the covariant derivative is defined using the Levi-Civita connection, and it amounts to the assumption that this is the only connection. (We ignore here the ambiguity introduced into

this procedure by the fact that curvature-dependent terms can be inserted with impunity in the SR version of the physics.)

While WEP can be considered a basic element of GR itself (not quite though, because there are versions of the theory with torsion), SEP is an add-on that allows one to integrate the rest of physics, apart from gravity, into GR. One then gets GR+SEP. One guiding principle in the present book is that one could not do the other physics if one did not have SEP, or at least, that one would need some other procedure if SEP were dropped.

Other versions, or statements, of equivalence principles are discussed in Chap. 2. A general conclusion of the present book is that it is better not to formulate these principles in terms of what may or may not be distinguished by an observer. This is a hangover from the experimentally observed equality of inertial mass and passive gravitational mass and the mathematical consequence of this in the context of Newtonian gravitational theory that an observer cannot tell by local observations whether an object merely appears to be accelerating due to the motion of the observer or is in free fall under gravity.

Bondi and Gold's problem arises as follows [2]. One can find coordinates (semi-Euclidean or SE coordinates) adapted to the worldline of the uniformly accelerating charge (see Chap. 2). The Minkowski metric takes on a different but still static form [see (2.16) on p. 9]. One would like also to interpret this as the metric for a static, homogeneous gravitational field (SHGF) in the coordinate frame of a person supported against free fall in that field. (In fact, 'supported' means 'held stationary relative to those coordinates'.) One is thus applying general relativity here, even though the gravitational field happens to be non-tidal because the curvature is zero. One can do electromagnetism in this context only be applying SEP. It tells us that the EM fields of the supported charge in the SHGF involve EM radiation terms, because they are mathematically identical in every respect with the EM fields of the uniformly accelerating charge in a Minkowski spacetime with no gravitational effects, not even non-tidal ones. Bondi and Gold's problem with this is that, for them, a charge that is stationary relative to coordinates in which the metric has the static form [see p. 51 and (13.1) on p. 127] cannot radiate.

If the latter were true, one option might be to say that SEP should be dropped. This is the threat to the equivalence principle mentioned in the title of the present book. The view adopted in this book is that one should keep SEP and hence accept that a charge that is stationary relative to coordinates in which the metric has the static form will actually sometimes radiate. The reasons for this choice are as follows:

- If one dropped SEP, one would have to have some other way of formulating all the bits of physics other than gravity in the context of GR. But what would it be?
- There is no particular reason to think that a charge that is stationary relative to coordinates in which the metric has the static form will not radiate. Stationarity and staticity are coordinate-dependent notions. Furthermore, GR+SEP introduces an interaction between gravitational effects and electromagnetic fields

in the minimal extension of Maxwell's equations (MEME) so the situation with regard to EM effects is quite different to the situation in SR.

This is of course just a theoretical stance, as is the whole discussion in this debate. It would be nice to have experimental evidence of what really happens out there.

Bondi and Gold's solution to this problem is not to drop SEP, but to discard the whole scenario of the SHGF as unphysical, then to point out that, in their interpretation of the equivalence principle, in a real spacetime with tidal effects, i.e., nonzero curvature, there would be no conflict (see p. 54 ff.). What makes the SHGF unphysical is just that it is purely non-tidal so that one has global inertial frames. Once one turns to curved spacetimes, the idea is that at any distance over which one could affirm the observation of EM radiation, the presence of a gravitational field would be revealed by its inhomogeneity, and the electromagnetic effects do not then have to be the same in the two cases. In other words, in a typical curved spacetime, the inertial frames are only inertial locally, so one expects the minimally extended Maxwell equations (MEME) to make at least a slightly different prediction to Maxwell's equations in a flat spacetime with no gravitational effects (not even non-tidal ones). This is supposed to save the charge from having to radiate in the case where it is stationary relative to coordinates in which the metric is static.

This solution is rejected in the present book for the following reasons:

- The SHGF undoubtedly is unphysical, but it should be a good approximation in certain contexts. The notion of approximation is a key issue when discussing this. Even real gravitational effects may be homogeneous enough for inhomogeneities not to be detectable at distances where the EM fields are detectable. Bondi and Gold believe that one must survey spacetime over some minimal region in order to determine the presence of EM radiation, and their argument is that real gravitational effects will always have revealed their inhomogeneity over such a region. The view in the present book is that there must be situations in which EM radiation effects will be sufficiently detectable over regions in which gravitational effects are still homogeneous.
- All the talk about what is detectable is irrelevant to the question of whether WEP and SEP apply. The fact that radiation effects are not detectable in some zone does not mean that there are no radiation effects. Such talk refers to the measurement accuracy of observers, something which improves all the time and cannot possibly be an element of fundamental theory. The word 'local' is not needed in the mathematical formulation of WEP and SEP. The point here is that either there is radiation or there is not, and it makes no difference whether this or that observer can detect it.
- There is in fact no particular reason to think that a charge that is stationary relative to coordinates in which the metric has the static form will not radiate. In Bondi and Gold's qualitative formulation, they forget that what is static in one coordinate system will not be in another, because staticity is relative, i.e., it is a coordinate-dependent notion. Furthermore, GR+SEP introduces an interaction between gravitational effects and EM fields which does not exist in SR, so intuitive arguments based on SR are inadequate here.

- We have here an application of SEP *par excellence* in the sense that local inertial frames turn out to be globally inertial. If SEP did not work here, why should it work in the general, more subtle context of a curved spacetime?

If Bondi and Gold really intend to uphold the idea that a charge that is stationary relative to coordinates in which the metric has the static form will not radiate, they really do need the mistaken notion that EM radiation can only be detected by observations made over some minimal region of spacetime within which gravitational effects will always have revealed their inhomogeneity. But even then, it is still not at all clear that this inhomogeneity would be sufficient to mean that MEME would not predict radiation over that region. This would also be essential to save their point of view, because it is not sufficient just to be unable to detect radiation over a given region. They are saying that, in some circumstances, there actually is no radiation.

This is perhaps where one sees the trouble caused by versions of the equivalence principle that speak about whether two situations can be distinguished. One at once assumes that one is talking about whether some observer or other could measure any differences. But that is not at all the idea of SEP. How could it be a serious theoretical consideration? The accuracies with which we can measure EM fields may depend on techniques that are quite different from those used to measure gravitational effects, and in any case, they are both purely technical questions. The strong equivalence principle provides an exact means of transposing pre-GR physics to GR, although whether this leads to a good theory or not remains open, of course.

The problem with statements referring to the indistinguishability of two situations is that they are almost never strictly true, either in Newtonian gravitational theory or in GR. This is why they have to refer to local observations. Real gravitational fields are always inhomogeneous and if one has sufficiently accurate observations, they can be distinguished from effects due to the motion of an observer by observations made over any neighbourhood of the observer, no matter how small. Furthermore, these observations may concern only the motion of neutral test particles and pay no attention whatever to other physical effects, such as electromagnetic effects. For effects due to the motion of an observer to be confused with gravitational effects in GR, the gravitational effects have to be non-tidal, i.e., with zero curvature, so the strict application of such a principle, insofar as there is one, concerns only this zero curvature case. Although WEP and SEP may be loosely stated with this kind of reference to local observations, these are only loose statements, because the mathematical formulation is exact, as befits fundamental theory, whether the theory successfully corresponds to reality or not.

To sum up, in order to save their idea that a charge that is stationary relative to coordinates in which the metric components are static cannot radiate, Bondi and Gold are having to assume that EM radiation can only be detected by observations made over some minimal region of spacetime within which gravitational effects will always have revealed their inhomogeneity; but even if this were true, it would only rescue a loose version of the strong equivalence principle, while the exact theory GR+SEP which pays no lip service to questions of local measurements or what some observer may or may not be able to distinguish quite clearly requires charges that are stationary relative to coordinates in which the metric components are static to radiate in some circumstances. The example of the SHGF provides as clear an example as one could hope to find, since it is an application of SEP *par excellence* likely to reveal the tendencies of the theory, and since it is in any case a good approximation in many situations where one could clearly determine the presence of an EM radiation field.

Fulton and Rohrlich agree with the solution provided by Bondi and Gold [3]. The main achievement of their paper is a simpler derivation of the EM fields due to the uniformly accelerating charge in Minkowski spacetime with no gravitational effects, and a demonstration that there is EM radiation for an inertial observer in this context. But it is argued in the present book that they were confused, not only about the difference between SR and GR (see Chap. 10), but also about the equivalence principle, and as a result did not really understand Bondi and Gold's problem (see Chap. 9). However, one has the same confusion over the role of equivalence principles that refer to local measurements and the question of whether EM radiation can be measured close to the source. Parrott deals with this issue in [7, Sect. 5]: EM radiation can be measured arbitrarily close to the source, if one has sufficiently accurate measurement techniques. Indeed, this is strongly suggested by the results due to Fulton and Rohrlich (see p. 62), something they do not appear to have noticed.

Chapter 5 discusses Fulton and Rohrlich's calculation of a radiation rate associated with any given source point on the charge worldline, and Chap. 6 compares this with the more detailed version due to Parrott [13], in order to clear up a slight ambiguity in Fulton and Rohrlich's statement in Chap. 7. It is in Chap. 7 that we spell out the reasons why one does not need to be far from the source to measure the radiation rate (at least in theory, which is all that matters here). Chapter 8 considers Fulton and Rohrlich's view of how energy can be conserved in the uniform acceleration of a charged particle given that it radiates EM energy but there is zero radiation reaction according to the Lorentz–Dirac equation. It is concluded that their view is not convincing and that one ought to be more critical of the Lorentz–Dirac equation.

Chapter 9 discusses Fulton and Rohrlich's assessment of how all this relates to the equivalence principle. Unfortunately, there is no clear statement of the principle they are referring to, a characteristic of all the papers cited here. However, they appear to be using some version that applies to SR, so even though Fulton and Rohrlich claim to agree with Bondi and Gold's proposed solution to their problem, there must have been some confusion and it seems unlikely that Fulton and Rohrlich really grasped Bondi and Gold's problem. In any case, the above criticism of Bondi and Gold's proposed solution would then apply to Fulton and Rohrlich.

Chapter 10 covers one of the key points made in the present book: SR and GR make different predictions, in fact totally opposite predictions, about whether freely falling or supported charges will radiate. It is one of the simplest chapters, but it is relevant because it would seem that the views expressed by some of the authors discussed here are coloured by a special relativistic intuition, rather than a general relativistic one. The main point is that GR+SEP introduces an interaction between gravitational effects and other physical fields, like EM fields. This is why one may still have reason to expect a charge that is stationary relative to coordinates in which the metric has the static form to radiate in certain cases. Indeed, the case of the SHGF strongly suggests that this is what will generally happen.

Chapter 11 asks how it is that, if uniformly accelerating charges in flat spacetime (without gravitational effects) do radiate, the radiation reaction derived from energy–momentum conservation turns out to be zero? It approaches the question by looking at the derivation of the Lorentz–Dirac equation, because the radiation reaction terms turn up there. Parrott's derivation in [13] is compared with Dirac's original 1938 derivation in [14]. The discussion is descriptive. It explains how two very different inertial mass renormalisations are required to arrive at the Lorentz–Dirac equation and points out (see pp. 104 and 109) that both DeWitt and Brehme [4] and Hobbs [12] miss one of these in their extension of the Lorentz–Dirac equation to curved spacetimes, probably because they reproduce the reasoning due to Dirac.

Parrott's point in his derivation was to show that the boundary conditions on the charge motion do make a difference. There is hope for the equation when the acceleration is zero outside some bounded proper time range (scattered particle), but it becomes much less obvious in the highly idealized case of hyperbolic motion (eternal uniform acceleration). Note also that one does not have $a(-\infty) = a(\infty)$ in the latter case, so one cannot save the derivation in this way (see p. 294). This means that the Lorentz–Dirac equation cannot be applied in some cases, and this may seem to save us from having to take the zero radiation reaction seriously in the case of hyperbolic motion.

However, this radiation reaction is corroborated by self-force calculations [16] which assume the charge to have some spatial extension in any spacelike hypersurface (see pp. 105 and 110). So it does look as though one has to take the zero radiation reaction seriously in the case of hyperbolic motion, and this in turn raises the question of how energy–momentum can be conserved for this scenario. Boulware's discussion of energy flow may give some comfort in this regard, but it seems to depend on the hyperbolic motion having been eternal so that (idealistic) distribution fields build up on certain null surfaces (see Sect. 15.12).

Chapter 12 briefly describes the paper by DeWitt and Brehme [4], but only insofar as it may be relevant to the issues discussed here. This paper is highly recommended for the mathematical techniques it introduces. Their aim is to generalise the Lorentz–Dirac equation to curved spacetimes, with the result (12.1) on p. 107. One discovers that free fall is different for a charged particle and for a neutral particle, basically because the former produces EM fields that affect its own motion. In this process, the curvature of spacetime along the charge worldline directly enters the proposed equation of motion.

DeWitt and Brehme seem confused about the relevance this has for 'the' equivalence principle, and unfortunately do not state their principle clearly. Their discussion seems to be coloured by a special relativistic prejudice. What is clear though is that they agree with Bondi and Gold that a charge that is stationary relative to coordinates in which the metric is static should not radiate. Interestingly, they do not directly check this with the formulas they generate for the EM fields produced by a charged particle. They also think that there should be electrogravitic bremsstrahlung, i.e., a charged particle falling freely in a curved spacetime should radiate. However, the idea put forward in the present book is that, to a first approximation, GR+SEP contradicts both of these claims. This idea is based on a very simple observation: according to GR+SEP, which is the theory used by DeWitt and Brehme to establish all their results, there are (normal) coordinates near any chosen event on the charge worldline in which electromagnetism looks approximately as it would in flat spacetime.

The approximation here is just the approximation that one must remain close enough to the worldline for the connection coefficients to remain small enough. These grow roughly linearly with the coordinate separation from the relevant event, the constant of proportionality being roughly the value of the curvature at this spacetime locality, but in the normal coordinates for the chosen event on the worldline, the connection coefficients have been made equal to zero at the chosen event on the charge worldline. How far one can move from the chosen event on the worldline without the connection coefficients growing too big thus depends on the value of the curvature there. In the neighbourhood of this event, we are hoping to examine the EM fields for emergent radiative terms, and the strength of these depends principally on the charge of the particle. So what decides whether the connection coefficients are too big is basically the charge, source of EM effects.

The perfect situation for seeing the kind of thing that will happen in a first approximation, i.e., when the charge is strong and the curvature weak, is a situation in which the curvature is actually zero, and this is effectively a case like an SHGF. We know in this case that (12.1) just becomes the Lorentz–Dirac equation, and that the stationary motion relative to SE coordinates is a solution for the same external force that would achieve this for a neutral particle. Although there is no radiation reaction, it is established that there is radiation when the particle is charged. We also know that free fall in the usual sense of geodesic motion is a solution when there are no external forces, and that there is no radiation here. This is why the case of the SHGF can be taken as an indicator of what can be considered a first approximation in cases of small curvature.

The contribution from DeWitt and Brehme is just to show how this approximation gets perturbed when the curvature starts to grow. Of course, with curvature, interacting as it does directly with the EM fields generated by the charge, and with those fields acting back on the charge, this particle will no longer follow a geodesic, and that in turn will affect the fields it generates. These are the curvature-generated effects, which mean that a charge that is stationary relative to coordinates in which the metric has static form will not just radiate in a first approximation that is curvature-independent, but will also radiate curvature-dependent terms; and likewise which mean that a charge left to its own devices, without any external field, even though it roughly follows a geodesic and does not radiate in the first approximation, will nevertheless radiate curvature-dependent terms.

But is it possible that curvature effects could dominate the first-order expectation implied by GR+SEP (see p. 92 at the end of Chap. 10)? For example, could the radiation effect expected for a supported charge in a real (non-homogeneous) gravitational field be somehow cancelled by curvature effects? One can already guess the answer: there is no reason why it should, because EM effects depend on the strength of their sources, while gravitational effects like curvature depend on the strength of their sources, and these strengths are physically independent. Of course, the effects might perfectly cancel to give no radiation in this case, but what is evident is that they will not generally do so.

Chapter 13 pays further attention to the question of whether a charge that is stationary relative to coordinates in which the metric has static form will radiate. It is based on an argument exposed by Parrott in [7, Appendix 3]. Parrott assumes that the EM field components of the retarded field due to the charge are independent of the time coordinate, which may appear to be tantamount to assuming that there is no radiation (but see below). Even then, assuming that the retarded field solution is unique, one only shows that the magnetic field components and hence the Poynting vector are identically zero.

The catch here, which is not discussed enough in the literature, is that one ought to ask in what sense the E_i are really components of an electric field as we know it, relative to these coordinates. Because the EM field tensor is a two-form (antisymmetric, covariant, and rank two), its component matrix can always be written in the form

$$
(F_{\mu\nu}) = \begin{pmatrix} 0 & E_1 & E_2 & E_3 \\ -E_1 & 0 & B_3 & -B_2 \\ -E_2 & -B_3 & 0 & B_1 \\ -E_3 & B_2 & -B_1 & 0 \end{pmatrix},
$$

for some functions E_i and B_j on the spacetime, whatever coordinate system one chooses. Does this mean that $\mathbf{E}$ and $\mathbf{B}$ are electric and magnetic fields as we know them?

One could of course make this the definition, but it should be borne in mind that they only correspond to our usual understanding in a locally inertial frame, and then only locally, i.e., with decreasing reliability as one moves away from the spacetime event at which one has made the connection components equal to zero. This is the utility of the strong principle of equivalence: physics looks roughly as we know it in locally inertial frames. Without this principle, one could not even begin to interpret the constructions one has made on these manifolds. The weak and strong principles of equivalence taken together allow one to interpret motion relative to arbitrary coordinates, and also to interpret other fields defined relative to arbitrary coordinates.

It is worth remembering that one could say nothing about the link between the spacetime manifold and the spacetime observed without having principles like these. Without SEP, one would require some other principle that made the link. To understand F as given above, one needs to transform to normal coordinates at the event one is interested in and then look at the components of F. This problem is illustrated by the debate over whether the $\mathbf{E}$ we have found here is a Coulomb field, as discussed so decisively in [7] and relayed in Chap. 13. Of course, there is absolutely no reason to expect it to be a Coulomb field in any possible sense. In fact, in semi-Euclidean coordinates for an observer comoving with an accelerating charge in flat spacetime, the EM fields generated by the charge do not look like a Coulomb field (see p. 231), and even if they did, the thesis put forward here is that this would mean nothing in itself. What is more, they are obviously not going to look like a Coulomb field in the global inertial frame which happens to be available in this case.

Finally, with regard to the proof in Chap. 13, even the fact that the EM fields are independent of the time coordinate in such a context is not actually tantamount to assuming that there is no radiation, since we have in the uniformly accelerating SE coordinate system a relatively simple example where this happens, with zero magnetic field components and hence zero Poynting vector, but there is radiation, as explained in Sect. 16.8. What goes wrong is just that staticity is coordinate dependent: what is static relative to one coordinate system may not be static relative to another.

Chapter 14 discusses a paper by Mould [5] which claims to describe (entirely theoretically) a way of detecting whether point charges are radiating EM energy. This paper deals with precisely the kind of issue just raised: how would one link the theoretical energy flux with something that could be measured in the real world? Unfortunately, it is not easy to understand how Mould's detector works, because at the critical stage in the description of its construction, there are no diagrams. However, we discover that his detector has the following attributes:

- When comoving with a uniformly accelerating charge in a flat spacetime with no gravitational effects (not even non-tidal ones), or indeed, when comoving with a charge that is supported (stationary) relative to an SHGF, it does not register any radiation.
- When freely falling in an SHGF in which there is a stationary charge, it does register radiation.
- When the charge is freely falling in the SHGF and the detector is supported (stationary) relative to the SHGF, the detector registers radiation.

Here we clearly have a definition of radiation to suit the prejudices circulating in this debate. The last feature is perhaps the most striking, since it means (by the mathematical equivalence of the theories) that a stationary charge in a flat spacetime without even non-tidal gravitational effects will appear to radiate if one is accelerating. So we have the extraordinary idea that this detector would pick up photons from a Coulomb field in flat spacetime by accelerating through it. This kind of radiation is clearly a relative phenomenon, in fact a detector-dependent phenomenon.

It is worth reflecting a little further on this question: Can a Coulomb field in Minkowski spacetime look like a radiating field to an accelerating observer? To answer that, one ought to choose some frame for the latter, but no such frame is laid down by the theory. The situation here is different from SR, because when the observer is inertial, a preferred frame is in fact laid down by the theory. But in GR one is stuck with all frames, a paradoxically unlucky situation. In reality, the best thing the accelerating observer could do would be to get hold of the description relative to some (locally) inertial frame. The fact that one can find semi-Euclidean coordinates adapted to the motion of the accelerating observer means nothing. They have no particular relevance according to GR, or even according to GR+SEP. The theory says precisely that no coordinates have any particular relevance. There are no favoured coordinate systems. The best we can do, with the help of SEP, is to refer to a locally inertial system. For this reason, one ought to be suspicious of glib calculations in accelerating frames. That is one of the main themes of the present discussion.

However, it is not difficult to find arguments against this point of view and the debate here ought to continue. One obvious point is that it may be a good idea for some purposes to define radiation for accelerating observers, in the sense that one's detectors are bound to be accelerating in some situations. However, there does seem to be a risk that this definition depend on the coordinates chosen by that observer. The problem here is that there are no canonical coordinates for such an observer, although one can find a whole class of quasi-canonical coordinate systems with certain good properties (see Sect. 2.4). It is not obvious that such a concept would be useful. On the other hand, even locally inertial systems at a given event are not unique, and not only because there are infinitely many locally inertial systems related by Lorentz boosts. But one thing is certain: it is not necessary for the assessment of WEP and SEP, and if it is purely motivated by the desire to save an obsolete version of the equivalence principle which speaks about what can or cannot be distinguished then it serves no purpose.

Another point is that a real physical detector that always gives the right measurement results when moving inertially might behave as Mould claims when accelerating. This would give a physical reason for defining radiation as he suggests. It may be that the Mould detector does this. But one does need to consider the possibility that other real physical radiation detectors of a different construction, which always give the right measurement results when moving inertially, might give different results to Mould's detectors when accelerating. Presumably, Mould's thesis is that his theoretical basis for the detector described in [5] is general enough to be all-encompassing, so that this could not happen. But then, since this result is not necessary for the assessment of WEP and SEP, and not implied by them, these results would suggest a kind of conspiracy between electromagnetism and gravity that is really not required by any standard aspects of these theories.

The claim being made in the present book is not therefore a complete rejection of Mould's work, only a criticism of its relation to the question of equivalence principles. The conclusion here is that Mould's detector is unable to resolve Bondi and Gold's paradox and is in fact irrelevant to the whole issue.

The long Chap. 15 deals with the Boulware paper, often quoted as definitive with regard to the questions discussed here. Boulware does not claim that a charge that is stationary relative to coordinates in which the metric is static does not radiate, only that an observer who is also stationary relative to these coordinates will not be able to measure any radiation there is. We find that the radiation escaping over this observer's event horizon is not the only thing preventing her from measuring it, according to Boulware. The fact of being restricted in a certain way to a certain kind of data from the region of spacetime within that event horizon also plays an essential role for this author.

The reason why Boulware needs to demonstrate that the comoving observer cannot observe the radiation seems to come from a version of 'the' equivalence principle which says that two situations should be indistinguishable, but one can only hypothesise about this, because he does not explain why explicitly, just as there is once again no clear statement of the principle he applies. The loose statement he supplies says that a uniformly accelerating frame must be indistinguishable from a gravitational field. The hypothesis proposed here to explain why Boulware tries to show that an observer comoving with a uniformly accelerating charge cannot observe its radiation is as follows: Boulware still believes Bondi and Gold's claim that, in the case of a gravitational field, a charge that is stationary relative to coordinates in which the metric is static cannot radiate, and by the above indistinguishability (referred to loosely as 'the' equivalence principle), concludes that an observer co-accelerating with the uniformly accelerating charge (in flat spacetime with no gravitational effects) should not detect any radiation.

The problem is this: strictly speaking, the above indistinguishability actually implies that, if in the case of a gravitational field a charge that is stationary relative to coordinates in which the metric is static cannot radiate, then there really is no radiation for an observer co-accelerating with the uniformly accelerating charge. It is not just that the observer would be unable to detect it. What creates all the difficulties is of course the fact that the uniformly accelerating charge is known to radiate according to the theory. At least, it radiates in the inertial frame, and Boulware confirms that he accepts this. The present view is that radiation has to be defined relative to locally inertial frames where SEP allows a physical interpretation. (If one hopes to say that it does not radiate in non-inertial frames, then one has to define what is meant by radiation relative to such frames, in the way attempted by Mould.) So we still have Bondi and Gold's paradox, assuming that one can uphold their opinion that, in the case of a gravitational field, a charge that is stationary relative to coordinates in which the metric is static cannot radiate. In any case, Boulware does not resolve the original paradox by claiming that the co-accelerating observer would be unable to detect the radiation. He merely demonstrates once again the risk of using versions of 'the' equivalence principle that refer to indistinguishability.

Boulware's account is mathematically quite sophisticated, and a large part of Chap. 15 is concerned with spelling out the details of his approach, but it is suggested here that there are some very subtle problems with the interpretation. This must be a recurrent theme in theoretical physics these days, where mathematical descriptions have become so abstract and little time is wasted describing how they are supposed to relate to the real world. The problems in this specific case begin in the highly mathematised description of the static gravitational field, described here in Sect. 15.1. One obtains a metric but there is no attempt whatever to say how it should be interpreted physically, and Boulware duly falls into a trap that awaits anyone who has not noticed that, if there is a link between this metric and a real gravitational field, then that field is not just static but also homogeneous. As discussed in Sect. 15.3, he claims that a co-accelerating observer attributes the strange behaviour of light to the extremely strong gravitational field which produces a future event horizon. The problem here is to reconcile the extremely strong gravitational field with a static and homogeneous gravitational field in which there are no tidal effects (zero curvature).

So how does the observer explain this event horizon and this infinite redshift if she considers herself to be supported in some SHGF? Viewed from the global inertial frame, the event horizon of the uniformly accelerating observer is due to the fact that her speed approaches the speed of light asymptotically as $\tau \to \pm\infty$. The redshift is just a Doppler shift, when viewed in the inertial frame, getting bigger without limit as the particle speed approaches the speed of light. Section 15.6 presents another way of explaining the redshift which is intended to reveal the origins of Boulware's perhaps controversial interpretation. It is basically a problem of interpreting the semi-Euclidean coordinate system and it draws attention to one of the main questions in general relativity which is almost never discussed: how do we link the coordinate systems of the theory, and the corresponding components of tensors representing physical quantities, with things that can be measured in the real world?

Sections 15.3–15.5 discuss the geometry of the SE frame as a preparation for the redshift analysis in Sect. 15.6. The alternative redshift analysis, in which the manifold is described by SE coordinates, is just the kind of analysis used in cosmology, and a parallel is drawn with the Milne universe. In the cosmological view, where the metric attributes a hyperbolic geometry to the spatial hypersurfaces of the Milne universe, one has the same kind of cosmological redshift as for any other hyperbolic Robertson–Walker universe, but in the Minkowski view (and the Milne universe is the only RW universe that is actually a flat spacetime), the redshift is just the ordinary Doppler effect due to relative motion of source and observer.

So infinite redshifts can be explained in two different ways:

- Viewed from a global inertial frame, they are basically due to the fact that the origin of the SE frame asymptotically approaches the speed of light. If one considers a series of emitters $G_1, G_2, \dots, G_n$, etc., at fixed values $y_1^1, y_2^1, \dots y_n^1$, etc., of the SE coordinate, approaching the forbidden point where the SE metric is singular, the redshift observed by the uniformly accelerating observer O at the origin of the SE frame depends only on the fixed value y_n^1. The point is that, following the sequence of galaxies toward the singularity, the difference of speed between G_n and O at the event when G_n emits its signal and the event when O receives it is constant for a given emitter G_n, i.e., independent of when the emission event occurs, but that this difference of speed increases without limit as n increases, i.e., as G_n approaches the singularity.
- Viewed from the SE frame itself, one imagines a series of galaxies sitting fixed at values of y_{gal}^1 that approach the point $y^1 = -c^2/g$ where the metric components become singular in these coordinates, with these galaxies sending light signals to the observer at the origin of the SE frame. But in order to sit fixed at these values, the galaxies have to have a series of (uniform) 4-accelerations that tend to infinity, as can be seen from (15.36) on p. 176. And it is the fact that the factor $(1 + gy_{\mathrm{gal}}^1/c^2)$ goes to zero that makes the proper time interval (15.59) between emissions go to zero and leads to ever larger redshifts. But something has to provide these galaxies with these accelerations. It is pure assumption to suppose that they arise merely because each galaxy is held up against a gravitational field.

> In fact, there is no explanation for how they manage to move like this. There is no explanation for the way any of the other points at fixed values of y^1 should be able to move like this. One solution to the paradox is to observe that the spacetime is not necessarily populated by objects like this.

Here is a situation where it is crucial to pay careful attention to the question raised above, about how a coordinate view should be related to real observations.

This question of how we interpret a set of coordinates and a metric, leads to a more pressing issue in the case of an SHGF, which illustrates something quite fundamental about general relativity: If we are given a metric of semi-Euclidean form in which the origin of the coordinate system has uniform acceleration g, then we are dealing with a static and truly homogeneous gravitational field (an SHGF), but on what grounds could we conclude that the 'acceleration due to gravity' is everywhere equal to g? This is certainly one possible interpretation, but it is not the only one.

There remains a clear ambiguity in the application of GR due to the fact that this spacetime is flat. There can be no intrinsic way to extract the specific value of g that turns up in the chosen coordinates. Indeed, we know that, starting from the flat spacetime and choosing any uniformly accelerating observer, with any value of g, we can always find some coordinates for the spacetime in which g turns up in the metric. At the same time, we would like to say that it is the flatness of the metric that rules out the possibility of a strong gravitational field in some zones and a weaker one in others, because such a field would involve tidal forces that would show up in a nonzero curvature. Whatever the value of g we eventually accept, we would like to insist that it be the same everywhere in the spacetime. Referring to the discussion in Sect. 2.4, we might argue that, insofar as we can apply the usual approximations that link Einstein's equation with Poisson's equation for a gravitational potential, these approximations lead to a gravitational potential of the type gz, which implies a constant 'acceleration due to gravity' of g everywhere.

In the end it is not so worrying that one can merely change coordinates and obtain a different value of g, since one should be able to explain this as viewing the original situation from an accelerating frame. How much of the acceleration is considered to be gravitational and how much due to frame motion is an inevitable ambiguity in a theory which was partly inspired by the idea that any inertial acceleration can be confused with a gravitational acceleration. With this argument, it is only when one can examine the source of the gravity that one can make this breakdown. On the other hand, looking back at Sect. 2.4, we did in fact fail to identify an energy–momentum distribution that would generate these metrics via Einstein's equation (apart from the zero distribution).

So one might imagine the following line of argument that opposes the simple idea that the value of g should be the same everywhere in spacetime. As an observer in this world, it would be natural to consider the views of observers placed at constant proper distances. Each of these would come up with a different value of g, and without other landmarks in the spacetime, such as a source of gravity, nobody would know whose value of g to accept, and this despite the fact that they would all measure zero curvature. Democratically, one would have no way of distinguishing one particular value. Although this may seem a sorry state of affairs, it does look as though one could argue this way in the very ideal case of the SHGF.

In the present view it is nevertheless a poor strategy to accept the idea that the gravitational acceleration may be stronger in some places than others, whatever that may mean. It is better to consider that the SHGF is a good approximation to some real spacetime with well-defined gravitational sources, relative to which distances can be assessed. One then simply remembers that the fact that a four-acceleration a^μ is required to hold an object at some fixed coordinate position does not mean a priori that a^μ is a measure of the gravitational field. The curvature gives an indication of the gravitational field when there are tidal effects, but when the gravitational field is non-tidal, we have to look at the sources.

This view perhaps looks more convincing if we reconsider the semi-Euclidean coordinates set up by an observer with uniform acceleration in a Minkowski space-time without gravitational effects. Such an observer adopts the planes of simultaneity of instantaneously comoving inertial observers in order to attribute her own instantaneous proper time to spacelike hypersurfaces and the space coordinates of those inertial observers. This leads to a remarkable situation which is unique to the hyperbolic motion of a uniformly accelerating observer. When we consider the motion of a point with a fixed value of the spatial coordinate in the semi-Euclidean frame, it also turns out to be uniformly accelerating, and its worldline could be the worldline of another uniformly accelerating observer, but with a different value of the acceleration (and a different Minkowski spatial coordinate at Minkowski time zero). Furthermore, such an observer, located at this specific point relative to the first, would always agree with the first about the simultaneity of pairs of events. In other words, their planes of simultaneity coincide, although they attribute different times to them (their own proper times). Moreover, if the new observer set up her own semi-Euclidean coordinate system, with her own acceleration, it would be related to the first semi-Euclidean coordinate system by a simple transformation, merely scaling the time coordinate by a constant factor and shifting the spatial origin.

In this context, it would be easy to look at the worldlines of points at different values of the space coordinate ζ along the axis of acceleration in the coordinates with non-Minkowski metric and point out that each has a different uniform acceleration, claiming as a consequence that there is a non-uniform gravitational field which turns out to be infinite at certain values of ζ. The trouble with this is that the 4-accelerations of such points are irrelevant. Nobody ever said that such points represented actual motions, let alone the motions of bodies supported in a gravitational field. What the coordinates do relative to the Minkowski frame is irrelevant, because they are just coordinates, and one can do anything with coordinates. A better policy in this non-gravitational Minkowski spacetime is thus to refer to a global inertial frame to understand what is happening. Then in the SHGF case, referring in the same way to the happily available global inertial frame, one should select one uniformly accelerating observer as supported relative to whatever source one can observe, in the sense that her four-acceleration corresponds to the strength of the gravitational field.

In the fully ideal case where there is no source and the SHGF permeates the whole of spacetime, the theory of GR would appear to be mute. Such gravitational fields without tidal effects have zero Riemann tensor and that is all we can ascertain. In fact it is in this sense that the theory implements the well-advertised idea that a gravitational effect and an inertial acceleration cannot be distinguished, at least when the gravitational effect is static and spatially homogeneous. But this does not mean that different observers held fixed at different spatial coordinates should be allowed to disagree about the gravitational potential in the Newtonian approximation.

Section 15.2 also describes how Boulware relates his mathematically obtained static gravitational field with Minkowski spacetime, through four maps of the original spacetime to four different regions of Minkowski spacetime. He describes this as the analytic continuation of the accelerated frame from region I to the three other regions (see p. 163 for the definitions of the four regions). However, what one actually has here are four coordinate charts on Minkowski spacetime, whose images happen to be the image of the coordinate chart on the manifold Boulware found for the static gravitational field. There is no way that this can be construed as an analytic continuation of something, or even as the natural extension of the SE coordinates to regions I, II and III, as Boulware claims elsewhere. This is significant because what Boulware wants is a proof that there is no time-independent coordinate system for region II in which a uniformly accelerated charge in region I is at rest. But the fact that one has found coordinates on region II in which the Minkowski metric takes on a non-static form is a long way from proving this.

Later one discovers why Boulware needs this unproved theorem: radiation can be detected in region II. If one had a coordinate system covering regions I and II relative to which the charge were stationary and for which the metric components had the static form, one would be back on the dangerous territory espied by Bondi and Gold. The problem is that Boulware's maps do not give a coordinate system for the combined region. The form of the metric components is even discontinuous across the hypersurface $Z = 0$ separating regions I and II, where the metric is actually degenerate in this representation, and the proposed map from the union of regions I and II to $\mathbb{R}^4$ cannot be a coordinate chart in an atlas because it is many–one. If Boulware hoped to justify finding radiation in region II, his whole argument would appear to be hopelessly unfounded from a logical and mathematical standpoint.

Of course, the claim that there is no time-independent coordinate system for regions I and II in which a uniformly accelerated charge in region I is at rest may actually be true and it could be interesting to try to prove this for its own sake, but this has not been attempted here because in the present view the whole motivation for doing so cannot be justified.

Section 15.8 tries to identify some of the causes of Boulware's misinterpretations of the theory. It brings out once again the idea that he thinks there are real bodies at fixed spatial positions in the SE frame and that they have these motions merely by being supported against the gravitational field. It also discusses his claim that it is not possible to find a single static coordinate system in which bodies at rest at different points undergo the same proper acceleration, because two bodies experiencing the same proper acceleration do not maintain the same proper distance. This is actually false, if one is allowed to have curvature. Boulware merely omitted to mention that he was insisting on the spacetime being flat. At the same time, this suggests that he was not aware of the link. Even in his mathematical derivation of the metric for a static gravitational field, he assumes that the spacetime is flat without comment or justification. He is clearly concerned by the fact that the fixed spatial points in Rindler coordinates do not actually have the same 4-accelerations, because intuitively one would expect this model to describe a spatiotemporally constant gravitational field (precisely because the spacetime is flat). We note that he describes the gravitational field as static, but never mentions that it must also be homogeneous.

Section 15.9 obtains the EM field components in the Minkowski frame following Boulware. All the details are spelt out, especially the logic of the argument leading to the fields on the null cones at the origin, which are distributions, with a full check that they do satisfy all Maxwell's equations, in particular the source-free equations. The Minkowski components of the EM fields due to the uniformly accelerating charge are given on p. 216.

Section 15.10 discusses Boulware's interesting account of how the distribution terms in the EM fields arise on the null cones: he considers a charge that was initially free, i.e., it has not always undergone uniform acceleration, and carries out bigger and bigger Lorentz boosts which have the charge coming down the z axis toward the origin with ever greater speed in the initial constant-velocity phase. In the limit as the boost goes to infinity, one retrieves the EM fields of the eternally uniformly accelerating charge. The details of the Lorentz transformation process are spelt out in detail here. The idea was originally suggested by Bondi and Gold [2], but they chose a more sophisticated limiting process to explain the fields on the null cones. What is shown is that the delta function terms in the retarded fields on p. 216 can be thought of as arising in the above limit from the Coulomb field of the charge before it began its acceleration, but viewed from a fast-moving frame. In a way, this feature indicates how idealistic this situation is, not just because a uniform acceleration is already idealistic, but because of the symmetry in time and the fact that it started infinitely long ago and will continue forever.

The derivations of the EM fields are interesting in themselves, illustrating just how sophisticated the problem is, but the key issues for the present discussion appear in Sect. 15.11, where we analyse Boulware's interpretation of the fields. It is shown that the retarded fields obtained in region I are identical to the advanced fields, another oddity of the ideal scenario under investigation. It is also shown how the fields along the forward light cone of a point on the charge worldline can be analysed into Coulomb and radiation terms in the standard way, by their dependence on $1/r^2$ and $1/r$, respectively, where r is what Dirac called the invariant retarded distance [14], i.e., the radius at the field point of the light cone centered on the retarded point in the inertial frame for which the charge is instantaneously at rest at the retarded point. For what it is worth, one could interpret the field at any point in region I either as the Coulomb field plus outgoing radiation field of the charge at the retarded time, or as the Coulomb field plus ingoing radiation field of the charge at the advanced time.

There is a connection with Dirac's redefinition of the radiated field $F_{\text{rad}}^{\mu\nu}$ as the difference between the retarded and advanced fields [14]. This is clearly zero in region I for the uniformly accelerated charge. It shows why the Lorentz–Dirac equation for the charge motion, including an incident (external) EM field that drives the motion, does not include radiation reaction terms (see p. 235). For a general motion of a point charge, subtraction removes Coulomb terms and one obtains a field that is finite on the charge worldline. Another key thing about $F_{\text{rad}}^{\mu\nu}$ is that it does usually give just the usual radiation field from the retarded field $F_{\text{ret}}^{\mu\nu}$ in regions where one is likely to measure the radiation field.

The advantage of this new definition (15.243) on p. 235 is thus that it provides a definite value for the field of radiation throughout spacetime, in particular, giving a meaning to the radiation field close to the charge. The usual theory gives a field that is inextricably mixed up with the Coulomb field. The disadvantage is that this result also attributes a meaning to the radiation field before its time of emission, when it can have no physical significance. Dirac merely says that this is unavoidable if one is to have a well-defined radiation field near the charge. The disadvantage of the definition (15.243) in the present case of a hyperbolically moving charge is undoubtedly that it gives zero for the radiation field everywhere, whereas one can see from results like (15.240) on p. 233 that there is in fact a radiation field. However, it is indeed $F_{\text{rad}}^{\mu\nu}$ that occurs in the renormalised equation of motion (15.244) on p. 235, which thus reduces to the usual Lorentz force law without radiation reaction.

On p. 236, we come to what Boulware considers to be the crux of the matter, where we look seriously at the supposed radiated field in region I and try to decide whether it could be detected experimentally for this uniformly accelerated source charge. It is claimed that:

- no limit can be taken such that the radiation field is large compared with the Coulomb field,
- the field expressed in terms of the distance to the worldline of the charge does not drop off as $1/l$ but rather as $1/l^2$, where l is what Boulware calls the invariant distance from the field point to the worldline, viz., the distance in the instantaneously comoving inertial frame for the unique comoving observer who considers the field point to be simultaneous (see p. 238).

In fact, what we have in the first claim is not a bound on the radiated field, but a bound on the distance at which one may measure the radiated field whilst remaining within region I, i.e., a bound that is particularly relevant only to an observer comoving with the charge. We have to go into region II, above the plane $z = t$, to appreciate how the radiated field outgrows the Coulomb field. So there is radiation in region I, but it will be impossible to detect if one is looking for a field that rapidly outgrows the Coulomb field. The second claim expresses field components relative to one frame in terms of a distance measured in another and it is not clear that this would really confound our comoving observer.

One can see where this kind of argument is leading. One is trying to convince oneself that, even though there is a radiation field, it might be difficult for an observer comoving with the charge to find out about it. In a way, one is wishing that the radiation field were not there. However, difficult does not mean impossible.

As mentioned above, the present interpretation of what Boulware is trying to do is as follows. He believes Bondi and Gold's claim that, in the case of a static, homogeneous (non-tidal) gravitational field, a charge that is stationary relative to coordinates in which the metric is static cannot radiate. By application of a loosely stated version of the equivalence principle, the situation for an observer comoving with a uniformly accelerating charge and using the semi-Euclidean coordinate system (for which the Minkowski metric is static) must be indistinguishable. The word 'indistinguishable' is taken to allow some leeway over the presence or otherwise of a radiation field. Rather than meaning that there cannot be a radiation field in some objective sense, it is taken to mean that there might be a radiation field, but that the comoving observer must not be able to detect it.

As noted already, there are two difficulties with this stance:

- There is no particular reason to concord with Bondi and Gold on their opinion that, in the case of a static, homogeneous (non-tidal) gravitational field, a charge that is stationary relative to coordinates in which the metric is static cannot radiate.
- The word 'indistinguishable' cannot be allowed to give this leeway. Strict application of WEP and SEP tells us that there could not be radiation, in an objective, observer-independent sense, for a uniformly accelerating charge in a flat space-time without gravitational effects, if one accepts Bondi and Gold's opinion stated in the first point.

This therefore serves also to illustrate the dangers of using a version of the equivalence principle that says whether things can be distinguished or not.

Finally, there is another serious argument against Boulware's conclusions. An observer moving with the charge could presumably still carry out careful measurements and discover the form (15.240)–(15.242) of the fields on p. 233 in some neighbourhood. She would then deduce that there was radiation, even despite the results of Sect. 15.11.4. Once the fields are known in the neighbourhood of the source, presumably there is a unique analytic continuation of these fields, i.e., one could detect the fact that they were going to look like radiating fields further away. This is no problem for the view presented in the present book because we are not constrained by the rather arbitrary opinion voiced by Bondi and Gold that a charge that is stationary relative to coordinates in which the metric is static cannot radiate.

It is when Boulware discusses the situation in region II, that one realises all the strangeness of his arguments (see Sect. 15.11 on p. 229 ff). The first problem is that we do not appear to have an analytic continuation of the accelerated frame, whatever that might be, to region II. If we take the coordinate transformations from the Minkowski coordinates on regions I and II, as given by the inverses of the first two relations of (15.11), we find that each point of the (τ, Z, x, y) space maps to one point in each region and that no points map to the horizon $t = z$ separating the two regions. This can hardly be described as an analytic continuation of the semi-Euclidean coordinates in region I. Perhaps there is such a continuation, perhaps there is not. It is difficult to see how one could continue beyond the horizon $t = z$, since it is given by $Z = 0$ in these coordinates, and the metric components relative to these coordinates are then singular. But even supposing that one could contrive some coordinates that coincide with the SE coordinates in region I and extend also over region II, would the fact that the charge happens to be at rest in the spatial hypersurfaces of these coordinates (in region I) really allow us to make any deductions about what is happening in region II? Would those coordinates in region II really describe the view of some observer, viz., the observer moving with the charge? For that matter, do the SE coordinates in region I really describe the view of any observer? Coordinates have no a priori physical significance. We face this question again in Chap. 16, because it is shown that, when the definition of the Poynting vector for the fields is extended to the SE coordinates, this quantity is identically zero, and one wants to interpret this in terms of what the hyperbolically moving observer will observe.

At least it is clear what Boulware is trying to do here: he aims to show that there is radiation from the uniformly accelerating charge, that it cannot be detected in region I by an observer moving with the charge (in order to save Bondi and Gold's opinion that a charge that is stationary relative to coordinates in which the metric is static cannot radiate), but that it is really there anyway because it can be detected in region II by an observer moving with the charge (where there is no constraint due to Bondi and Gold's opinion because the metric turns out not to be static for this same observer in region II). This statement shows how odd the whole argument really is: the observer is not actually in region II but these are her coordinates. There is even the idea that she has no choice over her coordinates, because the analytic continuation of the SE coordinates to region II given in (15.11) (if indeed it could be interpreted in this way) is 'the' analytic continuation, and by implication unique. But who said the observer had to use the SE (Rindler) coordinates in region I? Surely she could use any coordinates she liked. One cannot draw conclusions because of coordinates.

Therefore the main criticism in this chapter is that there is no reason to think that the coordinates τ, Z, x, y proposed for region II are in any sense coordinates in which the hyperbolically moving charge is stationary, so the fact that the Minkowski metric in region II turns out to have non-static components relative to these coordinates is really irrelevant, even if one wanted to save the competing claims that the uniformly accelerating charge is actually radiating and a particle at rest in a static gravitational field cannot radiate, or cannot be seen to radiate by an observer using coordinates in which the particle is at rest and the metric components have static form.

Boulware makes a final point in his comments concerning region II, namely the observation that the fields in region II are invariant under a certain transformation of the data, viz., replacing the charge e by a charge $-e$ and having it follow the mirror-symmetrical worldline in region III. This can be seen by a suitable examination of the fields in Sect. 15.9.2. The conclusion from this is that no measurements restricted to region II can distinguish between the Coulomb plus outgoing radiation field of a uniformly accelerated charge e in region I and the Coulomb plus outgoing radiation field of a uniformly accelerated charge $-e$ in region III.

So observers in region II will proclaim the presence of the radiation but they will be unable to determine where it came from. Indeed, we have to look in other regions of spacetime to find the source. Boulware gives us three things that could signal the location of the charge in region I:

- the presence of the field in region I,
- the absence of the field in region III,
- the delta function field along the surface $z + t = 0$, rather than along the surface $z - t = 0$.

He appears to have overlooked another possibility: we could just look at the charge.

It is not at all clear what one is supposed to deduce from this. However, it does once again raise the question of what is meant by an observer 'in' region II, or even 'an observer'. Who are these people? Are they just sets of coordinates and components of tensors relative to these coordinates? Are they people who refuse to communicate or cannot communicate with other observers in other regions? And if they will not or cannot communicate, does that really matter for a theory that purports to model spacetime?

Looking back over Boulware's paper as a whole, it looks as though he has transformed the original problem raised by Bondi and Gold. Instead of seeing a difficulty for any version of the equivalence principle, he seems to have taken Bondi and Gold's idea that a charge that is stationary relative to coordinates in which the spacetime metric has static form cannot radiate and, noting that this would appear to be false in the case of an SHGF, he then proceeds to try to save it by showing that an observer moving with the charge would never be able to detect the radiation. Not only do the arguments used along the way fail, but the very principle of his attempt is untenable. Let us deal with the latter point first.

The problem is that it is really irrelevant what a comoving observer would be able to detect. The equivalence principles WEP and SEP that are essential to the whole theory, and which he abundantly uses, imply that since the uniformly accelerating charge does radiate, the stationary charge relative to the semi-Euclidean coordinates for a spacetime containing an SHGF must also radiate. So Bondi and Gold's idea is actually wrong, unless one rejects SEP or declares it inapplicable, as they rightly pointed out in their original paper [2]. Indeed, Boulware accepts that there is radiation and hence that Bondi and Gold's idea is wrong. He does not need to show that a comoving observer will not detect radiation, and indeed, it would make no difference whether she could detect it or not, since we agree that it is there.

Of course, it may still be an interesting question in its own right. However, Boulware's arguments are inadequate to show this. The first stage of the argument, that the comoving observer should only have access to data in region I, seems valid enough, since no information can ever reach this observer from the other region affected by EM fields sourced by the charge, viz., region II. It does make the whole attempt look a little odd, however, because one knows very well that radiation can be detected in region II and it seems rather strange to think that such a fundamental theory as this should be dependent in any way on the fact that one particular observer is partly isolated from most of spacetime by an event horizon.

In any case none of the three arguments put forward to show that the comoving observer cannot detect radiation within region I is actually valid:

- The first argument points out that the ratio $|\mathbf{E}_{rad}|/|\mathbf{E}_{Coulomb}|$ of the radiated field to the Coulomb field does not go as r in the region accessible to the comoving observer. This was discussed in Sect. 15.11.4, in particular (15.250) on p. 237. The problem is that the observer need in no sense depend on such a simplistic criterion as this to decide whether there is radiation or not. In particular, the observer does not have to examine the fields a long way from the charge worldline in order to determine how the fields are going to evolve.
- The second argument points out that

$$|\mathbf{E}_{rad}| \sim \frac{e}{4\pi}\frac{2\sin\theta}{l^2} \, ,$$

where l is called the invariant distance from the field point to the worldline. This is also discussed in Sect. 15.11.4, in particular (15.257) on p. 240. It is interpreted to mean that the radiated part of the field seems to vary like a Coulomb field with distance from the source. This argument comes to grief in the same way as the last. The operative word is 'seems', but there is no particular reason why the observer should be duped, especially if she can make accurate measurements nearby. Indeed, it seems highly probable that the observer could accurately predict the fields even into region II on the basis of sufficiently accurate measurement of the fields in the close neighbourhood of the charge worldline. This too reveals something of the oddity of the idea that the limitations on what one particular observer can measure might be relevant.
- The third argument, often cited as definitive proof that the comoving observer could not detect radiation, or even that there is no radiation, is the fact that the generalisation of the Poynting vector to the semi-Euclidean coordinate system is identically zero. This was shown in Sect. 15.12.1 on p. 246. However, no attempt is made to establish the physical meaning of this generalisation of the Poynting vector to coordinates other than inertial coordinates. In fact, it does not give an energy flow when integrated over a spacelike hypersurface.

The last item is discussed in greater detail in Chap. 16, in the light of Parrott's criticism [7].

To end this rather bleak conclusion, let us recall briefly once again the present explanation for what motivated Boulware to try to show that a comoving observer would not detect the radiation from the charge. The present idea is that Boulware believes Bondi and Gold's claim that, in the case of a static, homogeneous (non-tidal) gravitational field, a charge that is stationary relative to coordinates in which the metric is static cannot radiate. He then applies a version of the equivalence principle in which the situation for an observer comoving with a uniformly accelerating charge and using the semi-Euclidean coordinate system (for which the Minkowski metric has static form) must be indistinguishable. However, he takes the word 'indistinguishable' to mean that there might be a radiation field, but that the comoving observer must not be able to detect it. The view in this book is that, if one accepts Bondi and Gold's opinion that a charge that is stationary relative to coordinates in which the metric is static cannot radiate, in the case of a static, homogeneous (non-tidal) gravitational field, then strict application of WEP and SEP tells us that there could not be radiation in the case of a uniformly accelerating charge in a flat spacetime without gravitational effects. In the present view, this should be taken as a warning to those using a version of the equivalence principle that says whether things can or cannot be distinguished.

Chapter 16 discusses the paper [7] by Parrott. The main achievement is to throw doubt on Boulware's paper [6] by showing that a zero Poynting vector in some arbitrary (non-inertial) coordinate system does not mean that there is no EM radiation. Despite Parrott's rather critical discussion of Bondi and Gold's idea that a charge that is stationary relative to coordinates in which the metric has static form cannot radiate (relayed in Chap. 13), he does seem to concur with it. He does not seem to have noticed the real threat this poses to SEP in the case of an SHGF, and unfortunately he does not state what version of the equivalence principle he is referring to when he discusses this issue.

Since Parrott disposes of one of Boulware's arguments against EM radiation being detectable by a co-accelerating observer (the Poynting vector argument), it may even be that there is a self-contradiction, for here we have a case of a charge that is stationary relative to coordinates in which the metric has static form and which is actually radiating. As discussed above, however, this depends on the version of the equivalence principle he is using, because he may believe like Boulware that it is enough to show that the charge does not appear to be radiating to the comoving observer.

The present book supports Parrott's view that purely local experiments can distinguish radiating from non-radiating situations (at least in principle), in particular, experiments using small rockets to accelerate neutral or charged particles, but it does not support the claim in Parrott's abstract for [7] that such experiments would actually distinguish a stationary charged particle in a static gravitational field from an accelerated charged particle in a (gravity-free) Minkowski space. This is because, as pointed out by Bondi and Gold at the beginning of this debate [2], such a distinction between the two scenarios would contradict GR+SEP. However, the present book does support Parrott's view that Boulware misidentifies energy with his Poynting vector argument (see Sect. 16.8).

Any argument that says there is objectively no radiation from the charge that is stationary relative to the usual SE coordinates for the SHGF, or that there might be radiation but that a comoving observer would not observe it, or that the existence or otherwise of radiation can be usefully defined in a coordinate-dependent way in this context, is denied in the view presented in the present book. Parrott disposes of one of Boulware's arguments to support the idea that there is radiation but that a comoving observer would not observe it, but he then implicitly claims that the stationary charge relative to these standard SE coordinates for the SHGF will not radiate, because he expects his charged rocket to distinguish this particle from an accelerated charged particle in a gravity-free Minkowski space.

Parrott clearly agrees with the idea that one can say a priori without further discussion, as everyone seems to, that a charged particle held stationary relative to Schwarzschild coordinates in the Schwarzschild spacetime will not radiate. Why is there never any proof of this? What about the rocket holding a charge in position relative to the semi-Euclidean coordinates in an SHGF? By an application of SEP *par excellence* (because the local inertial frames happen to be global), this rocket will also have to do more work than one that was just holding a neutral particle. That is what the theory says anyway, because the mathematics is strictly identical to the case of the gravity-free Minkowski spacetime with the charged or neutral particles being uniformly accelerated. Then one does agree with Parrott that the idea of the little rocket doing the necessary acceleration in each case is a perfect local test of whether there is radiation. The disagreement is with the claim that one would distinguish the two cases. The theory (with SEP) definitely says that one would not.

The Schwarzschild spacetime is then another situation and a priori one has no reason to suppose that the charge held stationary relative to the Schwarzschild coordinates will not be found to radiate in the sense detectable by the little rocket. After all, they are only coordinates, even if they happen to be coordinates relative to which the metric assumes what we like to call a static form.

Section 16.1 discusses the way Killing vector fields can be used, in those spacetimes symmetric enough to possess any, to define conserved vector quantities from conserved energy–momentum tensors. Section 16.2 then discusses a Killing vector field in Minkowski spacetime that is associated with the symmetry of the Minkowski metric under Lorentz boosts. The point noted by Parrott is that the conserved vector field constructed from the electromagnetic energy–momentum tensor and the Lorentz boost Killing symmetry does not give the energy–momentum of the EM fields.

Section 16.3 shows how any static metric expressed in static form relative to some choice of coordinates has an obvious Killing vector field that one would naively consider to represent the invariance of the metric under 'time' translation. As always in relativity, 'time' is just a coordinate here. An obvious example is Schwarzschild spacetime in the usual coordinates. The Killing vector fields of this spacetime are discussed in Sect. 16.4. Given a conserved EM energy–momentum tensor, the question is whether these 'time' translation Killing symmetries give the energy–momentum of the EM fields, as suggested by Parrott. The view in the present book is that the fact of their being conserved is not reason enough to claim that they should correspond to what we usually call energy–momentum.

Sections 16.5 and 16.6 just lead back in a different way to the metric for the Rindler (SE) coordinate description of a gravity-free Minkowksi spacetime, or for a static, homogeneous gravitation field (SHGF), if it can be so interpreted. Section 16.7 then discusses this coordinate system as a description of spacetime that might be adopted by a uniformly accelerating observer in a gravity-free Minkowski spacetime, and poses the question as to whether this might be a natural or privileged frame for such an observer. The question asked is: why should the observer take these as her space coordinates? Of course, there are some interesting features, well known from our earlier study of semi-Euclidean coordinates. But surely, the question is whether one would set up such coordinates if one were accelerating. In other words, are these physically the same as accelerating coordinates, in some sense, for some operational way of setting up coordinates? Indeed, is there any privileged set of coordinates for an accelerating observer, as there is for an inertial observer?

This is an important question here, because a lot of the debate at issue here ought to have been about how one should interpret this general theory of relativity in specific cases, where one happens to have a coordinate system on a manifold, and a metric in component form. It is an easy matter to generate long calculations in general relativity, and even if the mathematics soon gets tough, the hard part of the theory is to link the results with what might be measured. Once one has raised the question of how to interpret coordinates and metric components, one finds that there is little reference to this in textbooks and papers, and that authors do fall into traps as a result.

To return to the relatively simple case of an accelerating observer in a flat space-time, where we may or may not consider that there is a non-tidal gravitational field, we are asking whether Rindler's coordinates, which are obviously closely related to our semi-Euclidean coordinates, are in fact a privileged set in some way. In fact it really looks as though they are not. When we set them up as in Chap. 2, we can see why. The fact that we lay down a set of conditions for the coordinates to be suitably adapted to the accelerating worldline shows just how artificial they actually are. Indeed, why should there be privileged coordinates for accelerating observers? There are such coordinates for inertially moving observers because it turns out that inertial motion is physically irrelevant. But accelerated motion makes a difference, because acceleration is not relative but absolute, for reasons we still do not understand.

At the end of Sect. 16.7, there is a brief discussion of the notion of rigidity in the context of accelerating solid objects, since it is often claimed that the Rindler coordinates specify a rigid accelerating elevator. The point here is once again to remind readers that even special relativity makes such claims much more difficult to uphold than their counterpart in pre-relativistic physics. In fact, it is not obvious that one could get any solid object to move according to Rindler's notion of rigidity [19]. Fortunately, and this is also an important point, there is no reason why one should consider that there are any bodies at fixed Rindler space coordinates.

Section 16.8 is the most important one in the discussion of Parrott's paper [7]. It is concerned with the interpretation of electric and magnetic fields, and the associated Poynting vector, when working in a non-inertial coordinate system. Because the EM field tensor is a two-form (antisymmetric, covariant, and rank two), its component matrix can always be written in the form

$$(F_{\mu\nu}) = \begin{pmatrix} 0 & E_1 & E_2 & E_3 \\ -E_1 & 0 & -B_3 & B_2 \\ -E_2 & B_3 & 0 & -B_1 \\ -E_3 & -B_2 & B_1 & 0 \end{pmatrix},$$

for some functions E_i and B_j on the spacetime, whatever coordinate system one chooses. But does this mean that **E** and **B** are electric and magnetic fields as we know them?

One could of course make this the definition, but it should be borne in mind that they only correspond to our usual understanding in a locally inertial frame, and then only locally, i.e., with decreasing reliability as one moves away from the spacetime event at which one has made the connection components equal to zero. This is the utility of the strong principle of equivalence: physics looks roughly as we know it in locally inertial frames. Without this principle, one could not even begin to interpret the constructions one has made on these manifolds. The weak principle of equivalence allows one to interpret motion relative to arbitrary coordinates, and the strong principle of equivalence allows one to interpret other fields defined relative to arbitrary coordinates. It is worth remembering that one could say nothing about the link between the spacetime manifold and the spacetime observed without having principles like these. Without SEP, one would require some other principle that made the link.

We return here to the Rindler elevator and imagine dividing the elevator walls into a large number of small coordinate patches with an observer stationed on each patch. The observers would measure the fields, calculate the corresponding energy–momentum tensor, and subsequently approximate (to arbitrary accuracy) the energy component of the integral giving the energy–momentum radiated through the walls of the elevator in a specified proper time interval (for the charge). The key point now is that this procedure is not the same as the following one: each of the observers calculates her own local energy outflow

$$\mathbf{n} \cdot \frac{\mathbf{E} \times \mathbf{B}}{4\pi} \Delta S \Delta \tau \ ,$$

where **n** is the outward unit normal vector to the wall in the observer's rest frame, ΔS is the area of her patch, $\Delta \tau$ the increment in her proper time, and **E** and **B** her electric and magnetic fields, respectively. Generally speaking, we could not attach any covariant meaning to the sum of these quantities over the different observers. This would amount to adding up quantities referring to different rest frames, which would not give energy as we know it. Furthermore, if we changed elevators to calculate the outflow of this quantity, e.g., having a bigger elevator, containing the first, then we would not come up with the same total: there is no reason to suppose that the observers on the larger elevator would obtain the same number for the energy radiation using this definition as those on the smaller one. Worse still, neither number would be expected to be related in any simple way to the additional energy required by a rocket for a charged versus uncharged payload.

In the procedure just described, the observers are not measuring what is usually called energy. They are measuring something else, a kind of pseudo-energy. It may be tempting to call it something like energy as measured in the elevator frame, but it is conceptually and experimentally distinct from the usual Minkowski energy. For arbitrary motion, it is not a conserved quantity and does not therefore even deserve the name 'energy'. For the special case of an elevator occupying constant

Rindler coordinates, it does happen to be independent of the elevator shape, because it is in fact zero for all such elevators, but this does not in itself imply that it is energy in the usual sense of the term. In fact Parrott shows that it is the conserved quantity constructed from the Killing vector field corresponding to the flow of the one-parameter family of Lorentz boosts. Two small changes are made to Parrott's proof, which appears to contain errors, but his final conclusion is fully supported here.

There is probably a general result lurking here, which says that a stationary charged particle in a static spacetime does not radiate the pseudo-energy defined as the conserved quantity corresponding to translation by the formal time coordinate in this spacetime. For we observe that the time coordinate provides a Killing vector field for the static metric, and this Killing vector field provides a conserved quantity in conjunction with the electromagnetic energy–momentum tensor.

Parrott agrees with Boulware that there is no radiation of the conserved quantity corresponding to the Lorentz boost Killing vector in Minkowski spacetime, but says that this is irrelevant to questions concerning physically observed radiation. Parrott also says [7, p. 13] that it is an experimental question whether it is Minkowski energy radiation or pseudo-energy radiation which corresponds to energy that must be furnished by the driving forces. He would settle the question by uniformly accelerating a large charge in a rocket and observing whether more fuel were required than for a neutral payload of the same mass. He would expect more fuel to be used, implying that Minkowski energy is the physically relevant energy.

However, there is no attempt anywhere to prove that the Boulware pseudo-energy has any connection with driving forces on the charge. It may have been surreptitiously assumed somewhere, but an actual demonstration is nevertheless absent. On the other hand, there are theoretical reasons for expecting an extra driving force to provide for Minkowski energy radiation. It is only this that would be tested by an experiment. That is to say, we never discuss how a driving force might be connected with pseudo-energy radiation, so there is no theoretical claim there to test by the proposed experiment. The experiment described would help us to understand the Lorentz–Dirac theory and the radiation formulas found in Minkowski coordinates.

Another question to raise at this point is the following: is there any direct connection between the fact that the radiation reaction terms in the Lorentz–Dirac (LD) equation are zero and the fact that the Poynting vector is zero in the semi-Euclidean (SE) frame? Of course, there is a connection: both facts occur only when the acceleration is uniform. However, we know that there is radiation in the Minkowski frame. So even though the LD radiation reaction is zero, we do not need to show that the SE observer will have zero Poynting vector. The two facts in the above question look more or less like two accidents that happen to occur when the acceleration is uniform.

Section 16.9 reconsiders the Schwarzschild spacetime, for which the metric has static components relative to the usual Schwarzschild coordinate system. As mentioned above, Parrott's conclusions with regard to the SHGF, disposing so effectively of one of the main arguments used to justify the idea that a stationary charge in a static spacetime will not radiate, seem to be at loggerheads with his more general agreement with Bondi and Gold's unproven principle: he considers that it takes no energy to hold a charged particle stationary in a static gravitational field, and he seems to agree implicitly that it will take energy to hold a charged particle 'stationary' relative to an SHGF, i.e., stationary relative to some SE coordinates for this spacetime.

It does indeed seem right to expect that a charged particle sitting still relative to a static gravitational field will not require an energy supply to do so, and nor will it do anything. This is certainly what one would say in Newtonian gravity theory, or special relativity theory with the obvious naive theory of gravity as a force field. A stationary particle is inertial in SR and so Maxwell's theory says there will be no radiation.

But in order to understand the physical implications of a scenario in general relativity theory, one is supposed (presumably) to look at what is happening in the locally inertial frame guaranteed by the weak equivalence principle (WEP) and deduce things about electromagnetic fields by applying the strong equivalence principle (SEP). If one did not apply SEP, one would have to find some other principle for relating the mathematics of the theory to physical observations. But when one looks at the static charge mentioned above as it would be described in a locally inertial frame, one finds that it is accelerating. There is then nothing obvious at all about the above conclusion. Indeed, the minimal extension of Maxwell's equations (MEME) to this context, which is an application of SEP, says that this charge will radiate. If energy radiates out, and we have seen that this could indeed be detected locally, within the given locally inertial frame, then it will have to be supplied by whatever is holding the charge up against the gravitational effects.

One then realises that just the fact that the conserved quantity corresponding to the Killing vector field ∂_t happens to be the only natural mathematical candidate for an energy is really not sufficient to make it what one would normally call energy physically. In contrast with a commonly held attitude in GR, mathematical convenience is not sufficient to be sure that one is doing physics, i.e., that one is getting the right relation with what is out there. It is surprising that Parrott accepts this, despite the general scepticism with which he greets the interpretation of Boulware's Poynting vector in the uniformly accelerating frame for Minkowski spacetime.

Of course, there is a major difference between the two scenarios: in the Minkowski spacetime there are several Killing vector fields that might be taken to give the energy, even if one of them undoubtedly should take precedence. But what could one do in Schwarzschild spacetime if one did not use ∂_t to build an energy quantity from energy–momentum tensors? There is nothing else, at least nothing else obvious. On the other hand, even in Schwarzschild spacetime, there are locally Minkowski frames and the view presented here is that our physical interpretation of the theory has to pass by them.

It must also be stressed that GR is very different from SR as regards gravitational effects. GR builds in an interaction between gravity and other fields, like electromagnetic fields, which just does not exist in special relativistic versions of these other fields. For example, light is affected by gravitational effects in GR, whereas it really is not in Newtonian gravity theory unless one attributes a mass to photons, but that is not then an interaction between EM fields and gravity. If a charge is stationary relative to the coordinates one happens to have chosen, one must remember that these are only coordinates. It is not stationary relative to the kind of coordinates one is supposed to use to understand the theory physically. And not only is the particle interacting with the gravitational effects via its own passive gravitational mass, but its EM fields are also interacting with the gravitational effects in this theory.

It seems that the case of the uniformly accelerating frame in Minkowski space-time amply illustrates this. If one can interpret this scenario as holding a charge stationary relative to an SHGF, one sees that the stationarity of the particle relative to SE coordinates, in which the appropriate (Minkowski) metric has the static form, means nothing with regard to the question of EM radiation. One has to view the situation from the locally (and in this case, globally) inertial frame, whereupon one finds that the charge is in fact accelerating, and that the minimal extension of Maxwell's theory clearly implies EM radiation. There may still be a question about whether the existence of this radiation is a subjective thing, in some sense, that can be observed by the locally inertial observer but not by one accelerating with the charge, but Parrott seems to have answered that with the conclusion that the existence of EM radiation should be considered as an objective phenomenon, i.e., if the locally inertial observer detects it, then so will the comoving observer, provided the latter looks at the right quantities locally and measures them carefully enough.

Some would say that the scenarios of a non-tidal gravitational field or an eternally uniformly accelerating charge are not physical and cannot be used to illustrate anything, but the present view in this book is quite the opposite: they provide an application of the equivalence principles WEP and SEP *par excellence* which allows one to be very clear about what is different in GR, as compared with SR. It is this clear case that warns us not to just go by the mathematics in a less intuitively obvious case like Schwarzschild spacetime.

So why should the conserved quantity constructed from ∂_t and the energy–momentum tensor of the EM fields be what we expect to call the EM energy, when t is just a convenient coordinate? Should the energy of the EM fields be conserved when these fields are interacting with another field? In other words, should we be able to find a conserved thing to correspond to the energy of the EM fields in Schwarzschild spacetime?

Chapter 17 returns to the Lorentz–Dirac equation in a different way and hopefully sorts out the question of how the radiation reaction can be zero for a uniformly accelerating charge when it is in fact radiating. The content of this chapter is entirely due to Parrott [7]. It is a striking result that the radiation reaction in the Lorentz–Dirac equation is zero in precisely those cases where the charge is accelerated as it would be by a static, homogeneous gravitational field in a special relativistic version of a gravitational theory (see Chap. 3). This adds to the surprising result that the retarded field generated by a charge in such motion has zero magnetic field and time-constant electric field in the semi-Euclidean coordinates of an observer moving with the charge (Sect. 15.11.1), as well as the fact that the semi-Euclidean metric for precisely this observer is static (Chap. 2).

But to return to the above proof, does this mean, as it appears, that there is no radiation reaction for a uniformly accelerated charged particle, i.e., that a rocket-driven uniformly accelerated charge requires no more energy from the rocket than an otherwise identical neutral particle, despite the fact that one gathers radiation energy in the charged case? We return to the idea of energy conservation. If the uniformly accelerated charged particle does indeed radiate energy into Minkowski spacetime, which can subsequently be collected and used, then this radiated energy must be provided by some decrease in energy of other parts of the system. In the case where the uniform acceleration is eternal, Fulton and Rohrlich say that it may come from the field energy, but they do not explain how. In this same case, Boulware discovered a flow of energy that focuses in on the charge from the hypersurface $z = -t$, which might answer Fulton and Rohrlich's question. But one can also consider this zero radiation reaction in a situation where the uniform acceleration occurs over a finite period of time. In any case, it is understood here that the uniformly accelerated charge will radiate as described, so that there would then be a problem for energy conservation.

Now the derivation of the Lorentz–Dirac equation is motivated by energy–momentum conservation: the change of energy–momentum of the particle over a given proper time interval should equal the energy–momentum provided by the external forces driving the particle minus the radiated energy–momentum. (We have to assume that infinite terms of a certain structure can be absorbed by mass renormalisation.) Despite this, and as Parrott discusses in [13, Sect. 4.3], the Lorentz–Dirac equation does not generally guarantee conservation of energy–momentum. One special case where it does is the situation when the particle is asymptotically free, i.e., its proper acceleration is zero in the infinite past and future. Conversely, it is not clear that the equation should apply to a particle which is not asymptotically free, such as a particle that is uniformly accelerated for all time. At any rate, the equation does not guarantee energy–momentum conservation in this case, by any of the reasoning that went into its derivation. In this sense, the fact that the radiation reaction term is zero implies nothing about the additional force which the rocket must provide for perpetually uniformly accelerated motion. But what about the asymptotically free case?

Parrott asks what happens in the case of a particle which is unaccelerated in the distant past, nudged into uniform acceleration, uniformly accelerated for a long time, then finally nudged back into an unaccelerated state. In this case, the Lorentz–Dirac equation does imply conservation of energy–momentum. However, the radiation reaction term is zero for the period of uniform acceleration, no matter how long it is. So if the radiation reaction term in the Lorentz–Dirac equation were the correct radiation reaction, one could say that all the energy that goes into radiation must therefore be provided at the beginning and end of the trip, while the particle is nudged into and out of its uniformly accelerated state. This is what Parrott analyses in his Appendix 2, and which is discussed in full in Chap. 17.

Note in passing that, apart from the nudging into and out of the accelerated state, the other, uniform acceleration does not incur a radiation reaction, whence one might think that it is the changing of the acceleration which leads to radiation. It would then not be a second time derivative but a third time derivative that counts here. However, the third derivative is nonzero even during the uniform acceleration, as can be seen from (17.2) on p. 290.

We are saying that, if we believe in the Lorentz–Dirac equation, we need to add a bit of energy to start the uniform acceleration, slightly more than if the rocket were uncharged, and thereafter the radiation is free, even if it persists for an arbitrarily long time and adds up to an arbitrarily large amount! Of course, one still needs energy to keep the charged rocket in uniform acceleration, but no more than if it were neutral, according to the radiation reaction term provided by the Lorentz–Dirac equation. In effect, we can borrow an arbitrarily large amount of radiated energy, which can in principle be collected and used by other observers in Minkowski spacetime, although it turns out that we do have to pay it back at the end of the trip, whenever the uniform acceleration is switched off. As Parrott says, this is hard to accept physically, and it seems to be a good reason for abandoning the Lorentz–Dirac equation.

The theoretical reasons for thinking that there is radiation from a uniformly accelerated charge are reconsidered, returning to the derivation of the Lorentz–Dirac equation as described in Chap. 11 and [13, Sect. 4.3]. In the case of an asymptotically free particle, it is established that

$$m_{\mathrm{ren}} u_{\mathrm{out}} - m_{\mathrm{ren}} u_{\mathrm{in}} = - \int_{\mathrm{side}} {}_{\perp}T \,, \tag{18.1}$$

where m_{ren} is the mass renormalised by an infinite contribution from the tube caps, but not yet completely renormalised. The right-hand side is an integral over the tube walls. Equation (18.1) is the condition for energy–momentum conservation. We seek an equation of motion that guarantees this.

We put (18.1) in differential form, viz.,

$$m_{\mathrm{ren}} \frac{\mathrm{d}u}{\mathrm{d}\tau} = - \lim_{\Delta\tau \to 0} \frac{1}{\Delta\tau} \lim_{r_0 \to 0} \int_{S_{r_0}(\tau,\tau+\Delta\tau)} {}_{\perp}T \,, \tag{18.2}$$

whose integral with respect to τ gives the limit as $r_0 \to 0$ of (18.1). We have to take the limit here, because the left-hand side of (18.2) cannot depend on r_0. We then calculate the integral over the piece of tube wall. This leads to the proposed equation of motion

$$\bar{m} \frac{\mathrm{d}u}{\mathrm{d}\tau} = q F_{\mathrm{ext}}(u) + \frac{2}{3} q^2 \langle a,a \rangle u \,, \tag{18.3}$$

where the mass has been renormalised yet again to give $\bar{m}$, due to more infinite terms (of the right structure) arising in the integral over the tube walls. This equation is impossible to satisfy unless $\langle a,a \rangle = 0$, because the first two terms are orthogonal to u and the last is parallel to it.

This is where we note that (18.2) implies (18.1), but not conversely. In other words, there may still be an equation of motion similar to (18.3) which implies energy–momentum conservation as expressed by (18.1). We therefore aim to see what would lead to the integrated form

$$\bar{m}u_{\text{out}} - \bar{m}u_{\text{in}} = q \int_{-\infty}^{\infty} F_{\text{ext}}\big(u(\tau)\big)\,d\tau + \frac{2}{3}q^2 \int_{-\infty}^{\infty} \langle a(\tau), a(\tau)\rangle u(\tau)\,d\tau \,. \tag{18.4}$$

The last integral can be rewritten as an integral of something else, i.e.,

$$\int_{-\infty}^{\infty} \langle a, a\rangle u\,d\tau = \int_{-\infty}^{\infty} \left(\frac{da}{d\tau} - \left\langle \frac{da}{d\tau}, u \right\rangle u \right) d\tau \,, \tag{18.5}$$

provided that we have $a(-\infty) = a(\infty)$, something we already assumed (in fact that both were zero) in order to deal with the integrals over the caps (the assumption of an asymptotically free particle). This is how we can say that the proposed equation of motion

$$\bar{m}\frac{du}{d\tau} = q F_{\text{ext}}(u) + \frac{2}{3}q^2 \left(\frac{da}{d\tau} - \left\langle \frac{da}{d\tau}, u \right\rangle u \right) \tag{18.6}$$

implies the equation (18.1) of energy–momentum conservation. Equation (18.6) is just another writing of the Lorentz–Dirac equation.

Now it is the integral of the energy–momentum tensor (or something derived from it) over the tube walls which leads to the idea of radiated energy–momentum. Parrott exposes the calculation in great detail in [13, Sect. 4.4], where he shows that (see also p. 67)

$$\int_{S_r(\tau_1, \tau_2)} {}^{\perp}T = \frac{q^2}{2r}\big[u(\tau_2) - u(\tau_1)\big] \quad \text{(mass renormalisation)}$$

$$-\frac{2}{3}q^2 \int_{\tau_1}^{\tau_2} u(\tau)\langle a(\tau), a(\tau)\rangle\,d\tau \quad \text{(radiation)}$$

$$-q \int_{\tau_1}^{\tau_2} \hat{F}_{\text{ext}}(u)\,d\tau \quad \text{(Lorentz force)} \,. \tag{18.7}$$

So we interpret the term

$$-\frac{2}{3}q^2 \int_{\tau_1}^{\tau_2} u(\tau)\langle a(\tau), a(\tau)\rangle\,d\tau \tag{18.8}$$

as energy–momentum radiation. It is because this term arises that we expect accelerating charged particles to radiate energy and momentum.

So what happens to this term in the case of a uniformly accelerating charge? Why do we say that this is nonzero, whilst the term it leads to (indirectly) in the Lorentz–Dirac equation is zero? Now the point is that the integral in (18.8) is between any two proper times, and we can check from Parrott's derivation [13, pp. 140–1] that

$$\int_{\tau_1}^{\tau_2} u(\tau)\langle a(\tau), a(\tau)\rangle\,d\tau = -a(\tau_2) + a(\tau_1) + \int_{\tau_1}^{\tau_2} \left(\frac{da}{d\tau} - \left\langle \frac{da}{d\tau}, u \right\rangle u \right) d\tau \,. \tag{18.9}$$

For uniform acceleration between these two proper times, this gives

$$\int_{\tau_1}^{\tau_2} u(\tau)\langle a(\tau), a(\tau)\rangle \, \mathrm{d}\tau = -a(\tau_2) + a(\tau_1) \neq 0 \,. \tag{18.10}$$

This brings out quite clearly why we still have grounds for claiming that there is radiation, whilst the proposed radiation reaction term is identically zero for this kind of motion.

The point is then that the acceleration changes during a period of uniform acceleration, because the word 'uniform' only refers to the constancy of the pseudolength $\langle a, a\rangle$. This is very clear from (16.20) on p. 260 or (17.2) on p. 290. Indeed, the condition of uniformity on the acceleration only requires $\mathrm{d}a/\mathrm{d}\tau$ to be orthogonal to a, not necessarily zero, and we have the explicit formula in (16.20) to confirm this.

Therefore, in this version, the uniformly accelerated charge is radiating energy–momentum at the rate implied by (18.10), and this despite the fact that the terms in the Lorentz–Dirac equation that are usually interpreted as describing the radiation reaction are actually zero. The Lorentz–Dirac equation only implies the more fundamental energy–momentum conservation relation over a proper time interval $[\tau_1, \tau_2]$ when the acceleration is the same at the beginning and end of the interval. One could even say that the Lorentz–Dirac equation is wrong if this is not the case.

Indeed, we can see from the very derivation of the Lorentz–Dirac equation why it makes the strange prediction mentioned earlier in the context of a particle that is unaccelerated in the distant past, nudged into uniform acceleration, uniformly accelerated for a long time, then finally nudged back into an unaccelerated state. In this case, the Lorentz–Dirac equation does imply conservation of energy–momentum. However, the radiation reaction term is zero for the period of uniform acceleration, no matter how long it is. All the radiation energy must therefore be provided at the beginning and end of the trip, while the particle is nudged into and out of its uniformly accelerated state.

As noted above, the acceleration itself, when uniform, does not incur a radiation reaction, at least according to the terms in the Lorentz–Dirac equation. It is the non-uniformity of the acceleration which does, while the rocket is nudging itself into a uniform acceleration in the present example. But once again, let us stress that one should not consider that it is not a second derivative but a third derivative that counts here, because the third derivative $\mathrm{d}a/\mathrm{d}\tau$ is nonzero in the uniform acceleration case, a point that the appellation 'uniform' tends to conceal. On the other hand, the whole question of higher derivatives does suggest that it might be worth considering a model in which the charge is non-pointlike: in that case, although it is a difficult calculation, we find a self-force on the charge that not only avoids the need for renormalisation, but potentially explains inertial mass, explains how the radiation reaction occurs, and includes higher order derivatives than just the acceleration.

So if we believe in the Lorentz–Dirac equation, we need to add a bit of energy to start the uniform acceleration, slightly more than if the rocket were uncharged, and thereafter the radiation will be free, even if it persists for an arbitrarily long time and adds up to an arbitrarily large amount. No more energy is needed to keep the charged rocket in uniform acceleration than if it were neutral, according to the radiation reaction term provided by the Lorentz–Dirac equation, which suggests that one could borrow an arbitrarily large amount of radiated energy, which could in principle be collected and used by other observers in Minkowski spacetime, with the proviso that one must pay it back at the end of the trip, whenever the uniform acceleration is switched off. Here is yet another reason for doubting the Lorentz–Dirac equation. Indeed, the equation of energy–momentum conservation arising directly from (17.9) seems to be a superior equation.

Chapter 17 establishes the equation of motion of a charged rocket on the basis of the Lorentz–Dirac equation in the following case: the rocket begins at rest in some Lorentz frame up to an initial time τ_i, before being nudged into uniform acceleration over a small proper time interval $[\tau_i, \tau_i + \delta]$, uniformly accelerated up to proper time $\tau = \tau_f - \delta$, nudged back into an unaccelerated state over the interval $[\tau_f - \delta, \tau_f]$, and continuing unaccelerated for $\tau > \tau_f$. At least, this is the first part of the motion considered.

It turns out that the excess energy required to accomplish the final deceleration of a charged rocket as compared with a neutral rocket (strictly speaking, it is not a deceleration because we allow the rocket to maintain its velocity at whatever value it has reached; what we have is a decrease in acceleration to zero), as measured in the final rest frame at $\tau = \tau_f$, is independent of the period of uniform acceleration. It would thus seem that, from the standpoint of the rocket pilot, only a modest excess amount of fuel must be burned to start the acceleration at the beginning of the trip and stop it at the end, with no excess fuel relative to an uncharged rocket required during the arbitrarily long period of uniform acceleration, during which a correspondingly arbitrary amount of radiation energy can be gathered.

Now some would suggest that one could allow the uniform acceleration to continue indefinitely without using any excess fuel as compared with the neutral rocket. They even suggest that perpetual uniform acceleration would require no extra fuel at all as compared with the situation for a neutral particle. If there were radiation energy from the charged rocket but not from the uncharged rocket, and if, as suggested for the latter scenario, no extra energy had to be put in to get this energy, one would obviously have a violation of energy conservation somewhere. One way out is to say that there might be radiation, but that it is not physically accessible, as proposed by Boulware, but rejected here.

Parrott counters this 'free energy' argument by re-examining the more realistic (sic) case of asymptotically free rockets, i.e., with constant velocity outside some finite period of uniform velocity, and showing that, not only does the Lorentz–Dirac equation predict something very strange in the case of the charged rocket, but it does put a limit on the duration of the uniform acceleration if one hopes eventually to remove the acceleration and allow the rocket to continue at constant velocity beyond some point in time. This is what is analysed mathematically in Chap. 17.

This shows how the Lorentz–Dirac equation saves itself from disaster in the case where it is supposed to work, i.e., when the motion is asymptotically free: if we do have to remove the uniform acceleration at some point, then that state of uniform acceleration cannot have been going on beyond some time determined by the initial rest mass of the charged rocket. Furthermore, even if the uniform acceleration has been going on for some time within the acceptable bound, one should have no doubt that the radiated energy does have to be paid for at the transition from the uniform acceleration to the uniform velocity state. The fact that only a modest amount of extra energy (as compared with the neutral rocket) is required to make this transition, as measured in the final frame of the rocket, does not mean that one is getting most of the radiated energy for nothing: the modest amount of energy used in the final frame, along with its associated momentum, can Lorentz transform into a large amount of energy in the initial rest frame at $\tau = \tau_i$, so energy conservation can indeed hold in the initial frame. After all, we know that the Lorentz–Dirac equation expresses energy conservation in an asymptotically free case like this.

Despite the decisive result that there is a limit on the duration of the uniform acceleration if one hopes eventually to remove the acceleration and allow the rocket to continue at constant velocity beyond some point in time, Parrott nevertheless extends his mathematical analysis in the following way: At time τ_f, we reverse the proper acceleration in a time-symmetric way, so that

$$a(\tau_f + \sigma) = -a(\tau_f - \sigma)\,, \quad \sigma > 0\,,$$

whence the rocket will eventually come back to rest in the initial frame at $\tau = 2\tau_f$. This was in answer to a more sophisticated challenge he mentions in [7], in which it was claimed that the excess fuel used over the whole interval $[\tau_i, 2\tau_f]$ (as compared with the neutral rocket) is modest and independent of the duration of the uniform acceleration. If this were possible, one could argue as follows. At the beginning and end of the trip, the rocket is at rest in the initial frame, so the energy of the radiation plus the exhaust should equal the rest mass loss of the rocket, i.e., the rest mass of the fuel used. If the loss of rest mass is finite and independent of the duration of the uniform acceleration, hence independent of the arbitrarily large energy radiation, we have a violation of energy conservation in the initial frame.

Since we have a situation here where we expect the Lorentz–Dirac equation to apply, and since it implies energy conservation when it applies, this would appear to cast a serious shadow over the proceedings. However, Parrott uses the Lorentz–Dirac equation to show that, with a fixed amount of initial fuel, one cannot obtain an arbitrarily long period of uniformly accelerated motion, i.e., τ_f cannot be arbitrarily large. Put another way, according to the Lorentz–Dirac equation, a charged rocket will run out of fuel if it uniformly accelerates for long enough, so the proposed time-symmetric motion is impossible with fixed initial fuel and arbitrarily large τ_f. In contrast, a uniformly accelerated uncharged rocket can accelerate forever, provided that all its mass can be consumed as fuel. So it looks as though the Lorentz–Dirac equation can save itself from what would here be tantamount to a self-contradiction.

All this would seem to suggest that there is a physical radiation reaction for a particle which is uniformly accelerated for a finite time even though the Lorentz–Dirac radiation reaction expression is identically zero during the period of uniform acceleration. However, if we believe the Lorentz–Dirac equation, we must accept the strange conclusion that all of this radiation reaction occurs at the beginning ($\tau \approx \tau_i$)

and end ($\tau \approx \tau_f$) of the trip, while the particle is being nudged into or out of its uniformly accelerated state.

Whether there is a radiation reaction for a perpetually uniformly accelerated particle depends on the definition of the radiation reaction. The radiation reaction term in the Lorentz–Dirac equation is identically zero, but there is no good physical reason to identify this term with the physically observed radiation reaction. As Parrott suggests, it seems more reasonable to obtain the answer for perpetual uniform acceleration as a limit of whatever answer is eventually accepted for uniform acceleration over a finite time. The Lorentz–Dirac equation probably has to be rejected in the process, if only because it accounts so strangely for the events of the finite-duration uniform acceleration in the above discussion.

One might say that the most revealing part of this last discussion occurs on p. 293, with regard to (17.11) and (17.12). The radiation through the tube walls between τ_1 and τ_2 is the left-hand side of (17.11), according to the careful calculation in Parrott's book. For uniform acceleration throughout the period $\tau \in [\tau_1, \tau_2]$, we obtain (17.12), which says that the radiation is numerically equal to the nonzero quantity $-a(\tau_2) + a(\tau_1)$. Now in the present context, we have arranged for $a(\tau_1) = a(\tau_2)$, but this does not mean that the radiation given by the right-hand side of (17.11) will be zero, because we do not have uniform acceleration throughout the interval $[\tau_1, \tau_2]$ when we put $\tau_1 = \tau_i$ and $\tau_2 = \tau_f$.

Chapter 19
Conclusion

Bondi and Gold's solution to their problem was to say that there is no such thing as a static homogeneous gravitational field throughout the whole of spacetime, so one did not have to worry about the fact that the generally accepted strong equivalence principle implies that a charge that is stationary relative to standard coordinates for such a spacetime, in which the metric components happen to be independent of the time coordinate, is predicted to radiate electromagnetic energy. Although this solution looks rather weak, it would seem that Bondi and Gold were the only ones among the cited authors who understood the seriousness of the threat to the strong equivalence principle, without which one has no obvious way of doing electromagnetism in a general relativistic framework.

What Bondi and Gold really objected to was the idea that a charge that is stationary relative to coordinates in which the metric components of a spacetime are static should be able to radiate electromagnetic energy. The present thesis is that it would have been better to reject this as a pre-general-relativistic prejudice and treat the unrealistic case of a static homogeneous gravitational field as an application *par excellence* of the strong equivalence principle. Indeed, the spacetime is flat, so the locally inertial frames that are always promised by the weak equivalence principle, which in turn always holds in torsion-free spacetimes, turn out to be global inertial frames. According to the strong equivalence principle, a charged particle sitting still relative to the usual coordinates for this spacetime will be observed to radiate electromagnetic energy as viewed from the globally available inertial frames in precisely the same way that an eternally uniformly accelerating charge in a flat spacetime with no gravitational effects, not even non-tidal ones, will be observed to radiate such energy.

This is a better solution because there is really no reason to expect a charge that is stationary relative to coordinates in which the metric components of a spacetime are static to be unable to radiate electromagnetic energy. Quite the contrary, in fact. Indeed, this application *par excellence* could have been used precisely to illustrate how general relativity differs from a naive model of gravity in the framework of special relativity. In the latter context, gravity is modelled by a force field. If the source of gravity is treated as stationary in an inertial frame and the charged particle is held still with respect to the source, Maxwell's theory predicts that it will not radiate electromagnetic energy. However, if it is allowed to fall toward the source, it will accelerate relative to the inertial frame, and Maxwell's theory predicts that it will radiate.

The situation in general relativity is very different for a fundamental reason: Maxwell's theory has to be adapted in order to fit it in with the idea that space-time is curved. The prescription provided by the strong equivalence principle is to replace coordinate derivatives of Maxwell's equations in flat spacetime by covariant derivatives relative to the Levi-Civita connection, because this is how the strong equivalence principle is implemented mathematically. One then finds that the electromagnetic fields interact with gravitational fields as expressed through the metric, something that just does not happen in the naive model of gravity described above for the special relativistic framework. A well known example is that the paths of light waves 'bend' when they pass by a massive object, or again, the wavelength of light expands with space when it passes through an expanding universe.

The upshot of this is that the electromagnetic fields of a stationary charge relative to some coordinates will generally interact with the gravitational field, even if the metric happens to have the static form in that coordinate frame, and that this will a priori introduce radiation terms into the electromagnetic fields. In fact, the application *par excellence* of the strong equivalence principle to the static homogeneous gravitational field serves as a perfect illustration, unrealistic though it may be, because it points precisely to what will happen in an approximate way in more realistic spacetimes. Consider, for example, the Schwarzschild spacetime. A charge that is stationary relative to the usual coordinates will be seen to accelerate relative to a locally inertial coordinate system. Insofar as this gravitational field can be approximated by a static homogeneous gravitational field in some small enough neighbourhood, the idealistic application of the strong equivalence principle rejected by Bondi and Gold is telling the locally inertial observer to expect electromagnetic radiation.

So just ruling out the static homogeneous gravitational field as unrealistic and keeping the strong equivalence principle as it is in all other cases is not in fact a solution to Bondi and Gold's problem. This principle implies that a charge that is stationary relative to the usual coordinates for Schwarzschild spacetime, in which the metric takes the usual static form, is in fact going to radiate electromagnetic energy. Note in passing that we are not just talking about the sophisticated curvature-dependent effects unearthed by DeWitt and Brehme when they extended the Lorentz–Dirac equation to curved spacetimes. No sophisticated analysis is required to get these radiation terms: the charge is accelerating relative to the locally inertial frame in which one can apply flat spacetime Maxwell theory to a first approximation.

There is another less physical but probably equally fundamental reason for not accepting Bondi and Gold's general principle that a charge that is stationary relative to coordinates in which the metric components of a spacetime are static cannot radiate electromagnetic energy. This is the relativity of staticity: although one can declare a metric to be static in a coordinate-independent way by saying that there exist coordinates in which it assumes a static form, the idea of a particle being static relative to a coordinate system cannot have any coordinate-independent import. This is a death blow for the above principle when one examines it in the static homogeneous gravitational field, which must in turn be a sufficient approximation to a great many curved spacetimes over some small enough region: a stationary charge relative to the usual coordinates for the static homogeneous gravitational field is in fact uniformly accelerating relative to the locally, and in this case globally inertial frames.

If these arguments can be accepted, then the conclusions of all the papers cited relative to this issue have to be discarded. The discussion here usually has two facets: a first part in which one tries to get a precise idea of what equivalence principle the authors are talking about, and a second part in which one tries to understand the way they react to it. Since none of the authors cited give a precise statement of the equivalence principle, the first part is a challenge in its own right. However, it is clear that they all have in mind a version which talks about whether an observer would be able to distinguish two different physical scenarios. This version is criticised here as an attempt to hang on to an obsolete pre-general-relativistic equivalence principle. It may be that this was the norm historically speaking. The present book defines and applies the now standard versions commonly known as the weak and strong equivalence principles.

The weak equivalence principle follows from the very construction of general relativity when one declares that the metric alone determines the structure of the spacetime, so that the connection has to be the Levi-Civita connection (assumption of zero torsion). For then, given any event in spacetime, and any level of measurement accuracy, one can find coordinates in some neighbourhood of that event such that the metric is Minkowskian, up to that level of measurement accuracy, throughout the neighbourhood, and the connection is zero. In other words, there are always locally inertial frames. The strong equivalence principle tells us that we can adapt any theory in flat spacetime to a general relativistic version by replacing coordinate derivatives by covariant derivatives with respect to the Levi-Civita connection. (There is a proviso that one could vastly complicate this step by introducing curvature-dependent terms, with impunity in the flat spacetime where the curvature is zero.)

The name of 'principle' is probably a misnomer in the case of the weak equivalence principle. However, the view adopted here is that this result, along with the strong add-on principle, are precisely what enable us to link theory with measurement. Indeed, one of the themes here is that not enough is said about this link in otherwise didactic textbooks. It is easy to generate mathematics of great intricacy in general relativity, but it is no easy matter to interpret the results. The weak and strong equivalence principles are essential, not just in formulating theories, e.g., of electromagnetism, in the case of the strong principle, but also in making this last step.

A related issue that is much discussed here, and illustrated by the difficulties that arise as a consequence, is the problem of interpreting results when they are not expressed relative to locally inertial coordinate systems. The spacetime of the static homogeneous gravitational field, or the coordinates usually advocated for an eternally uniformly accelerating observer in a flat spacetime with no gravitational effects, not even non-tidal ones, serve to illustrate this beautifully. This question alone justifies consideration of these highly idealistic models. Some of the qualitative errors of judgement that seem to ensue are quite remarkable, e.g., the idea that the singularity in the famous Rindler frame might actually correspond to something like a black hole in its interpretation as the spacetime of a static homogeneous gravitational field, or again, the idea that one can treat the Poynting vector relative to such coordinates as specifying a flow of electromagnetic energy and momentum.

The system of semi-Euclidean (Rindler) coordinates usually proposed for an observer moving with an acceleration relative to some inertial frame (in the context of special relativity) is examined in some detail. For example, it is shown that the (Minkowski) metric will only have a static form for eternal uniform acceleration. It is also shown that, if the observer carries a charged particle, it will not appear to produce a magnetic field and also that the resulting electric field will be independent of the time coordinate. Although this extension of the notions of magnetic and electric field to non-inertial coordinate frames is regarded critically, it is a remarkable and otherwise uninterpreted fact that they should satisfy such simple conditions for precisely the observer motion that corresponds to free fall in a static homogeneous gravitational field in a naive special relativistic model of gravity.

Naturally, the cited papers contain much more than just an assessment of some outdated equivalence principle and would still be worth reading for the rest of their content. Other major questions discussed here are the validity of the Lorentz–Dirac equation, but also the issue of what constitutes electromagnetic radiation and whether it is local enough to be observed in arbitrarily small coordinate systems, or even whether it should be a frame-dependent concept. These are more difficult and technical questions than those concerning the equivalence principle. The views adopted in the present discussion are as follows.

Firstly, the Lorentz–Dirac equation is not always valid, a generally held conclusion. There are situations where one can see this from the derivation, because one makes assumptions about the charge motion that are not always fulfilled. The eternally uniformly accelerating charge is a case in point. But even in cases where one might expect it to work, e.g., asymptotically free motion, because it does then imply energy conservation, it still tells us strange things. The discussion here largely follows Parrott. With regard to the other questions mentioned, it is considered here (following Parrott once more) that radiation can be measured as locally as one would like, at least in principle, and that its existence should not be considered as a frame-dependent concept. The second point will undoubtedly be a bone of contention.

There is some mention of the motion and fields of spatially extended charge distributions, but without detailed calculations. If all particles are in fact spatially extended rather than mathematical points, as might reasonably be expected, it is well known that one can explain in part why they have inertia. Indeed, this is one of the two widely accepted mechanisms one appeals to in particle physics, along with the Higgs mechanism. The point particle approach is condemned to renormalise here, even in pre-quantum theory. One can also understand why accelerating charges should radiate electromagnetic energy, there being no explanation in the point particle model. Although these models would appear to corroborate the Lorentz–Dirac equation, not an argument in their favour, they deserve much more attention, even in the context of general relativity, because one can understand why the self-force contribution of a particle to its inertial mass should equal the self-force contribution to its passive gravitational mass. (See [23] for a discussion of these issues and further references.)

The whole content of this book stems from the discomfort in hearing that there might be a difficulty in applying 'the' equivalence principle in some cases. Very likely this difficulty only actually concerns a version of the principle that is not at all crucial for general relativity as it is usually formulated. If it did really concern the strong principle, as Bondi and Gold suggest, then we would be in trouble. This is why the present approach is to reject another of their hypotheses. At the same time, whether the present theses are accepted or not, there must be lessons here for the way general relativity is taught and interpreted.

References

1. M. Born: Ann. Phys. Lpz. **30**, 39 (1909)
2. H. Bondi, T. Gold: The field of a uniformly accelerated charge, with special reference to the problem of gravitational acceleration, Proc. Roy. Soc. A **229**, 416–424 (1955)
3. T. Fulton, F. Rohrlich: Classical radiation from a uniformly accelerated charge, Annals of Physics **9**, 499–517 (1960)
4. B.S. DeWitt, R.W. Brehme: Radiation damping in a gravitational field, Annals of Physics **9**, 220–259 (1960)
5. R.A. Mould: Internal absorption properties of accelerating detectors, Annals of Physics **27**, 1–12 (1964)
6. D. Boulware: Radiation from a uniformly accelerated charge, Annals of Physics **124**, 169–188 (1980)
7. S. Parrott: Radiation from a uniformly accelerated charge and the equivalence principle, available from `arXiv:gr-qc/9303025 v8` (2001)
8. G.A. Schott: *Electromagnetic Radiation*, Cambridge University Press, Cambridge (1912) pp. 63–69
9. W. Pauli: Encycl. Math. Wiss. **19**, 648, 653 (1920)
10. A. Einstein: *The Principle of Relativity*, Dover, New York (1923) pp. 99–100
11. M. Friedman: *Foundations of Space–Time Theories*, Princeton University Press, NJ (1983)
12. J.M. Hobbs: A vierbein formalism of radiation damping, Annals of Physics **47**, 141–165 (1968)
13. S. Parrott: *Relativistic Electrodynamics and Differential Geometry*, Springer, New York (1987)
14. P.A.M. Dirac: Classical theories of radiating electrons, Proc. Roy. Soc. London A **167**, 148 (1938)
15. R.P. Feynman, R.B. Leighton, M. Sands: *The Feynman Lectures on Physics*, Vol. II, Addison-Wesley, Reading, MA (1964)
16. A.D. Yaghjian: *Relativistic Dynamics of a Charged Sphere*, Lecture Notes in Physics 686, Springer-Verlag, New York (2006)
17. H.-D. Zeh: *The Physical Basis of the Direction of Time*, 5th edn., Springer-Verlag, Heidelberg (2007)
18. J.S. Bell: *Speakable and Unspeakable in Quantum Mechanics*, 2nd edn., Cambridge University Press, Cambridge (2004), Chap. 9
19. W. Rindler: *Introduction to Special Relativity*, Oxford University Press, New York (1982)
20. M. Göckeler, T. Schücker: *Differential Geometry, Gauge Theories, and Gravity*, Cambridge University Press, Cambridge (1987)
21. H.R. Brown: *Physical Relativity: Space-Time Structure from a Dynamical Perspective*, Clarendon Press, Oxford (2005)

22. F. Rohrlich: *Classical Charged Particles*, 3rd edn., World Scientific Publishing, Singapore (2007)
23. V. Petkov: *Relativity and the Nature of Spacetime*, Frontiers Series, Springer-Verlag, Berlin, Heidelberg (2005)
24. B.S. DeWitt: *Stanford Lectures on Relativity*, Springer-Verlag, Berlin, Heidelberg (to be published)

Index

Fundamental Theories of Physics

Series Editor: Alwyn van der Merwe, University of Denver, USA

1. M. Sachs: *General Relativity and Matter.* A Spinor Field Theory from Fermis to Light-Years. With a Foreword by C. Kilmister. 1982 ISBN 90-277-1381-2
2. G.H. Duffey: *A Development of Quantum Mechanics.* Based on Symmetry Considerations. 1985 ISBN 90-277-1587-4
3. S. Diner, D. Fargue, G. Lochak and F. Selleri (eds.): *The Wave-Particle Dualism.* A Tribute to Louis de Broglie on his 90th Birthday. 1984 ISBN 90-277-1664-1
4. E. Prugovečki: *Stochastic Quantum Mechanics and Quantum Spacetime.* A Consistent Unification of Relativity and Quantum Theory based on Stochastic Spaces. 1984; 2nd printing 1986 ISBN 90-277-1617-X
5. D. Hestenes and G. Sobczyk: *Clifford Algebra to Geometric Calculus.* A Unified Language for Mathematics and Physics. 1984 ISBN 90-277-1673-0; Pb (1987) 90-277-2561-6
6. P. Exner: *Open Quantum Systems and Feynman Integrals.* 1985 ISBN 90-277-1678-1
7. L. Mayants: *The Enigma of Probability and Physics.* 1984 ISBN 90-277-1674-9
8. E. Tocaci: *Relativistic Mechanics, Time and Inertia.* Translated from Romanian. Edited and with a Foreword by C.W. Kilmister. 1985 ISBN 90-277-1769-9
9. B. Bertotti, F. de Felice and A. Pascolini (eds.): *General Relativity and Gravitation.* Proceedings of the 10th International Conference (Padova, Italy, 1983). 1984 ISBN 90-277-1819-9
10. G. Tarozzi and A. van der Merwe (eds.): *Open Questions in Quantum Physics.* 1985 ISBN 90-277-1853-9
11. J.V. Narlikar and T. Padmanabhan: *Gravity, Gauge Theories and Quantum Cosmology.* 1986 ISBN 90-277-1948-9
12. G.S. Asanov: *Finsler Geometry, Relativity and Gauge Theories.* 1985 ISBN 90-277-1960-8
13. K. Namsrai: *Nonlocal Quantum Field Theory and Stochastic Quantum Mechanics.* 1986 ISBN 90-277-2001-0
14. C. Ray Smith and W.T. Grandy, Jr. (eds.): *Maximum-Entropy and Bayesian Methods in Inverse Problems.* Proceedings of the 1st and 2nd International Workshop (Laramie, Wyoming, USA). 1985 ISBN 90-277-2074-6
15. D. Hestenes: *New Foundations for Classical Mechanics.* 1986 ISBN 90-277-2090-8; Pb (1987) 90-277-2526-8
16. S.J. Prokhovnik: *Light in Einstein's Universe.* The Role of Energy in Cosmology and Relativity. 1985 ISBN 90-277-2093-2
17. Y.S. Kim and M.E. Noz: *Theory and Applications of the Poincaré Group.* 1986 ISBN 90-277-2141-6
18. M. Sachs: *Quantum Mechanics from General Relativity.* An Approximation for a Theory of Inertia. 1986 ISBN 90-277-2247-1
19. W.T. Grandy, Jr.: *Foundations of Statistical Mechanics.* Vol. I: *Equilibrium Theory.* 1987 ISBN 90-277-2489-X
20. H.-H von Borzeszkowski and H.-J. Treder: *The Meaning of Quantum Gravity.* 1988 ISBN 90-277-2518-7
21. C. Ray Smith and G.J. Erickson (eds.): *Maximum-Entropy and Bayesian Spectral Analysis and Estimation Problems.* Proceedings of the 3rd International Workshop (Laramie, Wyoming, USA, 1983). 1987 ISBN 90-277-2579-9
22. A.O. Barut and A. van der Merwe (eds.): *Selected Scientific Papers of Alfred Landé.* [*1888-1975*]. 1988 ISBN 90-277-2594-2

Fundamental Theories of Physics

23. W.T. Grandy, Jr.: *Foundations of Statistical Mechanics.* Vol. II: *Nonequilibrium Phenomena.* 1988
ISBN 90-277-2649-3

24. E.I. Bitsakis and C.A. Nicolaides (eds.): *The Concept of Probability.* Proceedings of the Delphi Conference (Delphi, Greece, 1987). 1989
ISBN 90-277-2679-5

25. A. van der Merwe, F. Selleri and G. Tarozzi (eds.): *Microphysical Reality and Quantum Formalism, Vol. 1.* Proceedings of the International Conference (Urbino, Italy, 1985). 1988
ISBN 90-277-2683-3

26. A. van der Merwe, F. Selleri and G. Tarozzi (eds.): *Microphysical Reality and Quantum Formalism, Vol. 2.* Proceedings of the International Conference (Urbino, Italy, 1985). 1988
ISBN 90-277-2684-1

27. I.D. Novikov and V.P. Frolov: *Physics of Black Holes.* 1989
ISBN 90-277-2685-X

28. G. Tarozzi and A. van der Merwe (eds.): *The Nature of Quantum Paradoxes.* Italian Studies in the Foundations and Philosophy of Modern Physics. 1988
ISBN 90-277-2703-1

29. B.R. Iyer, N. Mukunda and C.V. Vishveshwara (eds.): *Gravitation, Gauge Theories and the Early Universe.* 1989
ISBN 90-277-2710-4

30. H. Mark and L. Wood (eds.): *Energy in Physics, War and Peace.* A Festschrift celebrating Edward Teller's 80th Birthday. 1988
ISBN 90-277-2775-9

31. G.J. Erickson and C.R. Smith (eds.): *Maximum-Entropy and Bayesian Methods in Science and Engineering.* Vol. I: *Foundations.* 1988
ISBN 90-277-2793-7

32. G.J. Erickson and C.R. Smith (eds.): *Maximum-Entropy and Bayesian Methods in Science and Engineering.* Vol. II: *Applications.* 1988
ISBN 90-277-2794-5

33. M.E. Noz and Y.S. Kim (eds.): *Special Relativity and Quantum Theory.* A Collection of Papers on the Poincaré Group. 1988
ISBN 90-277-2799-6

34. I.Yu. Kobzarev and Yu.I. Manin: *Elementary Particles. Mathematics, Physics and Philosophy.* 1989
ISBN 0-7923-0098-X

35. F. Selleri: *Quantum Paradoxes and Physical Reality.* 1990
ISBN 0-7923-0253-2

36. J. Skilling (ed.): *Maximum-Entropy and Bayesian Methods.* Proceedings of the 8th International Workshop (Cambridge, UK, 1988). 1989
ISBN 0-7923-0224-9

37. M. Kafatos (ed.): *Bell's Theorem, Quantum Theory and Conceptions of the Universe.* 1989
ISBN 0-7923-0496-9

38. Yu.A. Izyumov and V.N. Syromyatnikov: *Phase Transitions and Crystal Symmetry.* 1990
ISBN 0-7923-0542-6

39. P.F. Fougère (ed.): *Maximum-Entropy and Bayesian Methods.* Proceedings of the 9th International Workshop (Dartmouth, Massachusetts, USA, 1989). 1990
ISBN 0-7923-0928-6

40. L. de Broglie: *Heisenberg's Uncertainties and the Probabilistic Interpretation of Wave Mechanics.* With Critical Notes of the Author. 1990
ISBN 0-7923-0929-4

41. W.T. Grandy, Jr.: *Relativistic Quantum Mechanics of Leptons and Fields.* 1991
ISBN 0-7923-1049-7

42. Yu.L. Klimontovich: *Turbulent Motion and the Structure of Chaos.* A New Approach to the Statistical Theory of Open Systems. 1991
ISBN 0-7923-1114-0

43. W.T. Grandy, Jr. and L.H. Schick (eds.): *Maximum-Entropy and Bayesian Methods.* Proceedings of the 10th International Workshop (Laramie, Wyoming, USA, 1990). 1991
ISBN 0-7923-1140-X

44. P. Pták and S. Pulmannová: *Orthomodular Structures as Quantum Logics.* Intrinsic Properties, State Space and Probabilistic Topics. 1991
ISBN 0-7923-1207-4

45. D. Hestenes and A. Weingartshofer (eds.): *The Electron.* New Theory and Experiment. 1991
ISBN 0-7923-1356-9

Fundamental Theories of Physics

Fundamental Theories of Physics

94.   V. Dietrich, K. Habetha and G. Jank (eds.): *Clifford Algebras and Their Application in Mathematical Physics.* Aachen 1996. 1998   ISBN 0-7923-5037-5

95.   J.P. Blaizot, X. Campi and M. Ploszajczak (eds.): *Nuclear Matter in Different Phases and Transitions.* 1999   ISBN 0-7923-5660-8

96.   V.P. Frolov and I.D. Novikov: *Black Hole Physics.* Basic Concepts and New Developments. 1998   ISBN 0-7923-5145-2; Pb 0-7923-5146

97.   G. Hunter, S. Jeffers and J-P. Vigier (eds.): *Causality and Locality in Modern Physics.* 1998   ISBN 0-7923-5227-0

98.   G.J. Erickson, J.T. Rychert and C.R. Smith (eds.): *Maximum Entropy and Bayesian Methods.* 1998   ISBN 0-7923-5047-2

99.   D. Hestenes: *New Foundations for Classical Mechanics (Second Edition).* 1999   ISBN 0-7923-5302-1; Pb ISBN 0-7923-5514-8

100. B.R. Iyer and B. Bhawal (eds.): *Black Holes, Gravitational Radiation and the Universe.* Essays in Honor of C. V. Vishveshwara. 1999   ISBN 0-7923-5308-0

101. P.L. Antonelli and T.J. Zastawniak: *Fundamentals of Finslerian Diffusion with Applications.* 1998   ISBN 0-7923-5511-3

102. H. Atmanspacher, A. Amann and U. Müller-Herold: *On Quanta, Mind and Matter Hans Primas in Context.* 1999   ISBN 0-7923-5696-9

103. M.A. Trump and W.C. Schieve: *Classical Relativistic Many-Body Dynamics.* 1999   ISBN 0-7923-5737-X

104. A.I. Maimistov and A.M. Basharov: *Nonlinear Optical Waves.* 1999   ISBN 0-7923-5752-3

105. W. von der Linden, V. Dose, R. Fischer and R. Preuss (eds.): *Maximum Entropy and Bayesian Methods Garching, Germany 1998.* 1999   ISBN 0-7923-5766-3

106. M.W. Evans: *The Enigmatic Photon Volume 5: O(3) Electrodynamics.* 1999   ISBN 0-7923-5792-2

107. G.N. Afanasiev: *Topological Effects in Quantum Mecvhanics.* 1999   ISBN 0-7923-5800-7

108. V. Devanathan: *Angular Momentum Techniques in Quantum Mechanics.* 1999   ISBN 0-7923-5866-X

109. P.L. Antonelli (ed.): *Finslerian Geometries A Meeting of Minds.* 1999   ISBN 0-7923-6115-6

110. M.B. Mensky: *Quantum Measurements and Decoherence Models and Phenomenology.* 2000   ISBN 0-7923-6227-6

111. B. Coecke, D. Moore and A. Wilce (eds.): *Current Research in Operation Quantum Logic.* Algebras, Categories, Languages. 2000   ISBN 0-7923-6258-6

112. G. Jumarie: *Maximum Entropy, Information Without Probability and Complex Fractals.* Classical and Quantum Approach. 2000   ISBN 0-7923-6330-2

113. B. Fain: *Irreversibilities in Quantum Mechanics.* 2000   ISBN 0-7923-6581-X

114. T. Borne, G. Lochak and H. Stumpf: *Nonperturbative Quantum Field Theory and the Structure of Matter.* 2001   ISBN 0-7923-6803-7

115. J. Keller: *Theory of the Electron.* A Theory of Matter from START. 2001   ISBN 0-7923-6819-3

116. M. Rivas: *Kinematical Theory of Spinning Particles.* Classical and Quantum Mechanical Formalism of Elementary Particles. 2001   ISBN 0-7923-6824-X

117. A.A. Ungar: *Beyond the Einstein Addition Law and its Gyroscopic Thomas Precession.* The Theory of Gyrogroups and Gyrovector Spaces. 2001   ISBN 0-7923-6909-2

118. R. Miron, D. Hrimiuc, H. Shimada and S.V. Sabau: *The Geometry of Hamilton and Lagrange Spaces.* 2001   ISBN 0-7923-6926-2

Fundamental Theories of Physics

119. M. Pavšič: *The Landscape of Theoretical Physics: A Global View.* From Point Particles to the Brane World and Beyond in Search of a Unifying Principle. 2001   ISBN 0-7923-7006-6
120. R.M. Santilli: *Foundations of Hadronic Chemistry.* With Applications to New Clean Energies and Fuels. 2001   ISBN 1-4020-0087-1
121. S. Fujita and S. Godoy: *Theory of High Temperature Superconductivity.* 2001
     ISBN 1-4020-0149-5
122. R. Luzzi, A.R. Vasconcellos and J. Galvão Ramos: *Predictive Statitical Mechanics.* A Nonequilibrium Ensemble Formalism. 2002   ISBN 1-4020-0482-6
123. V.V. Kulish: *Hierarchical Methods.* Hierarchy and Hierarchical Asymptotic Methods in Electrodynamics, Volume 1. 2002   ISBN 1-4020-0757-4; Set: 1-4020-0758-2
124. B.C. Eu: *Generalized Thermodynamics.* Thermodynamics of Irreversible Processes and Generalized Hydrodynamics. 2002   ISBN 1-4020-0788-4
125. A. Mourachkine: *High-Temperature Superconductivity in Cuprates.* The Nonlinear Mechanism and Tunneling Measurements. 2002   ISBN 1-4020-0810-4
126. R.L. Amoroso, G. Hunter, M. Kafatos and J.-P. Vigier (eds.): *Gravitation and Cosmology: From the Hubble Radius to the Planck Scale.* Proceedings of a Symposium in Honour of the 80th Birthday of Jean-Pierre Vigier. 2002   ISBN 1-4020-0885-6
127. W.M. de Muynck: *Foundations of Quantum Mechanics, an Empiricist Approach.* 2002
     ISBN 1-4020-0932-1
128. V.V. Kulish: *Hierarchical Methods.* Undulative Electrodynamical Systems, Volume 2. 2002
     ISBN 1-4020-0968-2; Set: 1-4020-0758-2
129. M. Mugur-Schächter and A. van der Merwe (eds.): *Quantum Mechanics, Mathematics, Cognition and Action.* Proposals for a Formalized Epistemology. 2002   ISBN 1-4020-1120-2
130. P. Bandyopadhyay: *Geometry, Topology and Quantum Field Theory.* 2003
     ISBN 1-4020-1414-7
131. V. Garzó and A. Santos: *Kinetic Theory of Gases in Shear Flows.* Nonlinear Transport. 2003
     ISBN 1-4020-1436-8
132. R. Miron: *The Geometry of Higher-Order Hamilton Spaces.* Applications to Hamiltonian Mechanics. 2003   ISBN 1-4020-1574-7
133. S. Esposito, E. Majorana Jr., A. van der Merwe and E. Recami (eds.): *Ettore Majorana: Notes on Theoretical Physics.* 2003   ISBN 1-4020-1649-2
134. J. Hamhalter. *Quantum Measure Theory.* 2003   ISBN 1-4020-1714-6
135. G. Rizzi and M.L. Ruggiero: *Relativity in Rotating Frames.* Relativistic Physics in Rotating Reference Frames. 2004   ISBN 1-4020-1805-3
136. L. Kantorovich: *Quantum Theory of the Solid State: an Introduction.* 2004
     ISBN 1-4020-1821-5
137. A. Ghatak and S. Lokanathan: *Quantum Mechanics: Theory and Applications.* 2004
     ISBN 1-4020-1850-9
138. A. Khrennikov: *Information Dynamics in Cognitive, Psychological, Social, and Anomalous Phenomena.* 2004   ISBN 1-4020-1868-1
139. V. Faraoni: *Cosmology in Scalar-Tensor Gravity.* 2004   ISBN 1-4020-1988-2
140. P.P. Teodorescu and N.-A. P. Nicorovici: *Applications of the Theory of Groups in Mechanics and Physics.* 2004   ISBN 1-4020-2046-5
141. G. Munteanu: *Complex Spaces in Finsler, Lagrange and Hamilton Geometries.* 2004
     ISBN 1-4020-2205-0

Fundamental Theories of Physics

142. G.N. Afanasiev: *Vavilov-Cherenkov and Synchrotron Radiation*. Foundations and Applications. 2004
ISBN 1-4020-2410-X
143. L. Munteanu and S. Donescu: *Introduction to Soliton Theory: Applications to Mechanics*. 2004
ISBN 1-4020-2576-9
144. M.Yu. Khlopov and S.G. Rubin: *Cosmological Pattern of Microphysics in the Inflationary Universe*. 2004
ISBN 1-4020-2649-8
145. J. Vanderlinde: *Classical Electromagnetic Theory*. 2004
ISBN 1-4020-2699-4
146. V. Čápek and D.P. Sheehan: *Challenges to the Second Law of Thermodynamics*. Theory and Experiment. 2005
ISBN 1-4020-3015-0
147. B.G. Sidharth: *The Universe of Fluctuations*. The Architecture of Spacetime and the Universe. 2005
ISBN 1-4020-3785-6
148. R.W. Carroll: *Fluctuations, Information, Gravity and the Quantum Potential*. 2005
ISBN 1-4020-4003-2
149. B.G. Sidharth: *A Century of Ideas*. Personal Perspectives from a Selection of the Greatest Minds of the Twentieth Century. 2007.
ISBN 1-4020-4359-7
150. S.-H. Dong: *Factorization Method in Quantum Mechanics*. 2007.
ISBN 1-4020-5795-4
151. R.M. Santilli: *Isodual Theory of Antimatter with applications to Antigravity, Grand Unification and Cosmology*. 2006
ISBN 1-4020-4517-4
152. A. Plotnitsky: *Reading Bohr: Physics and Philosophy*. 2006
ISBN 1-4020-5253-7
153. V. Petkov: *Relativity and the Dimensionality of the World*. Planned 2007.
ISBN to be announced
154. H.O. Cordes: *Precisely Predictable Dirac Observables*. 2006
ISBN 1-4020-5168-9
155. C.F. von Weizsäcker: *The Structure of Physics*. Edited, revised and enlarged by Thomas Görnitz and Holger Lyre. 2006
ISBN 1-4020-5234-0
156. S.V. Adamenko, F. Selleri and A. van der Merwe (eds.): *Controlled Nucleosynthesis*. Breakthroughs in Experiment and Theory. 2007
ISBN 978-1-4020-5873-8
157. F. Cardone and R. Mignani: *Deformed Spacetime*. Geometrizing Interactions in Four and Five Dimensions. 2007
ISBN 978-1-4020-6282-7
158. S.N. Lyle: *Uniformly Accelerating Charged Particles*. A Threat to the Equivalence Principle. 2008
ISBN 978-3-540-68469-5

Printing: Krips bv, Meppel, The Netherlands
Binding: Stürtz, Würzburg, Germany